T0136989

Algorithms for Intelligent Systems

Series Editors

Jagdish Chand Bansal, Department of Mathematics, South Asian University,
New Delhi, Delhi, India

Kusum Deep, Department of Mathematics, Indian Institute of Technology Roorkee,
Roorkee, Uttarakhand, India

Atulya K. Nagar, Department of Mathematics and Computer Science,
Liverpool Hope University, Liverpool, UK

This book series publishes research on the analysis and development of algorithms for intelligent systems with their applications to various real world problems. It covers research related to autonomous agents, multi-agent systems, behavioral modeling, reinforcement learning, game theory, mechanism design, machine learning, meta-heuristic search, optimization, planning and scheduling, artificial neural networks, evolutionary computation, swarm intelligence and other algorithms for intelligent systems.

The book series includes recent advancements, modification and applications of the artificial neural networks, evolutionary computation, swarm intelligence, artificial immune systems, fuzzy system, autonomous and multi agent systems, machine learning and other intelligent systems related areas. The material will be beneficial for the graduate students, post-graduate students as well as the researchers who want a broader view of advances in algorithms for intelligent systems. The contents will also be useful to the researchers from other fields who have no knowledge of the power of intelligent systems, e.g. the researchers in the field of bioinformatics, biochemists, mechanical and chemical engineers, economists, musicians and medical practitioners.

The series publishes monographs, edited volumes, advanced textbooks and selected proceedings.

More information about this series at http://www.springer.com/series/16171

Om Prakash Verma · Sudipta Roy ·
Subhash Chandra Pandey ·
Mamta Mittal
Editors

Advancement of Machine Intelligence in Interactive Medical Image Analysis

 Springer

Editors
Om Prakash Verma
G. B. Pant Government Engineering College
Okhla, New Delhi, India

Sudipta Roy
Washington University in St. Louis
St. Louis, MO, USA

Subhash Chandra Pandey
Birla Institute of Technology, Mesra
(Patna Campus)
Patna, Bihar, India

Mamta Mittal
Department of Computer Science
and Engineering
G. B. Pant Government Engineering College
New Delhi, Delhi, India

ISSN 2524-7565   ISSN 2524-7573   (electronic)
Algorithms for Intelligent Systems
ISBN 978-981-15-1102-8   ISBN 978-981-15-1100-4   (eBook)
https://doi.org/10.1007/978-981-15-1100-4

This Springer imprint is published by the registered company Springer Nature Singapore Pte Ltd.
The registered company address is: 152 Beach Road, #21-01/04 Gateway East, Singapore 189721, Singapore

Preface

Medical imaging issues are so complex owing to high importance of correct diagnosis and treatment of diseases in healthcare systems. Recent research efforts have been devoted to processing and analyzing medical images to extract meaningful information such as volume, shape, motion of organs, to detect abnormalities, and to quantify changes in follow-up studies. Medical image analysis is diverse, and the large amount of information introduced through the hybrid systems requires the next generation of image quantification that needs to be addressed. This book addresses the issues and describes the current advanced method in interactive medical image analysis.

The book is organized into 15 chapters.

Chapter 1, "Pragmatic Medical Image Analysis and Deep Learning: An Emerging Trend," presents current movements of medical image analysis using deep learning. So far, deep learning has shown some good outcomes in health care, but deep learning depends on large number of training data. The chapter discusses the latest issues and solution to medical image analysis using deep learning. Finally, it concludes with future of more open, novel, and innovative prospects of medical image analysis.

Chapter 2, "Aspect of Big Data in Medical Imaging to Extract the Hidden Information Using HIPI in HDFS Environment," presents the utilization of big data analytics to process the massive amount of medical imaging data in clusters of commodity hardware. This chapter also discusses the pixel count of medical images on distributed environment.

Chapter 3, "Image Segmentation Using Deep Learning Techniques," presents a brief overview of 2D and 3D image segmentation and feature extraction. This chapter also suggests that image pre-processing such as bias field correction, image registration, skull stripping, and intensity normalization is the necessary intermediate step to prepare the data. The significance of deep learning and its various techniques including FCN, ParseNet, PSPNet, DeepLab, DeepLabv3, DeepLabv3+, and UNET have been discussed.

Chapter 4, "Application of Machine Intelligence in Digital Pathology: Identification of Falciparum Malaria in Thin Blood Smear Image," proposes a new algorithm that uses a combination of both supervised and unsupervised learning methods as well as rule-based methods, for faster and effective segmentation of parasite-infected cells. However, the method does not take into consideration other forms of malarial parasite infections caused by different species.

Chapter 5, "Efficient ANN Algorithms for Sleep Apnea Detection Using Transform Methods," presents a low-power and high-performance method of night-time recordings of ECG signal to note the deviations of ECG parameters in terms of its amplitudes and time segments. The method is a combination of steps: noise removal, obstructive sleep apnea (OSA) QRS complex detection, extraction of features based on QRS complex reference, and classification of features using artificial neural network (ANN) for sleep apnea detection. The classification using ANN trained with scaled conjugate gradient (SCG) has emerged as the best possible solution.

Chapter 6, "Medical Image Processing in Detection of Abdomen Diseases," emphasizes the clinical aspects in healthcare imaging and emerging trends with some future directions in research. This chapter summarizes various abdominal diseases like tumor, cysts, and stone (calculi), their texture, and gray-level properties. Various diagnostic systems of different diseases like liver tumor, breast cancer from CT images, and their performance were also surveyed in this chapter.

Chapter 7, "Multi-reduct Rough Set Classifier for Computer-Aided Diagnosis in Medical Data," discusses the theoretical aspects of rough set and methods for dimensionality reduction. The idea of multi-reduction rough set classifier has also been discussed. The classifiers are constructed based on these reductions, and the outcome of these classifiers is integrated for the final classification called as ensemble of classifiers. Ensemble methods give the perfect classifiers by combining the less perfect ones.

Chapter 8, "A New Approach of Intuitionistic Fuzzy Membership Matrix in Medical Diagnosis with Application," presents a modified model for medical diagnosis to deal with some complicated features of medical issues using intuitionistic fuzzy membership matrix. The results of the numerical examples provide the best diagnostic conclusions. Therefore, this is a new way of performing medical diagnosis with a much wider concept.

Chapter 9, "Image Analysis and Automation of Data Processing in Assessment of Dental X-ray (OPG) Using MATLAB and Excel VBA," presents the role of science and technology to elevate the patient care by enhancing diagnosis, treatment plan, and the procedures by making it less prone to human errors. This chapter is used for identifying the stages of tooth development to assess the age of an individual. This phenomenon can be accomplished by training the machine to identify the images and to compute the values using artificial intelligence which in turn enable a novice person in dentistry to successfully and accurately identify the age from a given radiography.

Chapter 10, "Detecting Bone Fracture Using Transfer Learning," presents an application of transfer learning (TL) to detect open bone fracture using limited set of images. TL is a kind of self-evolving deep learning (DL) technique. In each iteration of TL, the model is improvised based on the knowledge transferred from the previously learned task. The transfer learning gives significantly comparable results than training model from scratch or even better in some cases. But it is also vulnerable to overfitting due to less training data, and in some cases, poor pre-processing leads to poor classification.

Chapter 11, "GAN-Based Novel Approach for Data Augmentation with Improved Disease Classification," implements a deep convolutional generative adversarial network architecture that can be used to create synthetic chest X-ray images for three classes (infiltration, atelectasis, and no findings). We proposed a novel schema for using online generative adversarial network model effectively. This approach of online augmentation using GAN model-based generated images helped in achieving improved accuracy on the test dataset. This has been demonstrated using three different models, particularly model 3 has produced reasonably good results for the disease classification.

Chapter 12, "Automated Glaucoma Type Identification Using Machine Learning or Deep Learning Techniques," presents an automatic detection of glaucoma using machine learning model that extracts features and classifies healthy or glaucoma-affected images. Extraction technique is used for the classification of healthy and glaucoma-affected images using deep learning technique. The image-processing technique for segmenting the ROI from the retinal fundus image and the identification of different important features that present in retinal images are also discussed.

Chapter 13, "Glaucoma Detection from Retinal Fundus Images Using RNFL Texture Analysis," focuses on peri-papillary atrophy analysis (PPAA) detection using retina nerve fiber layer (RNFL) analysis from fundus image of retina. Since the fundus image is less expensive than the OCT image, the analysis process is highly acceptable from various sectors of society. The experimental results reported the benefits of the classifier worked on 14 statistical features.

Chapter 14, "Artificial Intelligence Based Glaucoma Detection," provides an overview of various artificial intelligence-based techniques for the use of glaucoma detection and introduces various traditional machine learning-based techniques, followed by deep learning-based techniques including transfer learning and ensemble learning-based techniques used for automated glaucoma detection, along with their advantages and drawbacks.

Chapter 15, "Security Issues of Internet of Things in Health-Care Sector: An Analytical Approach," presents an analytic view of existing key security issues pertaining to IoT in health care. In this chapter, pre-occupied architecture which connects physical world to the network has been discussed. IoT is being the upcoming technology, and still it is in its early phase of development. Indeed, it seems that it will encompass a wide spectrum in coming future.

The editors are very thankful to all the members of Springer India Private Limited, especially Aninda Bose, for the given opportunity to edit this book.

New Delhi, India   Om Prakash Verma
Saint Louis, USA   Sudipta Roy
Patna, India   Subhash Chandra Pandey
New Delhi, India   Mamta Mittal

About This Book

Medical imaging offers useful information on patients' medical conditions and clues to the cause of their symptoms and diseases. Research efforts have been devoted to processing and analyzing medical images to extract meaningful information such as volume, shape, motion of organs, to detect abnormalities and to quantify changes in follow-up studies. Computerized image analysis aims at automated extraction of object features that play a fundamental role in understanding the image content. Machine intelligence technique usage in medicine is useful in storage, data retrieval, and optimal use of information analysis for decision making in solving problems. Medical imaging issues are so complex owing to high importance of correct diagnosis and treatment of diseases in healthcare systems. To face this challenge, commonly known image pre-processing steps such as image fusion, segmentation, feature extraction, inter-subject registration, longitudinal/temporal studies, texture, and classification need to have incorporated intelligent techniques. Typical application fields are computer-assisted diagnosis, image-guided therapy, treatment monitoring and planning, 2D and 3D modeling, big data integration, as well as rehabilitation and home care. The major technical advancements and research findings in the field of machine intelligence pertaining to medical image analysis has also been incorporated.

This research book focuses on the innovative research ideas and integrating intelligence to provide the comprehensive overview of the state-of-the-art computational intelligence research and technologies in medical decision making based on medical images. The recent technologies and studies have reached the practical level and become available in clinical practices in hospitals rapidly such as computational intelligence in computer-aided diagnosis, biological image analysis, and computer-aided surgery and therapy. Areas of interest in this proposed book are the highest quality and original work that contribute to the basic science of processing, analyzing, and utilizing all the aspects of advance computational intelligence in medical decision making based on medical images.

Contents

About the Editors

Dr. Om Prakash Verma was born in September 1966. Prof. Verma received B.E. degree in Electronics and Communication Engineering from Malaviya National Institute of Technology, Jaipur, India, M. Tech. degree in Communication and Radar Engineering from Indian Institute of Technology (IIT), Delhi, India and PhD degree in the area of Soft Computing and Image Processing from University of Delhi, Delhi, India. He joined Department of Electronics & Communication Engineering, Delhi Technological University (formerly Delhi College of Engineering) in 1998. Prof. Verma is a Professor of Delhi Technological University Delhi and currently working as Principal, GB Pant Govt. Engineering College Okhla Delhi since April 2017. He is a man of vision and firm commitment and resolve to nurture academic and professional excellence in Institution to which he has associate himself during his 28 years long professional career.

Sudipta Roy has received his Ph.D. in Computer Science & Engineering from the Department of Computer Science and Engineering, University of Calcutta, on 2018. Currently he is associated with PRTT Lab at Washington University in St. Louis, USA. He hold an US patent in medical image processing and filed an Indian patent in smart agricultural system. He is serving as an Associate Editor of IEEE Access, IEEE and International Journal of Computer Vision and Image Processing (IJCVIP), IGI Global Journal. He has more than five years of experience in teaching, and research. His fields of research interests are biomedical image analysis, image processing, steganography, artificial intelligence, big data analysis, machine learning and big data technologies.

Dr. Subhash Chandra Pandey is working as Assistant Professor at Birla Institute of Technology, Mesra, Ranchi (patna campus). He has completed B.Sc from the University of Allahabad, UP, India. Further, he completed the AMIE (Mechanical Engg.) from the Institution of Engineers (India) and subsequently M.Tech (Computer Engineering) from Motilal Nehru National Institute of technology, UP, India. Moreover, he has done his Ph.d from Dr. APJ Abdul Kalam Technical

University, Lucknow, UP, India. His current area of interest is Data mining. Soft computing, Bio-inspired computation, AI, machine intelligence, Philosophical aspects of machine cognition and IoT.

Dr. Mamta Mittal is graduated in Computer Engineering from Kurukshetra University Kurukshetra in 2001 and received Masters' degree (Honors) in Computer Engineering from YMCA, Faridabad. Her Ph.D. is from Thapar University Patiala in Computer Engineering and rich experience of more than 16 years. Presently, working at G.B. PANT Government Engineering College, Okhla, New Delhi (under Government of NCT Delhi) and supervising Ph.D. candidates of GGSIPU, New Delhi. She is working on DST approved Project "Development of IoT based hybrid navigation module for mid-sized autonomous vehicles". She has published many SCI/SCIE/Scopus indexed papers and Book Editor of renowned publishers.

Chapter 1
Pragmatic Medical Image Analysis and Deep Learning: An Emerging Trend

Pranjal Pandey, Smita Pallavi and Subhash Chandra Pandey

1 Introduction

Perhaps, it is of the utmost importance to process and analyze the health-related information and computed investigation of health-related images is substantial for many purposes such as automated decision-making, automated surgery and therapy, and anatomy-related medical research [1]. It is the latest trend to apply the techniques pertaining to artificial intelligence (AI) for this pursuit. Indeed, AI-based techniques are widely being used for the purposes of information analysis, data retrieval, and in decision-making. Further, it is also used for the purposes of problem-solving within the realm of improvement of healthcare systems so that effective treatment can be provided [2]. In earlier days, healthcare data were comparatively small compared to present scenario. As a matter of fact, owing the gigantic advancement within the purview of image acquisition devices, there is a dire need of the techniques which can analyze even the big data. Undoubtedly, image analysis is an arduous and interesting task. Perhaps, swift augmentation in medical images and succinct-related models requires substantial efforts by medical expert. Moreover, analyses of theses medial images are subjective and likely to be affected by natural human error. Furthermore, these images possess dissident views among different specialists. That is why, machine learning (ML) techniques are of paramount importance to automate

P. Pandey
Department of Electronics and Communication Engineering, Indraprastha Institute of Information Technology, Delhi, Okhla Industrial Estate, Phase 3, New Delhi, India
e-mail: pranjalpandey200@gmail.com

S. Pallavi · S. C. Pandey (✉)
Department of Computer Science and Engineering, Birla Institute of Technology, Mesra, Ranchi (Patna Campus), Patna, Bihar, India
e-mail: subh63@yahoo.co.in

S. Pallavi
e-mail: smita.pallavi@bitmesra.ac.in

© Springer Nature Singapore Pte Ltd. 2020
O. P. Verma et al. (eds.), *Advancement of Machine Intelligence in Interactive Medical Image Analysis*, Algorithms for Intelligent Systems,
https://doi.org/10.1007/978-981-15-1100-4_1

the analysis process. However, conventional AI techniques are not adequate to tackle complicated problems. From extensive literature survey, it has been observed that ML and AI have developed swiftly in recent past. Indeed, various methods of ML have displayed substantial impetus in the healthcare domain such as automated image processing, segmentation, etc. Further, machine learning techniques have also been used to extract the knowledge in an effective and efficient way from the medical images. The ML and AI enable the physician to diagnose and forecast precisely and quickly the diseases so that it can be rectified within the stipulated time. Moreover, these methods improve the efficacy of physician to comprehend the way to investigate the common diversion which spearheads the ailment. These approaches include traditional algorithms as given below:

- Support vector machine (SVM),
- NN,
- KNN,
- Recurrent neural network (RNN),
- Long short-term memory (LSTM),
- Extreme learning model (ELM),
- Generative adversarial networks (GANs),
- Deep learning methods such as convolutional neural network (CNN), etc.

Further, algorithms pertaining to machine intelligence can enhance the efficacy [3]. The accurate recognition of natural attributes from medical images has significant impact in precise inference. Attempts that have been made in recent past augment the efficiency of machine intelligence algorithms. In this realm, the prominent and indispensable feature is the implementation of hybridized AI techniques [4].

In machine learning algorithm, segmentation is often very decisive for obtaining more accurate results in subsequent phases such as feature extraction and image measurement. However, it has been observed that manual segmentation is considerably time-taking. Moreover, segmentation of many scans is also not pragmatically feasible. In addition, the quantity of images required to analyze the disease is also growing exponentially due to technological advancement. Thus, one can infer that the manual segmentation is not reasonable in healthcare diagnostic and it can also not produce ingenious human interpretation. Thus, intelligent tools are of paramount importance for the computed segmentation. In [4–18], in-depth information has been rendered pertaining to recent segmentation techniques of medical image. Moreover, some categorizations of these methods are also given in [4–18]. The lung's nodules segmentation is shown in Fig. 1. Perhaps, computerized disease identification from medical images has shown considerable precision since decades, but novel and innovative advancements in ML have emerged a new horizon in the deep learning-based medical imaging processing.

Indeed, hybridization of high-performance computing (HPC) with AI can cater the need to provide the efficacy to deal with big medical image data for the purpose of precise and effective diagnosis. Further, in this quest, deep learning helps in enormous ways. It provides the way to choose and extract attributes and it also fabricates new ones. Furthermore, deep learning identifies the disease besides the predictive

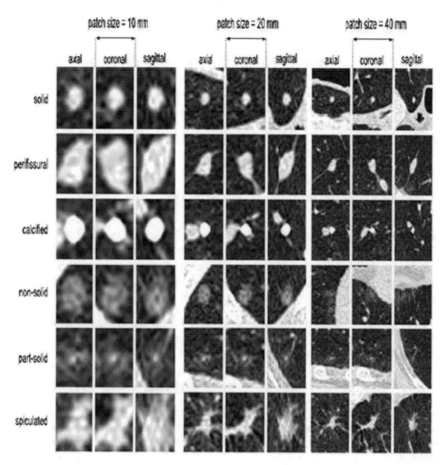

Fig. 1 Lung's nodules segmentation [42]

measures. In addition, deep learning imparts pragmatic prediction models so that physician could perform effectively. Indeed, there are certain pitfalls in traditional algorithm such as traditional algorithms are confined to process the natural images in their raw form and require more processing time. Moreover, traditional algorithms are based on expert knowledge and require substantial time in feature tuning. In contrast, hybridization of HPC and AI can perform with raw data and renders the automatic feature learning with faster pace. Moreover, combination of HPC and AI can demonstrate the learning of multiple levels of abstraction and representation. Besides these benefits, combination of HPC and AI can show automatic extraction of information from huge set of images.

Deep learning (DL)-based algorithms render speed and efficient performance within different domains of AI-related tasks. The aim of this chapter is to provide the in-depth and lucid view of deep learning architecture and comparative analysis of deep learning over the machine learning in the realm of medical image analysis and processing.

The remaining chapter is structured in the following manner. Section 2 briefly represents the rudiments of medical image analysis. Section 3 explains architectural view of NN and DL. Further, flamboyant features of DL over ML in medical image engineering (MIE) are given in Sect. 4. Furthermore, Sect. 5 renders the structural nuances of CNN. Section 6 critically analyzed the issue "will DL be pragmatic in medical imaging?" Finally, chapter is concluded in Sect. 7.

2   Rudiments of Medical Image Analysis

The anatomical structure plays vital role in the medical image analysis particularly in the quantitative analysis such as the volume measurement. The quantitative information is obtained through the process of image analysis. Figure 2 represents different steps required in the image processing. In general, different steps of image processing systems are as follows:

Step 1: The representation of concerned region is obtained by apparatus such as MRI.
Step 2: Then it is processed by the image processing unit.
Step 3: Feature extraction and feature selection are performed by the medical image analysis tools.
Step 4: Finally, the output data is inputted to the high-level processing unit of medical image processing system.

Indeed, computed analysis of medical images required number of image processing methods including preprocessing. Different preprocessing operations are as follows:

a. Noise removal,
b. Image enhancement,
c. Edge detection, etc.

After the completion of these primitive steps, the image is prepared for the purposes of analysis. Further, the image in ready state is inputted to the medical image analysis system. Subsequently, the mining of the region of interest (ROI) is performed during the phenomenon of segmentation. Furthermore, after the computation of region of interest, the depiction and dimensioning are worked out. In addition, features extraction is carried out to recognize and classify the region of interest. Indeed, machine learning approach-based computed image processing tools are the central tenet to advance the superiority of image analysis and understanding. Deep learning is a widely used method that imparts excellent accuracy. Undoubtedly, it created an unprecedented horizon in the field of MIA. Viability of DL within the purview of medical domain encompasses wide range of applications. However, researchers are lagging behind to convert all such data into the useful information. Next section will describe the architecture of neural network and deep learning. State-of-the-art

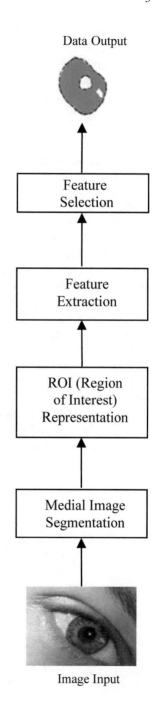

Data Output

Feature
Selection

Feature
Extraction

ROI (Region
of Interest)
Representation

Medial Image
Segmentation

Image Input

techniques of DL in the realm of MIA are also illustrated in subsequent sections. However, the discussion is only introductory. It renders a glimpse of wide-ranging DL impetus in MIA image processing sector particularly in health care.

3 Architectural View of NN and DL

Structurally and conceptually, artificial neural networks (ANNs) mimic the functionality of human biological nervous system. Literature survey reveals the fact that perceptron can be treated as one of the initial human brain-based neural networks (NN). Structurally, it comprises input layer which is connected with output layer. This neural network performs well for the linearly separable classification problem. In contrast, in order to sole the more complicated problem one or more layer is inserted in between the input and output layers, i.e., hidden layers. NN entails mutually interrelated and interconnected neurons. The input neurons receive input data, complete some sort of processing on these data, and subsequently transfer existing output to the approaching layer. The common structural details of NN are shown in Fig. 3. Finally, the NN displays the output which can further be inputted to the next layer. The sole purpose of hidden layer is to tackle the intricate problem efficiently.

Deep learning neural network (DNN) is an indispensable new cost-effective technique to train the network. Indeed, the weight learning is slow in DNN. However, extra layers in DNN facilitate the symphony of features in a bottom-up manner, i.e., from beneath layers to the apex layer. This symphony of features creates the efficacy to maneuver the intricate data.

In fact, DL is of paramount importance and is emerging trend in automated applications. It has been observed from the extensive literature survey that deep learning has substantially been advanced since 2013. In present scenario, there is plethora

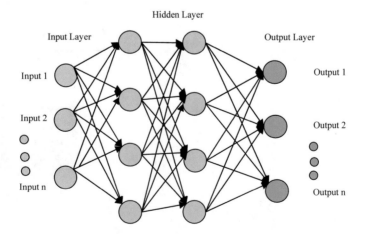

Fig. 3 Neural network architecture

Fig. 4 Interdependency of artificial intelligence, machine learning, and deep learning

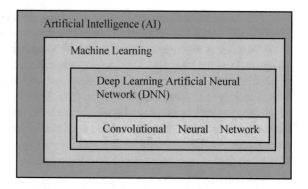

of DNN-based computer vision applications. Undoubtedly, these applications are well even in comparison of human beings. The DNN-based techniques can recognize blood and tumor cancer accurately from MRI scans. Obviously, DNN is an advancement of ANN that comprises more numbers of hidden layers.

The quantum of extended hidden layers renders abstraction of high-order and superior image analysis. Precisely, we can say that it is a widely applicable and accepted method owing to its incredible performance. Basically, DNN entails various layers of neurons in a well-structured and hierarchical fashion and thus rendering a hierarchical attribute illustration. The extended quantity of layers shows huge modeling capacity.

It is worthy to mention that DNN can fundamentally learn entire potential relationship after training even in the case of huge knowledge database. That is why, DNN can make intelligent predictions (Fig. 4).

Machine learning imparts the efficacy to the machine to learn while it is devoid of exhaustive program. In typical ML, human expertise is needed to select such features that illustrate the representation of visual data in an optimum fashion. Furthermore, DL is subclass of ML and is an efficient technique. In fact, deep learning does not require feature selection. Instead, deep learning learns which features are best suited for classification of data on its own accord [19]. Indeed, ANN is the foundation stone of the majority of deep learning techniques. CNN is a subclass of deep learning artificial neural network. In initial phase, CNN was implemented for classification and recognition of handwritten digits but it stepped forward substantially in 2012 in the ImageNet test [20]. CNN performs its functionality by the process of convolution. In the convolution phenomenon, point-wise multiplication of small matrices of weight vectors takes place across the whole image and thus generates the neuronal maps. The phenomenon of matrix convolution repeatedly takes place and subsequently generates the new hidden layers of neural maps. Further, optimization of the network is worked out on the basis of loss estimation. Moreover, backpropagation is also applied on the network thus obtained.

From extensive literature survey, it has been revealed that the deep learning image is widely used for the varying purpose in healthcare sector and it is an emerging area of research from last few years. The realm of deep learning entails some other streams also such as identification and detection of lung nodules as benign and malignant from the CT scans [21–29], classification of lung syndrome from CT scans [30]. Deep learning is also used successfully for the identification of lymph nodes into benign and malignant from CT [31]. It is pertinent to mention that new advancement in machine learning and particularly in deep learning will cause a succinct impact on human society in coming future. Further, the role of data science within the purview of radiology is indispensable and is considerable for the advancement of AI [32, 33]. However, the prospect is still unveiled because the uprising is still at the verge of commencement. Undoubtedly, it is imperative and pertinent from the viewpoint of medical institutions to investigate and comprehend the new technologies to abreast the emerging challenges.

4 Flamboyant Features of DL over ML in MIE

Accurate and precise analysis of disease is a function of some parameters such as image acquisition and image interpretation. However, the automated image interpretation is still in its initial phase and is being applied extensively in the domain of computer vision. The conventional ML techniques substantially depend on proficient attributes and identification needs structure attributes to be worked out. Further, owing to the wide-ranging discrepancies in the patient data, consistency of conventional machine learning approaches is not reliable. Perhaps, these are the reason evolution of machine learning techniques so that machine learning can cater requirement in the case of complex and big data.

In recent past, many image acquisition procedures and techniques have substantially been developed such as CT and MRI scans, X-ray, etc. These devices possess high degree of resolution. Indeed, like common steps of image engineering [5, 6], medical image analysis also entails following stages:

- Initial stage,
- Middle stage, and
- High stage (Table 1).

Low-level medical image processing entails basic operations and the inputs and outputs are medical images. The basic operations generally performed in low-level processing are as follows:

a. Noise reduction,
b. Edge detection,
c. Contrast enhancement, and
d. Sharpening of medical images.

Middle stage performs the tasks such as the following:

a. Segmentation and
b. Description.

Table 1 Comparative study of different DL models [43]

Types of network	Details of network	PROS	CONS
Deep neural network (DNN)	1. Number of layers is greater than 2 2. Generally implemented for classification and regression	1. Excellent accuracy	1. Nontrivial learning phenomenon 2. Learning rate is slow
Convolutional neural network (CNN)	1. Excellent performance for 2D data 2. Convolutional filters convert 2D into 3D	1. Excellent performance 2. Learning rate is fast	1. Plethora of data is needed for classification
Recurrent neural network (RNN)	1. Renders sequential learning 2. Sharing of weights take place across all steps and neurons	1. Sequential event learning is possible 2. Can perform time dependency model 3. Great accuracy	1. Gradient vanishing 2. Big datasets are needed
Deep conventional extreme learning machine (DC-ELM)	1. Implemented Gaussian probability function		
Deep Boltzmann machine (DBM)	1. Foundation stone of this model is Boltzmann phenomenon 2. Hidden layers encompass unidirectional connectivity	1. Imparts excellent inference	1. Parameter optimization cannot take place in case of big data
Deep belief network (DBN)	1. Unidirectional connectivity 2. Used for supervised as well as unsupervised learning 3. Hidden layers are treated as visible for the next layer	1. Greedy approach is followed in every layer	1. Training process is costly
Deep Auto-encoder (DA)	1. Solely for unsupervised learning 2. Often used for dimensionality reduction 3. The number of inputs and outputs is equal	1. No requirements of labeled data	1. Pretraining step is required 2. Training may adversely influenced from vanishing

Segmentation performs the partitioning of an image into region of interest (ROI). Further, in description tasks, the objects are transformed into a convenient form so that it can be effectively processed and identified by the machine. In description tasks, the inputs are generally the medical images and the outputs of description task are the worked out features. Further, high stage analysis incorporates following tasks:

a. Inference from collective identification from objects.
b. Vision-related cognitive functioning.

In the first task, the assembled identified objects are analyzed and hidden features are sensed. Further, in the second task, these features are interpreted and analyzed so that disease can be identified precisely and accurately.

In present scenario, deep learning has created an impetus in different walks of medical image analysis. It is undoubtedly an efficient and supervised machine learning method. This technique entails the phenomenon of DNN which is a modification of NN. Unlike neural network, deep neural network renders considerably high approximation of biological brain. This high approximation is obtained due to advanced architecture used in deep neural network as compared to simple neural network.

5   Structural Nuances of CNN

Feedforward NN is basically used for the purposes of image analysis. However, the connectivity of neurons of one layer with every neuron of the subsequent layer renders it strongly inefficient. On the other hand, the domain knowledge is often used for ingenious pruning of the neurons connections and it also displays the improved performance. In contrast, CNN is a specific type of ANN. CNN possesses the characteristics of the preservation of spatial associations of the data. Moreover, the preservation of the spatial association of the data requires only limited numbers of connections among the layers. First, a grid structure is implemented to organize the input of the CNN which is subsequently provided for layers which in turn preserve the spatial association of the data. In CNN, each layer performs the process on a little region of the preceding layer as shown in Fig. 5. CNN possesses the efficacy of excellent input data representation and it is substantially effective for the orientation-related tasks of the images. It is pertinent to mention that CNN entails several convolutional as well as activation layers as depicted in Fig. 5.

In addition, CNN also incorporates pooling layers. The training of CNN is performed with the help of backpropagation as well as using gradient descent like a NN. Mainly, CNN comprises following layers:

- Convolutional layers,
- Activation layers, and
- Pooling.

The activations received from preceding layers are convolved in convolutional layers with the help of small filters. These small filters are parameterized. It is worthy

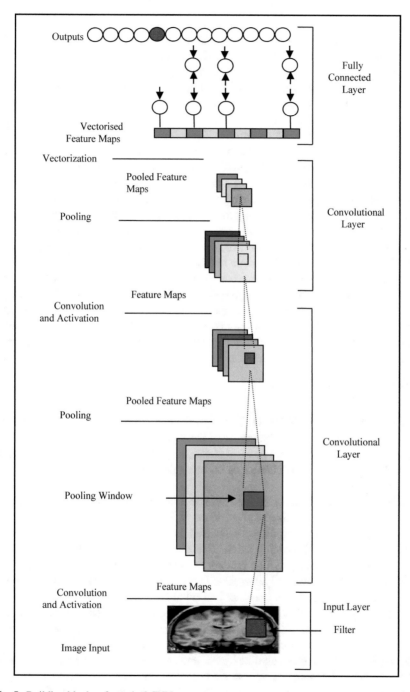

Fig. 5 Building blocks of a typical CNN

to mention that exact weights are shared throughout the entire input field by every filter. It implicates that variations are equal at each layer. In this way, the numbers of weights required to be learned are considerably reduced. The impetus behind this weight sharing is due to the fact that a particular feature occurring in one portion of the image might occur in other portions. Lucidly, we can say that if a filter under consideration is able to detect vertical lines, then it can be effectively implemented to detect the vertical lines wherever they exist.

Further, the output of convolutional layer, i.e., feature maps is inputted to the nonlinear activation functions. This phenomenon enables the CNN to approximate any sort of nonlinear function. In general, the activation function is taken as resolved linear units [34, 35]. The feature maps thus obtained are vital for further processing. Now, the feature maps are pooled in a pooling layer. Furthermore, a small grid region is inputted to perform the pooling process. Subsequent after the pooling process a single number is obtained for each grid region. Often, the pooling process can be classified as (a) max pooling and (b) average pooling. The max pooling entails max function for obtaining the single number for each grid region, whereas average pooling implements the average function for the same. It has been observed that subtle variation in the inputted image may cause subtle perturbation in the activation maps. That is why, this layer causes some translational invariance. In [36], it has been mentioned that exclusion of pooling layers can reduce the architectural complexity of the network without affecting its performance. Some extra important components of many recent CNNs incorporate the following features:

- Dropout regularization and
- Batch normalization.

Indeed, dropout regularization is an excellent concept that enhanced the performance of CNNs. In this technique, various models are averaged in an ensemble manner which comparatively provide increased performance than single model. In [37], this technique is considered as an averaging method based on probabilistic modeling of NN.

The second component, i.e., batch normalization, in general, follows the activation layers and it yields normalized activation maps. In this phenomenon, the mean value is subtracted from the normalized value and then the result is divided by the standard deviation for each training batch. The batch normalization creates the speedy training of the network and also renders the parameter initialization easy [38]. In order to create an innovative and modified version of CNN architectures, these components are assimilated in a complex and mutually connected manner, or one can even implement any other viable and effective processes. It should be noted that architectural design of CNN for any specific task entails many indispensable factors such as the way out to accomplish the task, the way data is inputted to the network, optimal utilization of computational cost, and memory consumption. However, in primitive days of recent DL, very naïve combinations were used, as in Lenet [39] and AlexNet [40]. Further, the architectures of CNN become quite complicated with the advancement in technology. Table 2 [41] describes brief overview of explicitly accessible code for ML pertaining to MI.

Table 2 Brief enumeration of explicitly accessible code for ML pertaining to MI [41]

Name	References	Link references
NiftyNet	[44, 45]	[46]
DLTK	[47]	[48]
DeepMedic	[49]	[50]
U-Net	[51]	[52]
V-net	[53]	[54]
SegNet	[55]	[56]
Brain lesion synthesis using GANs	[57]	[58]
GANCS:	[59]	[60]
Deep MRI reconstruction	[61]	[62]
Graph CNN	[63]	[64]

6 Will Deep Learning Be Pragmatic in Medical Imaging?

Some researchers believe that implication of deep learning technique may become the most troublesome approach since the advent of digital imaging. It is intuitive to ponder that deep learning technique might replace the involvement of human beings in the realm of medical image analysis within next few years as most of the analysis will be observed by intelligent machines. Even, intelligent machines may accomplish the tasks of disease prediction, selection of medicines, and decisions regarding the treatment procedures. Perhaps, deep learning is being used in different domains of medical science but it is intuitive to consider that structural and functional study of eye played crucial role in the research domain of DL. Further, pathological issues and cancerous disease have gained substantial implications of deep learning and researches have obtained satisfactory accuracy in these fields.

It is intuitive to visualize that the future of deep learning in medical image processing seems to not viably occur in near future as compared to other state-of-the-art medical image processing applications. Perhaps, it is because of the complexities included in this approach. From the authors' perspective, it is still a fascinating wild-goose chase. Researchers are striving strenuously; however, there are various hurdles and bottlenecking that are debilitating their progresses. There are so many aspects which need a lot of considerations such as

- Non-availability of dataset,
- Confidentiality and associated legalities,
- Committed therapeutic professionals,
- Non-standardization of data and ML techniques, etc.

In Table 3 [41], a brief enumeration of MI datasets and repositories is given. Further, a brief enumeration of competitions pertaining to MI is enumerated in Table 4 [41].

Table 3 A brief enumeration of MI datasets and repositories [41]

Dataset name	Link references
OpenNeuro	[65]
UK Biobank	[66]
TCIA	[67]
ABIDE	[68]
ADNI	[69]

Table 4 A brief enumeration of competitions pertaining to MI [41]

Competition name	Objective	Link references
Grand-challenges	Biomedical image analysis	[70]
RSNA	Analysis of chest radiography	[71]
HVSMR 2016	Analysis of blood pool	[72]
ISLES 2018	Segmentation of stroke lesions	[73]
BraTS 2018	Segmentation of brain tumors	[74]
CAMELYON17	Classification of breast cancer	[75]
ISIC 2018	Analysis of skin-related disease	[76]
Kaggle's 2018	Fast treatment	[77]
Kaggle's 2017	Diagnosis of lung cancer	[78]
Kaggle's 2016	Diagnosis of heart-related disease	[79]
MURA	Identification bone abnormality	[80]

7 Conclusions

Recent developments in ML render the substantial needs of MIA to augment the computer-aided diagnosis. In recent few years, DL has emerged as a pivotal technique for the automation of our different walks of life. Perhaps, this technique rendered tenacious and succinct impetus in this pursuit. Moreover, DL ameliorates the process of MIA as compared to conventional ML algorithms. However, the DL is creeping ahead slowly in the domain of health care as compared to other approaches. In this chapter, authors ingeniously portray different aspects of DL in the purview of MIA. In last section, authors discussed the bottlenecking emerging forth in pragmatic application of DL in medical image analysis. Visualizing the glittering side of ML, it seems that in recent future, ML will supersede and replace the human beings in medical applications and particularly in the process of diagnosis. However, there are many arduous challenges which still need to be resolved. One unanswerable issue in this pursuit is the non-availability of standard data. Therefore, we are still facing the scarcity of adequate training data and how big data could be implemented in medical domain is still implicit as on date. Contrary to this, recent modifications on other techniques display better result with increase in the size of the data. So far, DL has shown positive contemplation; however, health care is an extremely delicate and sensitive sector so we should conceptualize other effective techniques simultaneously because it seems that DL alone cannot result in an efficient way. Lastly, we conclude that certainly future will open more novel and innovative prospects for the pursuit of medical image analysis.

References

1. Bankman I (2009) Handbook of medical image processing and analysis, 2nd ed. Elsevier, pp 71–257 (2009)
2. Melonakos J (2009) Geodesic tractography segmentation for directional medical image analysis. PhD dissertation, Department of Electrical and Computer Engineering, Georgia Institute of Technology University (2009)
3. Harris G, Andreasen NC, Cizadlo T, Bailey JM, Bockholt HJ, Magnotta VA, Arndt S (1999) Improving tissue classification in MRI: a three-dimensional multispectral discriminant analysis method with automated training class selection. J Comput Assist Tomogr 23(1):144–154
4. Pham DL, Xu C, Princo JL (2000) A survey on current methods in medical image segmentation. Annu Rev Biomed Eng 2:315–337
5. Yu-Jin Z (2006) Advances in image and video segmentation. IEEE IRM Press
6. Zhang YJ (2002) Image engineering and related publications. Int J Image Graphics 2(3):441–452
7. Clarke LP, Velthuizen RP, Camacho MA, Heine JJ, Vaidyanathan M, Hall LO, Thatcher RW, Silbiger ML (1995) MRI segmentation: methods and applications. Magn Reson Imag 13:343–368
8. Leondes CT (2005) Medical imaging systems technology: methods in cardiovascular and brain. Singapore
9. Yang D, Zheng J, Nofal A, Deasy J, Naqa IME (2009) Techniques and software tool for 3D multimodality medical image segmentation. J Radiat Oncol Inform 1(1)
10. Olabarriag SD, Smeulders AWM (2001) Interaction in the segmentation of medical images: a survey. Med Image Anal (Elsevier) 127–142
11. Withey DJ, Koles ZJ (2007) Three generations of medical image segmentation-methods, available software. Int J Bioelectromagnetism 9(2)
12. Pitiot A, Delingette H, Thompson PM (2005) Automated image segmentation: issues and applications. In: Leondes CT (ed) Chapter 6 of Medical imaging systems technology: Methods in general anatomy, vol 3, pp 195–241
13. Sharma N, Aggarwal LM (2010) Automated medical image segmentation techniques. J Med Phys 3–14 [Online]. http://www.ncbi.nlm.nih.gov/pmc/articles/PMC2825001/
14. Nayak H, Amini MM, Bibalan PT, Bacon N (2009) Medical image segmentation-report
15. Neufeld E, Samaras T, Chavannes N, Kuster N (2006) Robust highly detailed medical image segmentation [Online]. http://citeseerx.ist.psu.edu/viewdoc/download?doi=10.1.1.116.4080&rep=rep1&type=pdf
16. Nayak H, Amini MM, Bibalan PT, Bacon N (2006) Medical image segmentation-report
17. Souplet JC (2007) Medical image navigation and research tool by INRIA (MedINRIA)

18. Pan Z, Lu J (2007) A Bayes-based region-growing algorithm for medical image segmentation. IEEE Trans 9(4):32–38
19. Chartrand G, Cheng PM, Vorontsov E et al (2017) Deep learning: a primer for radiologists. Radiographics 37:2113–2131
20. Krizhevsky A, Sutskever I, Hinton GE (2012) ImageNet classification with deep convolutional neural networks. In: Proceedings of the 25th international conference on neural information processing systems, vol 1, Lake Tahoe, Nevada. Curran Associates Inc., pp 1097–1105
21. Ciompi F, de Hoop B, van Riel SJ et al (2015) Automatic classification of pulmonary peri-fissural nodules in computed tomography using an ensemble of 2D views and a convolutional neural network out-of-the-box. Med Image Anal 26:195–202
22. Nibali A, He Z, Wollersheim D (2017) Pulmonary nodule classification with deep residual networks. Int J Comput Assist Radiol Surg 12:1799–1808
23. Song Q, Zhao L, Luo X et al (2017) Using deep learning for classification of lung nodules on computed tomography images. J Healthc Eng 8314–740
24. Ciompi F, Chung K, van Riel SJ et al (2017) Towards automatic pulmonary nodule management in lung cancer screening with deep learning. Sci Rep 7:464–479
25. Hua KL, Hsu CH, Hidayati SC et al (2015) Computer-aided classification of lung nodules on computed tomography images via deep learning technique. Onco Targets Ther 8:2015–2022
26. Sun W, Zheng B, Qian W (2017) Automatic feature learning using multichannel ROI based on deep structured algorithms for computerized lung cancer diagnosis. Comput Biol Med 89–530
27. Wang H, Zhao T, Li LC et al (2017) A hybrid CNN feature model for pulmonary nodule malignancy risk differentiation. J Xray Sci Technol
28. Wang C, Elazab A, Wu J et al (2017) Lung nodule classification using deep feature fusion in chest radiography. Comput Med Imaging Graph 57:10–18
29. Setio AA, Traverso A, de Bel T et al (2017) Validation, comparison, and combination of algorithms for automatic detection of pulmonary nodules in computed tomography images: the LUNA16 challenge. Med Image Anal 42:1–13
30. Anthimopoulos M, Christodoulidis S, Ebner L et al (2016) Lung pattern classification for interstitial lung diseases using a deep convolutional neural network. IEEE Trans Med Imaging 35:1207–1216
31. Wang H, Zhou Z, Li Y et al (2017) Comparison of machine learning methods for classifying mediastinal lymph node metastasis of non-small cell lung cancer from (18)F-FDG PET/CT images. EJNMMI Res 7–11
32. Kohli M, Prevedello LM, Filice RW et al (2017) Implementing machine learning in radiology practice and research. AJR Am J Roentgenol 208:754–760
33. Dreyer KJ, Geis JR (2017) When machines think: radiology's next frontier. Radiology 285:713–718
34. Clevert D-A, Unterthiner T, Hochreiter S (2015) Fast and accurate deep network learning by exponential linear units (ELUs). arXiv:1511.07289
35. He K, Zhang X, Ren S, Sun J (2015) Delving deep into rectifiers: surpassing human-level performance on image net classification. In: Proceedings of the IEEE international conference on computer vision, pp 1026–1034
36. Springenberg JT, Dosovitskiy A, Brox T, Riedmiller M (2014) Striving for simplicity: the all convolutional net. arXiv:1412.6806
37. Srivastava N, Hinton G, Krizhevsky A, Sutskever I, Salakhutdinov R (2014) Dropout: a simple way to prevent neural networks from over-fitting. J Mach Learn Res 15:1929–1958
38. Ioffe S, Szegedy C (2015) Batch normalization: accelerating deep network training by reducing internal covariate shift. In: International conference on machine learning, pp 448–456
39. LeCun Y, Bottou L, Bengio Y, Hanner P (1998) Gradient-based learning applied to document recognition. Proc IEEE 86:2278–2324
40. Krizhevsky A, Sutskever I, Hinton GE (2012) ImageNet classification with deep convolutional neural networks. In: Pereira F, Burges CJC, Bottou L, Weinberger KQ (eds) Advances in neural information processing systems, vol 25. Curran Associates, Inc., pp 1097–1105

41. Lundervold AS, Lundervold A (2018) An overview of deep learning in medical imaging focusing on MRI. Zeitschrift fur medizinische Physik. https://doi.org/10.1016/j.zemedi.2018. 11.00
42. Cui Z, Yang J, Qiao Y (2016) Brain MRI segmentation with patch-based CNN approach. In: IEEE control conference (CCC), pp 7026–7031
43. Razzak MI, Naz S, Zaib A (2017) Deep learning for medical image processing: overview, challenges and the future. Lect Notes Comput Vis Biomechanic (Springer) 26:323–350
44. Gibson E, Li W, Sudre C, Fidon L, Shakir DI, Wang G, Eaton-Rosen Z, Gray R, Doel T, Hu Y, Whyntie T, Nachev P, Modat M, Barratt DC, Ourselin S, Cardoso MJ, Vercauteren T (2018) NiftyNet: a deep-learning platform for medical imaging. Comput Methods Programs Biomed 158:113–122
45. Li W, Wang G, Fidon L, Ourselin S, Cardoso MJ, Vercauteren T (2017) On the compactness, efficiency, and representation of 3d convolutional networks: brain parcellation as a pretext task. In: International conference on information processing in medical imaging (IPMI)
46. http://niftynet.io
47. Pawlowski N, Ktena SI, Lee MC, Kainz B, Rueckert D, Glocker B, Rajchl M (2017) DLTK: state of the art reference implementations for deep learning on medical images. arXiv:1711. 06853
48. https://github.com/DLTK/DLTK
49. Kamnitsas K, Ledig C, Newcombe VF, Simpson JP, Kane AD, Menon DK, Rueckert D, Glocker B (2017) Efficient multi-scale 3D CNN with fully connected CRF for accurate brain lesion segmentation. Med Image Anal 36:61–78
50. https://github.com/Kamnitsask/deepmedic
51. Ronneberger O, Fischer P, Brox T (2015) U-net: convolutional networks for biomedical image segmentation. In: Medical image computing and computer-assisted intervention (MICCAI). LNCS, vol 9351. Springer, pp 234–241, arXiv:1505.04597 [cs.CV]
52. https://lmb.informatik.uni-freiburg.de/people/ronneber/u-net
53. Milletari F, Navab N, Ahmadi S-A (2016) V-net: fully convolutional neural networks for volumetric medical image segmentation. In: Fourth international conference on 3D vision (3DV). IEEE, pp 565–571
54. https://github.com/faustomilletari/VNet
55. Badrinarayanan V, Kendall A, Cipolla R (2017) Seg-Net: a deep convolutional encoder-decoder architecture for image segmentation. IEEE Trans Pattern Anal Machine Intelligence
56. https://mi.eng.cam.ac.uk/projects/segnet
57. Shin H-C, Tenenholtz NA, Rogers JK, Schwarz CG, Senjem ML, Gunter JL, Andriole KP, Michalski M (2018) Medical image synthesis for data augmentation and anonymization using generative adversarial networks. In: International workshop on simulation and synthesis in medical imaging. Springer, pp 1–11 (2018)
58. https://github.com/khcs/brain-synthesis-lesion-segmentation
59. Mardani M, Gong E, Cheng JY, Vasanawala S, Zaharchuk G, Alley M, Thakur N, Han S, Dally W, Pauly JM et al (2017) Deep generative adversarial networks for compressed sensing automates MRI. arXiv:1706.00051
60. https://github.com/gongenhao/GANCS
61. Schlemper J, Caballero J, Hajnal JV, Price AN, Rueckert D (2018) A deep cascade of convolutional neural networks for dynamic MR image reconstruction. IEEE Trans Med Imaging 37:491–503
62. https://github.com/js3611/Deep-MRI-Reconstruction
63. Parisot S, Ktena SI, Ferrante E, Lee M, Moreno RG, Glocker B, Rueckert D (2017) Spectral graph convolutions for population-based disease prediction. In: International conference on medical image computing and computer-assisted intervention. Springer, pp 177–185
64. https://github.com/parisots/population-gcn
65. https://openneuro.org
66. http://www.ukbiobank.ac.uk/
67. http://www.cancerimagingarchive.net

68. http://fcon_1000.projects.nitrc.org/
69. http://adni.loni.usc.edu/
70. https://grand-challenge.org
71. https://www.kaggle.com/c/rsna-pneumonia-detection-challenge
72. http://segchd.csail.mit.edu/
73. http://www.isles-challenge.org/
74. http://www.med.upenn.edu/sbia/brats2018.html
75. https://camelyon17.grand-challenge.org/Home
76. https://challenge2018.isic-archive.com/
77. https://www.kaggle.com/c/data-science-bowl-2018
78. https://www.kaggle.com/c/data-science-bowl-2017
79. https://www.kaggle.com/c/second-annual-data-science-bowl
80. https://stanfordmlgroup.github.io/competitions/mura/

Chapter 2
Aspect of Big Data in Medical Imaging to Extract the Hidden Information Using HIPI in HDFS Environment

Yogesh Kumar Gupta

1 Introduction

Big data attribute about data abundance specifically engendered every day from gregarious media sites, sensors, satellites, and so forth. As this data is accumulated from varied heterogeneous sources, the unstructured form of data explicitly has the need of special coding skills. We are unable to store big data using conventional method because of its form and there are special implements to store this type of data. Intricacy in handling the big data additionally increases with the rising in its volume [1]. Big data has several characteristics, which are additionally kenned as V's of big data. Here, we are presenting 7 V's of immensely colossal data: volume—specified as plethora of the data, velocity—refers to the celerity of data production, variety—refers to the miscellaneous category of data, veracity—refers to the meaningfulness of the data in all phase; validity—points out toward the correctness of the data with deference to certain application, volatility—refers to the duration for which data is utilizable for us, and value—is generally discussed with veneration to the outcome of the big data processing [2].

1.1 Big Data Sources and Its Formats

The big data can be classified into the following three categories:

(1) Structured data: The data produced from several articles and journals, Customer Relationship Management (CRM), business applications such as retail, finance,

Y. K. Gupta (✉)
Department of CS, Banasthali Vidyapith, Vanasthali, Rajasthan, India
e-mail: gyogesh@banasthali.in

© Springer Nature Singapore Pte Ltd. 2020
O. P. Verma et al. (eds.), *Advancement of Machine Intelligence in Interactive Medical Image Analysis*, Algorithms for Intelligent Systems, https://doi.org/10.1007/978-981-15-1100-4_2

Fig. 1 Characteristics of
large volume of data [2]

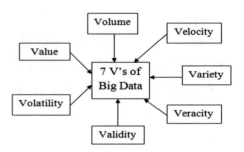

bioinformatics, and other such traditional databases in various sources such as
RDBMS, OLAP, and data warehousing [3].
(2) Semi-structured data such as Rich Site Summary (RSS) or XML formatted data,
    HTML, CSV, and RDF data [3].
(3) Unstructured data: These data are created by the users such as trading markets
    data, medical imaging data, web forums, social media sites, user feedback,
    emails, comments, audios, images, videos, etc. or it may be created by machine
    such as various sensors data and weblogs [3] (Fig. 1).

1.2 History of Big Data from Various Sources

A recent study projected that every minute the users of email send more than 200
million emails, Twitter users produce 277,000 tweets, more than 4 million searching
queries collected by means of Google and Facebook users contribute to additional
2 million portion of substance and more than 350 GB of data is processed. The
approximated figure about the tremendous amount of data is that, it was estimated
nearly about 5 Exabyte's (EB) till 2003, 2.7 Zeta bytes (ZB) till 2012, and it is
expected to grow nearly about four times greater till 2016. Figure 2 shows the data in
Terabytes (TB) during the years 2001–2012. From the year 2005 to 2012, it appears

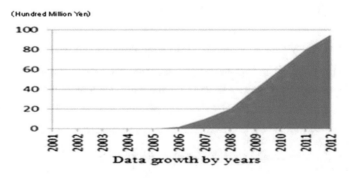

Fig. 2 Data in terabytes from 2001 to 2012 [4]

(Hundred Million Yen)

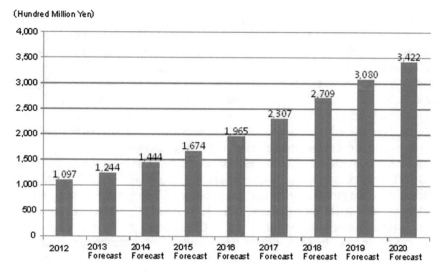

Fig. 3 Forecast of transition and size of big data analytics market from 2012 to 2020 [4]

from this graph that the amount of data exponentially grows within this period due to the significant contribution of big data analytics. Figure 3 shows the forecast of transition and size of big data market from the year 2012 to 2020 [4].

1.3   Big Data in Health Care

It is irresistible on account of its broadly explained aspects termed as degree, varieties of data and its pace, i.e., rate at which data is generated. The electronic healthcare data includes medical imaging generated through minimal invasive and noninvasive medical imaging modalities techniques such as PET scan, CT scan, X-ray, MRI, ultrasound, etc. diagnostic reports, pharmacy information for health insurance, medical research journals, laboratory experiment values, implants and surgery test results, medication information's, etc. which is prone to help the organizations to build a 360° view of each and every patient. The pace of Electronic Care Record (ECR) implementation exponentially increases in inpatient and outpatient aspects. To enhance the improvement in outcomes, reduction of costs, and excellence of care, this digitization of healthcare informatics is opening tremendous promising possibilities to help the payers, stakeholders, and providers. According to IBM report, big data analytics in health care may give its involvement in following subjects [5]:

- Genomic analytics,
- Evidence-based medicine,
- Scam analysis, and
- Patient contour analysis.

Big data contains many varieties of data; this data can be in the form of text, images, and other multimedia contents. In this paper, we have discussed about medical imaging data only. There are two categories of medical imaging modalities: invasive and noninvasive medical imaging. Invasive medical images are the one which are taken during the surgeries into a body cavity, whereas noninvasive images are such as CT scan, MRI, USG, and X-ray images. Most of the work has been done in the field of invasive imaging, and not in noninvasive imaging, and so, in this paper, we have taken only noninvasive images. In the field of health care, noninvasive medical imaging is widely utilized into final decade, and consequently here be an abundance of data available which were leftover unattended. This unattended data may have some utilizable information, which is able to salutary in front of future. It is extremely obligatory to the progression of images by means of the intention of extracting the hidden patterns and information as of them. Here, in order to extort this information, processing of image must be performed on these images [6, 7]. A consequence of processing of image to extort utilizable information from images and to avail some mathematical operations is done. These operations utilize any kind of signal processing. An image is provided as the input for this processing. Output of image processing can be some parameters or characteristics of the input image. Image processing and digital image processing are often used interchangeably. Image processing follows some steps to process the images: image acquisition, preprocessing, segmentation, feature extraction and feature cull, relegation, and performance evaluation. These are the very consequential step to extract the hidden information from the images [8].

We have sundry implements such as MATLAB to deal out images related to medical although within big data, here immense numbers of that images. Subsequently, not a feasible approach to deal out that images piecemeal with the avail of these implements. Big data make use of distributed methodical or analytical implements to deal out the abundance of data in expeditious celerity. These implements are capable of acclimated to deal out immensely colossal data images besides through the assimilation of some adscititious packages and libraries. Big data analytical implements work in distributed manner so even the plethora of data can be processed at a very expeditious rate. Hadoop is utilized within dispersed environment to deal out big data. Hadoop works on clusters of commodity hardware where each cluster contains a number of processing nodes [9, 10]. Immensely colossal data is divided and allocated to multiple nodes of Hadoop cluster, and then this data is processed to ascertain subsidiary information. There are two major components of Hadoop— MapReduce and HDFS (Hadoop distributed file system). HDFS is used to store the massive volume of data in distributed environment. It has two categories of data: metadata and cognate data to applications. Both types of data are stored on variant of nodes. MapReduce processes all the datasets stored within HDFS in parallel manner. It consists of several processors which perform parallel processing in Hadoop environment [11]. Hadoop Image Processing Interface (HIPI) is an implement of Hadoop which is utilized to deal out images in parallel fashion over dispersed environment. Images which are to be processed through HIPI are stored up in HDFS, and the concluding output of the processed image is as well stored within HDFS [12].

1.4 Hadoop Image Processing Interface (HIPI)

HIPI is a major framework to carry out processing of image on astronomically immense extent. This kind of framework is predicated upon Hadoop. It is considered to exertion with Hadoop, MapReduce due to its parallel dealing out module. By means of the avail of MapReduce, these abundances of images can be processed in parallel manner. HIPI runs among Hadoop and all its astronomical data such as image is stored up on the Hadoop distributed file system. This data is made available toward the MapReduce to efficiently process this data in order to find some paramount information [12]. It provides the facility to the user to perform sundry image processing algorithms having no deep cognizance of Hadoop framework. There are some objectives that are consummated by HIPI, which are as follows:

(a) HIPI is defined as an open-source framework which works on MapReduce.
(b) Through facilitate of HIPI, images can be stored in Hadoop in various file formats.
(c) HIPI processes images in parallel and distributed manner.

A utilize can be interoperated between multiple image processing libraries. Generally, HDFS is utilized to store up records on sundry nodes over Hadoop other than; it is arduous used for this dispersed file system to store up such an abundance of data such as image. In turn to store up these image files in the dispersed file system, a number of functions are required to be performed. An image needs to be stored in HIPI Image Bundle to store in the Hadoop environment. After that, these images are processed through HIPI.

1.5 Architecture of HIPI

HIPI defines three phases to deal with an image, such as

(a) Culling,
(b) Mapping, and
(c) Reducing.

In the very first phase, a HIB is made available to the framework wherever images are filtered as per the description presented in the program through the user. In the Map phase, all images with the purpose of carry on all the way through the Cull phase are made available sequentially for the enhancement of data locality. Subsequently, all the images are obtainable to the mapper separately. In this phase, key–value pair is stimulated. The outcome of the mapper phase is taken as the input intended for the reducer phase. In the direction of diminish, the exploitation bandwidth of network and pre-summative key–value pair of the third phase, i.e., Shuffle step is finished. At the end, all the task of reducer which were classified by user were carried out simultaneously in parallel way to acquire some outputs. By getting these outputs, all are collected to have final outcome from the framework, i.e., HIPI [13, 14] (Fig. 4).

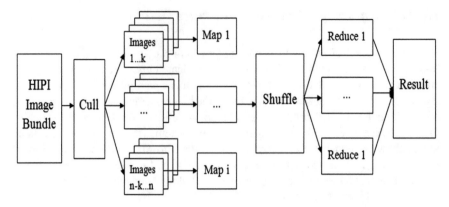

Fig. 4 HIPI architecture [13]

1.6   Medical Imaging Modalities

In medical science, the medical imaging modality is a technique that represents the internal structure of body for the analysis of medical and clinical intervention, besides creating the visual presentation of the functions of a body in the outline of various soft tissues and organs. Imaging modalities of medical looks for to expose interior structures concealed with the bones and skin, seeing that to make a diagnosis and show the ROI region corresponding to the disease. Medical imaging furthermore ascertains a database of normal and physiology to formulate it probable toward recognizing deformities. Even if imaging of detached organs and tissues be able to act upon for medical basis, the procedures are regularly well-thought-out part of pathology as an alternative to medical imaging. There are various types of invasive and noninvasive imaging modalities such as MRI, CT scan, X-ray, USG, etc. These modalities are elaborated further in detail.

1.6.1   Magnetic Resonance Imaging (MRI)

To record the internal structure and some portion of function in the body, a noninvasive medical imaging modality such as MRI is significantly used. MRI is a spectroscopic imaging technique based on nuclear magnetic resonance principle, and uses non-ionizing electromagnetic radiation. There are huge numbers of atoms of hydrogen in our body since the main ingredient in the body is water or lipid. The nucleus of a hydrogen atom is composed of positively charged protons. The protons produce MRI signals in the body. Patient's body is plant within a strong magnet field with a small gradient. There is a static magnetic field which is normally greater than 10^4 times and is powerful than that of earth magnetic field. Every atom has precession frequency along with their magnetic fields which is straightforwardly proportional to the magnetic field power. Radio Frequency (RF) pulse is applied to match the

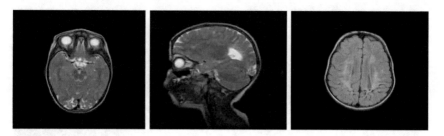

Fig. 5 Brain structure MRI image

precession frequency; some atoms soak up energy and change their direction. After that, these types of atoms reproduce the energy even as evolution to their novel orientation (relaxation). MRI makes use of radio frequency and strong magnetic field to construct high-quality cross-sectional, anatomical, and functional body images in any plane [15].

The main advantage of MRI in medical is that it is mainly used to provide in-depth analytic imaging of soft tissues in the body such as soft organs and cartilage tissues like brain and the heart. MRI is normally painless and does not use radiations, and so MRI scans are generally considered safe in pregnancy and for children. It allows defined and immediate finding of functional, molecular, and anatomical data. MRI shows some unique information which other techniques are not capable to show and has the great impact for cell tracking and breast tissue regeneration. On the other hand, MRI does have many drawbacks such as it has very tight space and is too much noisy. For some people, it may be claustrophobic and quite expensive. MRI is not for intraluminal abnormalities and there is a need of sedation to little children because they cannot remain still [15]. Some sample images of MRI are shown in Fig. 5.

1.6.2 Ultrasonography (USG)

Ultrasound imaging is very efficient medical imaging technique to ensure the blood circulation in heart and in blood vessels and to check the development of baby inside the uterus. It uses high-frequency sound pulses in the range of megahertz and their echoes to generate medical images. Transducer probe is used to pass the high-frequency sound pulses keen lying on the patient's body. After passing through a body, sound pulses hit a wall between soft tissues (e.g., between soft tissue and small structure). Many sound pulses find a wall and get reflected and other pulses wait until meeting another edge and then get reflected. Scattered sound pulses relayed to the ultrasound machine and calculate the space from probe to the tissue (5,005 ft/s or 1,540 m/s) and time of the echoes return (in terms of millionths of a second) [15].

Now the ultrasound machine demonstrates the space and echoes intensities on screen and displays an ultrasound image in 2D, 3D, and 4D. Many other ultrasounds are also in use such as Doppler ultrasound and 3D ultrasound imaging. Nowadays,

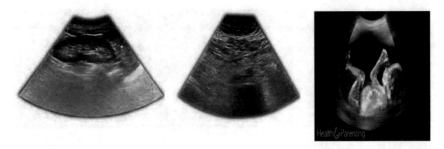

Fig. 6 USG images to detect fetus

ultrasound technology is readily present, inexpensive for the treatment of endometriomas, bladder lesions, rectovaginal septum, and ovarian endometriosis. Ultrasound has three major areas of applications such as obstetrics and gynecology, cardiology, urology, used to detect intra- and extra-luminal abnormalities, observing fetal health, checking the sex of the baby, view inside the heart to recognize the abnormal patterns, measuring the blood circulation in major blood vessels or heart, and finding kidney stones and cancer, respectively. Ultrasound can cause two major problems; first is that it enables increase of temperature and second is that ultrasound releases gases which get dissolved in the blood or tissues and it forms bubbles when gases go out [15]. Some sample images of USG are shown in Fig. 6.

1.6.3   X-ray Imaging

X-rays are a type of high-energy electromagnetic waves which can penetrate many body parts at varying levels such as fat, bones, tumors, and other body parts can absorb X-rays at different levels which will be reflected in the image of X-ray. For the creation of medical X-ray imaging, ionizing electromagnetic radiation travels through the body, and is absorbed by the body parts at different levels depending on the atomic number and density of the tissues, and it creates a profile. X-ray attenuation works more efficiently in bones rather than soft tissues. The produced image is simply a two-dimensional projection between the source of X-ray and film. X-rays consist of many medical applications such as chiropractic and dental. Radiographs are used to view the movement of body parts and are useful for blood vessels of heart and brain, and it enables to take the internal structure of stomach, intestines, and lungs, detect fracture in bones, etc. Mammography is used to detect the breast cancer and hysterosalpingogram is used for uterus and fallopian tubes. Due to the use of relatively high level of radiations, people suffer from many problems such as skin reddening, hair loss, tissue effects like cataracts, and also increased chances to have cancer later in life [15]. Some sample images of X-rays are shown in Fig. 7.

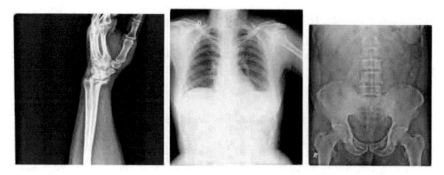

Fig. 7 X-ray images

1.6.4 Computed Tomography (CT scan)

CT scan is a diagnostic technology which consists of a rotating frame which has two parts; first is a tube of X-ray at one face and a detector at converse face of the frame. An image is acquired each time when the X-ray tube and detector completes one round of the patient body and many images will be collected from many angles. Every profile of X-ray beam is reconstructed by the computer to produce 2D image and then scanned. 3D CT scan can also be acquired with the help of spiral CT which is helpful in visualization of tumor in three dimensions. Recently, 4D CT scan is introduced to overcome the problem of respiratory movements. It produces temporal and spatial information about the organs. It is a painless and noninvasive method to diagnose the medical problems. In our body, it captures the picture of soft tissues, bones, and blood vessels. CT scan also provides detailed information about very minute abnormalities, even the body does not have the symptoms of it. CT scan provides detailed spatial information and a good picture of veins. CT scan releases relatively high radiations which have the risk of lung and breast cancer and mainly not recommended for pregnant women and has health issue for unborn babies and fetuses. It is not useful for intraluminal abnormalities [15]. Some sample images of CT scan are shown in Fig. 8.

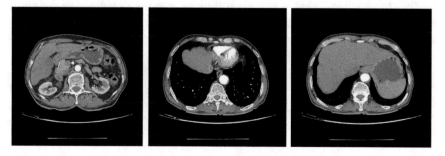

Fig. 8 CT scan images showing internal body structures

1.7   Medical Imaging Modalities Formats

The medical images come in various formats such as DICOM, PACS, etc. and each has its own specific interpretation.

1.7.1   Digital Imaging and Communication in Medicine (DICOM)

DICOM is undeniably a standard and is specifically utilized worldwide for storing, managing, transmitting, and printing information in medical imaging which incorporates several modalities of medical imaging, for instance, MRI, radiography, ultrasound imaging, CT scan, and radiation therapy [16]. DICOM standard has been developed by National Electrical Manufactures Association (NEMA) through a highlighting lying on medical imaging diagnostic. Mostly, image formats are being covered by DICOM standard which are appropriated in the direction of the medical informatics. A particular DICOM file encloses a header (i.e., patients record such as its name, image dimensions, etc.) and image data (which is in compressed or uncompressed form) [16]. These types of files can be switched among two entities with the aim of getting image in DICOM format of patient data.

DICOM standard supports interoperability by specifying the following:

1. For network communication.
2. Semantics and syntax of commands and related information that are able to exchange with protocols like TCP/IP.
3. For the communication of media, here, set of storage services of media.

It is applicable for both the networked environment and offline media environment.

1.7.2   Picture Archiving and Communication System (PACS)

It provides the short-term and long-term storage (i.e., economical storage), management, retrieval (i.e., convenient access to images from several imaging modalities), and distribution of medical images. It is comprised means of replacement of conventional radiological films based on medical images. By providing this facility, it facilitates the filmless clinical environment [17]. PACS is cost neutral with respect to conventional radiology. If economic saving is made, this is a bonus because it eliminates the operating cost of film processing and storage space among other things [17]. Many of the problems that were associated with film are resolved by PACS. More efficiently, the solution is provided by encouraging the immediate availability of patient data with imaging studies.

The main components of PACS are as follows:

1. Imaging modalities, for example, MRI, X-ray, CT scan, etc.
2. Highly secure network.
3. Workstations for handing out and interpreting images.
4. Files intended for storing and image retrieval with related reports and documents.

2 Literature Review

Wu et al. [1] enlighten the concept of big data as the perspectives of health care are conversed. Basavaprasad and Ravi [8] have discussed mechanism of image processing with some major applications. Mukherjee and Shaw [9] illustrate Hadoop to deal with this massive quantity of data. Ghazi and Gangodkar [10] discussed efficient working of Hadoop over cluster environment. Working plan of all the daemons of Hadoop has also discussed in this paper. Vidyasagar [11] has classified the various components and architecture of Hadoop. Vemula and Crick [18] delineate steps and procedures concerned in the image processing interface of Hadoop, namely, HIPI; in this paper, they also describe some filter algorithms and one of them is "Laplacian". Bhosale and Gadekar [19] define the concept of Hadoop for parallel data processing and also illustrated the architecture of HDFS and MR module. Barapatre et al. [14] personify the structural design and work plan of HIPI with MapReduce and HDFS also defines the "module" for better enhancement of performance within short amount of time. Mamulwar et al. [20] illustrate the new tools due to propagation of "Medical images" on the daily basis as conventional methods on single working node are sluggish and not capable to handle hasty creation of images. Consequently, "Hadoop-based medical image retrieval system" is scheduled to advance the ability of retrieving images. Yang et al. [21] point up the concept of storing and shearing large-scale medical images by defining a "MIFAS-Medical images File Accessing system" with the help of HDFS of Hadoop and through this system achieved reliability, fault tolerance, and scalability. Yuzhong and Lei [22] highlight the perception of big data which is categorized into structured or unstructured form that includes images, audios, and videos. Due to quick increment of data in massive amount, there is hadoop and image processing architecture of cloud which are discussed. Peter Augustine [23] personified the advantages of big data analytical tool such as Hadoop in the area of medical or healthcare centers where huge amount of data were generated to provide the finest economical services in the healthcare centers. Chris and Liu [24] describe the HIPI framework for large-scale image processing paradigm which has also discussed the benefits of HIPI to create and develop tools for dealing out of images. A main focus point of this paper is MapReduce, computer vision, and image processing. Muneto and Kunihiko [25] discussed the concept of Hadoop due to its ability of parallel data processing and also evaluate the performance by processing the images in correspondence with distributed environment. Timofei and Andrey [26] define that single computer cannot process collection of large amount of images since distributed environment is discussed. Use of openCV in Hadoop cluster is illustrated for image processing and also HIPI framework provides the capability to openCV for performing image processing over distributed environment. Patel et al. [12] had various techniques of processing images in parallel and distributed manner using HIPI just like MapReduce is used for data processing. Sweeney et al. [13] conversed with reference to HIPI, i.e., image processing interface of Hadoop and also enlightens the working stages of HIPI. Arsh et al. [27] illustrated various tools of processing of image over dispersed environment like OpenCV and HIPI. Fan and Bifet [28] described how data is increasing every day, because of increasing use of Google, Facebook, YouTube, Twitter, and other social networking sites. Hashem et al. [29] typify big data in three aspects: numerous data, this data cannot be classified into relational databases, and data is generated, captured, and processed in high speed. Categories to classify big data are discussed here. Dinakaran et al. [30] discussed various frameworks like OpenCV, HIPI, and HMIPr. Input/output tools are used to make clear internal representation of images.

3   Research Methodology

In our research, we acquired 500 noninvasive medical images such as CT scan, MRI, USG, and X-ray from secondary sources, that is, Internet. But, we used only 100 different images to store and process through distributed image processing interface that is called HIPI.

Every noninvasive medical imaging consists of two components: region of interest (ROI) and non-ROI. ROI of an image refers to the component of the image where a disease has occurred, such as the boundary of the tumor, etc. is included in it. All the radiologists work with ROI part of the image, which is genuinely only 10–20% part of the image. The remaining area is kenned as non-ROI part, which never focused but can have some subsidiary information for the future. In this research work, we have counted the pixel of the whole image which comprises both ROI and non-ROI part. The samples of several medical imaging modalities are shown in Fig. 9.

3.1   Experimental Setup

To store and process the medical images, we developed a cluster environment using Hadoop. The Hadoop infrastructure is made up of one cluster having four SlaveNodes with one MasterNode. Each node in a cluster environment is having the same hardware configuration: Intel® core™ i3-3210 CPU 540 @3.07 GHz, 4.00 GB RAM, 500 GB HDD. We used softwares such as Ubuntu 16.04 LTS, Openjdk Version_1.8.0_151, Hadoop-2.7.1, Gradle-2.10, HIPI 2.1.0.

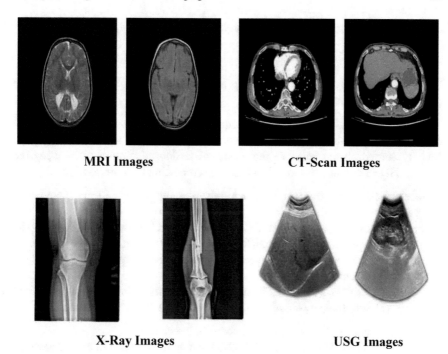

MRI Images   **CT-Scan Images**

X-Ray Images   **USG Images**

Fig. 9   Various noninvasive medical imaging modalities

3.2   Cluster Network Details

In a network of cluster, we used five computers with different IPs. The detail of the nodes in a cluster environment is shown in Table 1.

Table 1   Network detail of nodes in a cluster commodity hardware

Node name	Detailed
HadoopMaster	10.11.0.167
HadoopSlave1	10.11.0.168
HadoopSlave2	10.11.0.169
HadoopSlave3	10.11.0.170
HadoopSlave4	10.11.0.166

3.3   Implementation of Parallel Storing and Parallel Processing Environment

There is some list of steps to develop an environment for parallel storing and processing of medical imaging data.

Step1: Install the Hadoop with all prerequisites.
Step2: Configure the cluster environment having one master and four slave nodes.
Step3: Started all the nodes in the cluster of Hadoop environment by using the HDFS command start-all.sh and displayed using the command jps.

Master node plays the role of an administrator that means all the slave nodes work under the supervision of this. The proposed work uses one master node and four slave nodes. Master node contains five daemons such as DataNode, NameNode, ResourceManager, SecondaryNameNode, and NodeManager, while the slave node contains DataNode and NodeManager. The details of master node and slave node in cluster environment are shown in Fig. 10.

Thereafter, it uses webs URL mentioned below one by one in web browser to open the web interface of NameNode and DataNode in HDFS environment.

(1) http://localhost:8088/—shows the nodes in the cluster.
(2) http://localhost:50070/—shows the NameNode information in cluster environment.
(3) http://localhost:50075/—shows the DataNode.
(4) http://localhost:50090/—shows the SecondaryNameNode.

The web interfaces are shown in Figs. 11, 12, 13 and 14.

3.4   Proposed Model to Extract the Hidden Information from Medical Imaging Modalities Using HIPI in HDFS Environment

The purpose of this proposed model is to extract hidden information in the form of pixel count. Pixels are the most prominent addressable elements in an image, and are quantified in dpi. The number of pixels in an image refers to the amount of information. In medical science, radiologists face many quandaries as the images have very less number of pixels and the information is not feasibly extractable. With the avail of some implements and technologies used to analyze immensely colossal data, we can extract this hidden information from medical images and count the number of pixels of several images at the same period of time. There is some list of steps that perform the parallel storing and processing of medical images.

Fig. 10 Functioning daemons of master and slave nodes in cluster

3.4.1   Parallel Storing of Medical Images in Hadoop Environment

Step 1:   To store massive volume of medical images and create a HIPI Image Bundle (HIB) in HDFS environment.

Step 2:   After storing the images, we need to make a jar file so that we can view how the images are stored in HDFS.

Step 3:   The jar file is used to see the images in HDFS environment.

Fig. 11 Number of nodes in cluster environment

Fig. 12 NameNode information in cluster environment

Fig. 13   DataNodes information in cluster environment

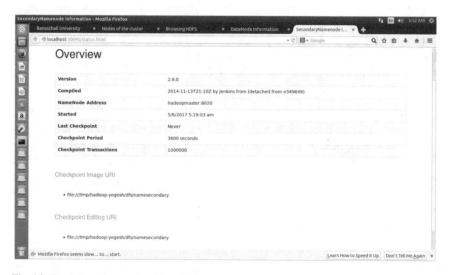

Fig. 14   Overview of secondary NameNode

3.4.2 Parallel Processing of Stored Medical Images in HDFS Environment

The proposed MapReduce algorithm is used to perform pixel count on the stored images. The algorithm contains the two most important classes: Mapper and Reducer. The working of mapper and reducer class step by step is as follows:

Mapper Class

Step1:   Check the value not equal to 0, width and height must be greater than 1 and band value must be 3.
Step 2: Add width and height to a variable.
Step 3: Now, create a loop and run till the value is less than height.
Step 4: Creating another loop and run till the value is less than weight.
Step 5: Add the R, G, and B values to the array.
Step 6: End Loop.
Step 7: Add the FloatImage value to a variable avg.
Step 8: Scale the avg variable value to get the result.
Step 9: Emit the result.

Reducer Class

Step 1: Add the FloatImage value to a variable avg.
Step 2: Traversing the values and add those values to avg value with an increment in total variable.
Step 3: Checking total value is greater than zero then
Step 4: Scale the avg variable value to average data.
Step 5: Add the average data values to result.
Step 6: Emit the result.

MapReduce algorithm is used for parallel data processing and analyzing massive amount of huge data which is stored in HDFS to handle automatic scheduling, communication, synchronization, and has the ability related with fault tolerance. In the mapping process, the input is massive amount of datasets which is divided into number of blocks and then each block is assigned to the Mapper as a form of key and value pairs. In the reducing process, the outputs coming from the Mapper are put into the Reducer and then processed and a final result is generated with the process of merging.

Input Medical Imaging Dataset

The acquired noninvasive medical images from secondary sources are shown in Fig. 15.

These above colored medical images are stored in Hadoop environment as an input to create a HIB file that will further be processed by the proposed model to get the total number of pixel count by executing the MapReduce program using HIPI in HDFS environment. It created two output files such as _SUCCESS and part-r-00000 which are shown in Fig. 16.

Fig. 15 Acquired medical images

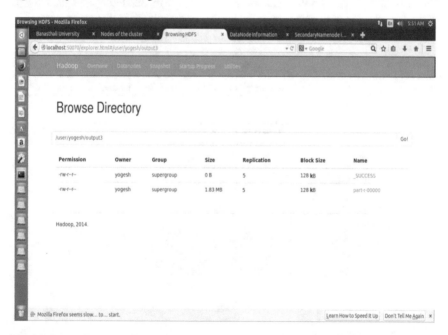

Fig. 16 Output files _SUCCESS and Part-r-00000

Fig. 17 Output of HIB file in the form of pixel count

The file Part-r-00000 contains the final output of execution. The results of the pixel count of image are shown in Fig. 17.

4 Results and Discussion

The average pixel values for the RGB colored image are shown in Table 2.

The analysis of results in the form of graph is shown corresponding to Table 2.

Figure 18 shows the average pixel values corresponding to red, green, and blue color that comprises both ROI and non-ROI part of the medical images, which assists the radiologist to know more about the hidden or meaningful information.

Table 2 Average pixel values of RGB

Color bands	R	G	B
Average pixel values	0.09386	0.09386	0.093823

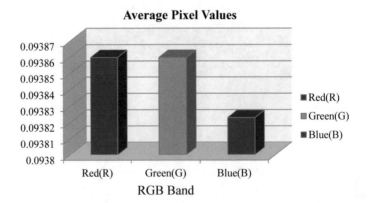

Fig. 18 Resultant average pixel values

5 Conclusions

An abundance of noninvasive medical images is present in big data. The processing of these images is very paramount from healthcare perspective. Radiologists face many quandaries due to the low pixel count of the images. Additionally, a sizably voluminous part of these medical images remains unattended even if there can be some paramount obnubilated information. In this paper, we have counted the number of pixels of the whole medical image containing both ROI and non-ROI part of the image. As big data contains a plethora of medical images, dealing with each image piecemeal will result in slow processing. So, here we have utilized big data analytical implements to process this massive amount of medical imaging data in clusters of commodity hardware. We have utilized HIPI, an implement of Hadoop, for the processing of these medical images in distributed cluster environment. We have presented the experiment and results cognate to our research.

References

1. Wu X et al (2013) Data mining with big data. https://doi.org/10.1109/tkde.2013.109. IEEE
2. Khan M et al (2014) Seven V's of big data understanding big data to extract value. In: Conference of the American Society for Engineering Education (ASEE Zone 1). 978-1-4799-5233-5/14/2014. IEEE
3. Gupta YK, Jha CK (2016) Study of big data with medical imaging communication. In: International conferences on communication and computing systems (ICCCS). CRC Press (Taylor & Francis group), pp 993–997
4. Gupta YK, Jha CK (2016) A review on the study of big data with comparison of various storage and computing tools and their relative capabilities. Int J Innov Eng Technol (IJIET) 7(1)
5. Raghupathi W, Raghupathi V (2014) Big data analytics in health: promise and potential. Health Inf Sci Syst (HISS). https://doi.org/10.1186/2047-2501-2-3
6. Zang R et al (2016) Big data for medical image analysis: a performance study. In: International parallel and distributed processing symposium workshops. https://doi.org/10.1109/ipdpsw.2016.61. IEEE
7. Wang L, Alexander CA (2015) Big data in medical applications and health care. Curr Res Med 6(1):1.8. https://doi.org/10.3844/amjsp
8. Basavaprasad B, Ravi M (2014) A study on the importance of image processing and its applications. In: National conference on recent innovations in engineering and technology. Int J Res Eng Technol 03(01). eISSN: 2319-1163|pISSN: 2321-7308
9. Mukherjee S, Shaw R (2016) Big data—concepts, applications, challenges and future scope. Int J Adv Res Comput Commun Eng 5(2)
10. Ghazi MR, Gangodkar D. Hadoop, MapReduce and HDFS: a developers perspective. In: International conference on intelligent computing, communication & convergence (ICCC). https://doi.org/10.1016/j.procs.2015.04.108
11. Vidyasagar SD (2013) A study on "Role of Hadoop in information technology era". Karnataka India 2(2). ISSN No 2277-8160
12. Patel HM et al (2015) Large scale image processing using distributed and parallel architecture. Int J Comput Sci Inf Technol 6(6):5531–5535. ISSN: 0975-9646.2015
13. Sweeney C et al. HIPI: a Hadoop image processing interface for image-based MapReduce tasks. Virginia

14. Barapatre HK et al (2015) Image processing using MapReduce with performance analysis. Int J Emerg Technol Innov Eng 1(4). ISSN: 2394-6598
15. Dubay S, Gupta YK, Soni D (2017) Role of big data in healthcare with non-invasive and minimal-invasive medical imaging modality. Int J Innov Res Comput Commun Eng (IJIRCCE) 5(3)
16. Rosslyn (2004) Digital imaging and communication in medicine (DICOM). National Electrical Manufactures Association (NEMA), Virginia
17. Hecht M. PACS—picture archiving and communication system. Vienna University of Technology, 0827459
18. Vemula S, Crick C (2015) Hadoop image processing framework. Int Congr Big Data. 978-1-4673-7278-7/15. https://doi.org/10.1109/bigdatacongress.2015.80. IEEE
19. Bhosale HS, Gadekar DP (2014) A review paper on big data and Hadoop. Int J Sci Res Publ 4(10). ISSN 2250-3153
20. Mamulwar AB et al (2015) A survey on medical image retrieval based on Hadoop. Int J Adv Res Comput Sci Softw Eng 5
21. Yang C-T et al (2015) Accessing medical image file with co-allocation HDFS in cloud. Futur Gener Comput Syst 61–73
22. Yuzhong Y, Lei H (2014) Large scale image processing research cloud. In: International conference on cloud computing GRIDs, and virtualization
23. Peter Augustine D (2014) Leveraging big data analytics and Hadoop in developing India's healthcare services. Int J Comput Appl 89:44–50
24. Chris S et al (2011) HIPI—a Hadoop image processing interface for image-based MapReduce tasks. Univ Virginia 2(1):1–5
25. Muneto Y, Kunihiko K (2012) Parallel image database processing with MapReduce and performance evaluation in pseudo distributed mode. Int J Electron Commer Stud 211–228
26. Timofei E, Andrey S (2015) Processing large amounts of images on Hadoop with OpenCV. In: CEUR workshop proceedings, vol 1513: Proceedings of the 1st Ural workshop on parallel, distributed, and cloud computing for young scientists (Ural-PDC), Yekaterinburg
27. Arsh S et al (2016) Distributed image processing using Hadoop and HIPI. In: International conference on advances in computing, communications and informatics (ICACCI), 21–24 Sept 2016. 978-1-5090-2029-4/16. IEEE
28. Fan W, Bifet A. Mining big data: current status, and forecast to the future. SIGKDD Expl 14(2)
29. Hashem I et al. The rise of "big data" on cloud computing: review and open research issues. http://dx.doi.org/10.1016/j.is.2014.07.006
30. Dhinakaran K et al (2016) Distributed image processing using HIPI. IJCTA (International Science Press) 9(12):5583–5589

Chapter 3
Image Segmentation Using Deep Learning Techniques in Medical Images

Mamta Mittal, Maanak Arora, Tushar Pandey and Lalit Mohan Goyal

1 Introduction

Modern-day computer vision technology has, by being developed on the roots of AI and deep learning techniques learned, changed and unfolded significantly in the past decade [1–3]. Now, it has applications in face recognition, image classification, picking out objects in pictures, video analysis, along with the processing of images in the robots and independently functioning vehicles (autonomous). Deep learning has the caliber of learning pattern(s) in its inputs so as to speculate and call out the classes of objects which contribute to the development of the entire object, which may be an image. The image segmentation algorithms and techniques nowadays use deep learning-based approaches to effectively understand the image structure at a level that seemed inconceivable only a decade ago [4]. They aim to develop the fact that exactly which real-world object a pixel is a constituent of, and hence the entire row and/or column of the image represents what. However, this is not necessary. The deep learning model also can be making notions and relations on clusters inside an image that need not be a specific row or column [5]. The prime deep learning frameworks that are employed for the computer vision or related tasks are Convolutional Neural Network (CNN), or some specific CNN architectures available, e.g., AlexNet, UNET, Inception, VGG, or ResNet [6].

M. Mittal · M. Arora (✉) · T. Pandey
Department of CSE, G. B. Pant Government Engineering College, Okhla, New Delhi, India
e-mail: maanakarora@gmail.com

L. M. Goyal
Department of CE, J. C. Bose University of Science and Technology, YMCA, Faridabad, India

© Springer Nature Singapore Pte Ltd. 2020
O. P. Verma et al. (eds.), *Advancement of Machine Intelligence in Interactive Medical Image Analysis*, Algorithms for Intelligent Systems,
https://doi.org/10.1007/978-981-15-1100-4_3

1.1 Image Segmentation and Its Essence

The process of image segmentation holds a vital role in the computer vision. It consists of a division of a given visual input into segments in order to rationalize the analysis of the image, as shown in Fig. 1a. After the system finishes computing over the images, the segments represent either the entire object or one of its elements depending upon the algorithm complexity. Such segments are made up of sets of pixels called "Super Pixels". The process of image segmentation classifies these pixels in comparatively sizeable components, thus eradicating the requirement to observe singular pixels as one unit, and as shown in Fig. 1b they deal in clusters of pixels known as classes or instances. Forming correct clusters is still a challenge [7, 8].

Generally, the image analysis has three major stages:

- Image classification: Categorizing the whole input image in a class like "cars", "girls", and "cheques".
- Object detection: Detect the object in the picture, form a quadrilateral (square or rectangle) around it, i.e., on a woman or on a cup. There may be multiple objects.

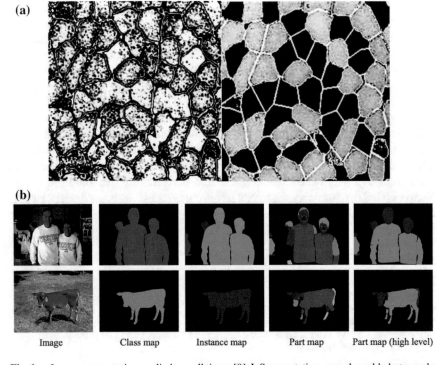

Fig. 1 **a** Image segmentation applied on cell tissue [9]. **b** Segmentation on real-world photographs

- Image segmentation: Pinning down the portions of input picture, along with interpreting that which object does the part belongs to. Segmentation lays down the basis for performing classification and object detection.

Further, segmentation process itself is divided into two wide divisions:

Semantic segmentation: Since a long time, this is termed as a process to allocate classes to each pixel in the given input image. Classify each pixel of the image in some purposeful and individual class of objects, such that one pixel gets just one class. Such classes are called as "Semantically interpretable", and along with this corresponds to object classes in the real world. As an example, all pixels covering a cat can be isolated and colored all brown. It is called "dense prediction" as it predicts a class for each pixel beforehand.

Instance segmentation: This segmentation tags every occurring instance or repetition of all subjects in a picture. This is different from semantic segmentation as this does not group the same kinds of the pixel into a single class. Instead, for example, if three women are there in the input image, instance segmentation tags every single woman as a unique instance. There are some more image segmentation methods which are used very commonly since a long time, but when compared to their other deep learning peers, sometimes these are a lot less efficient due to the use of immutable algorithms, along with the need of human proficiency and intrusion in the process [10–14].

- K-means clustering,
- Threshold,
- Histogram-based image segmentation, and
- Edge detection.

1.2   Need of Image Segmentation: Reason and Significance

In the reports of American Cancer Society from the year 2015, in the USA, approximately over 1600 humans were speculated to lose life to cancer on every passing day which corresponds to about 590,000 humans in a single year [15–17]. Even in today's age of technological advancements, cancer can be fatal if it is not identified at an early stage. If cancerous cells are detected as early as technology can and aware patients take necessary steps correctly, potentially millions of lives can be saved. The shape of the cancerous cells plays a pivotal role in governing the extent or severity of corresponding cancer. A common inference can effectively be drawn here that object detection will not be very helpful in this case. Only the bounding boxes will be generated and they will not help in identifying the shape of the cells. Image segmentation techniques prove to be helpful. They help to approach this problem in a more granular manner and get more meaningful results.

In this chapter, the authors delve into some very popular deep learning techniques used for image segmentation in medical image analysis. Differences among them are highlighted and their capabilities, limitations, along with advantages are discussed. To familiarize readers with the intricacy existent among segmentation in the medical sphere and address the challenges, the basic concepts of image segmentation are discussed first [18–21]. This involves the definition of the 2D and 3D images, description of an image segmentation problem, the image features, and the introduction of intensity distributions of sample images (medical images, etc.). Then the authors explain different preprocessing steps also consisting of image registration, bias field correction, and removal of the insignificant portions. The common validation problems are discussed in the chapters. Medical illustrations are included in the chapter to depict information wherever possible. Each topic contains relevant images for a proper explanation of the concepts.

2 Basic Concepts of Image Segmentation

This section explains the basic concepts of image segmentation used in this chapter using brain MRI scans as a subject [22].

2.1 2D and 3D Images

In a 2D space, an image could be effectively expressed as $I(i, j)$, a function, whereas inside a 3D space, it can be defined as a function $I(i, j, k)$, when $i = \{0 : M - 1\}$, $j = \{1 : N - 1\}$, along with $k = \{0 : D - 1\}$, denoting the space coordinates with M, N, and D being the lattice dimensions. Values of these functions are intensities, which are usually illustrated as gray values (i.e., 0–255) in MRI of the brain. When a 2D image is considered, an image can be classified as a set of fundamental image elements called pixels. Similarly, in 3D space, the fundamental components of an image are called voxels. Representation of pixels and voxels can be seen in Fig. 2.

The intensity values (0–255) and plane coordinates (i, j) uniquely specify the pixels, whereas voxels have unique specification of intensity values (0–255) and space coordinates (i, j, k), where i is row number, j is column number, and k is slice number in the volumetric arrangement. The average magnetic resonance characteristics that the considered corresponding element tissue exhibits generate a single numerical value that is assigned to that image element. Spatial resolution for the clarity of the image depends on the dimension of the elements taken into consideration. Figure 3 illustrates a voxel for the 3D brain volume and 2D MRI image of the same brain using a pixel for the representation of information. Also, voxel/pixel sizes do vary in different cases due to different imaging parameters, magnet strength of the MRI apparatus used, and the time allowed for entire image acquisition along with other factors. However, in conventional results provided by famed studies, voxels are often of a few millimeters, the most common being 1–2 mm large.

(a)

(b)

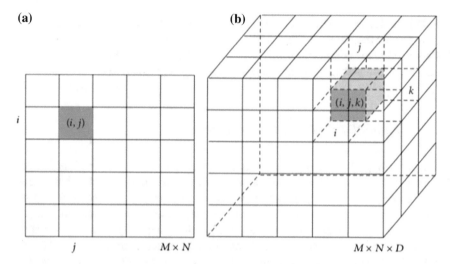

Fig. 2 Image elements in planar, volumetric representations. **a** Pixels in 2D space are represented with square lattice nodes. **b** Voxels in 3D space are represented with cubic lattice nodes

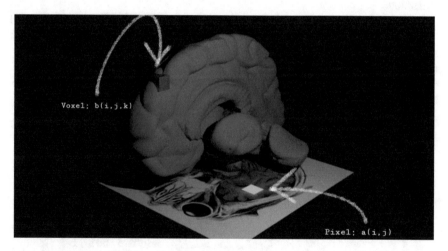

Fig. 3 Illustration of brain MRI elements. Square represents the pixel (i, j) in 2D MRI slice and cube represents image voxel (i, j, k) in 3D space

Even finer spatial resolutions can be obtained with longer scanning durations being allowed, but the increasing patient discomfort with passing time must be taken into consideration too. Allowed duration of image acquisition changes with the age groups involved too, as adults can withstand comparatively longer durations in the process in contrast to children who may get uncomfortable sooner.

As a result, adult brain MRI raw data acquisition is approximately 18–19 min often, whereas in pediatric MRI cases, the image acquisition varies in the limits of

5–15 min. After the raw data is acquired, the images are fed to a series of systems and there may be thousands of slices, which should be sorted [23]. Image segmentation is elaborated in the next section, where the authors talk about its objective as a process and the technical insights to how different types of segmentation are dealt with, etc.

2.2   Segmentation of an MRI Image

The prime objective of segmentation of the image is to mark clusters of distinct portions in it, which are visually distinct, homogeneous, and meaningful with respect to some computed properties or features, such as texture, gray level, or some other properties to facilitate easy image analysis (classification, object identification, and processing). Image segmentation is categorized into three major types, namely, threshold segmentation, region-based segmentation, and edge-based segmentation as shown in Fig. 4. MRI is an effective technique for noninvasive imaging for producing detailed images of portions inside the flesh, seemingly opaque to human eyes as it appears, but opens up like a transparent layer of jello to the magnetic resonant waves of the apparatus.

MRIs are readily preferred in the treatment of brain tumors for speculating and monitoring of patients from the inside. It can also be used to measure the tumor's size. Segmentation is a process of extracting information from an image and to group them into regions with similar characteristics. Structural brain MRI analysis includes the description and identification or classification of some particular components in human anatomy, the most important part being the classification of tissue. Each

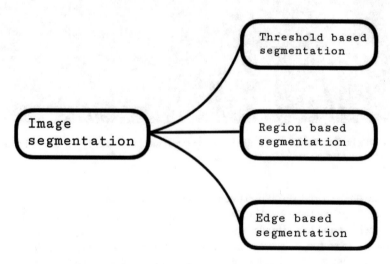

Fig. 4 Basic segmentation techniques

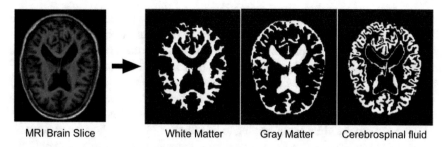

| MRI Brain Slice | White Matter | Gray Matter | Cerebrospinal fluid |

Fig. 5 Segmentation of MRI of brain shown alongside an original MRI result and the portions segmented shown in three different categories

region/portion in the picture is assigned a predefined class of tissue in the classification process. Image elements in brain MRI are classified into three primary tissue variants—White Matter (WM), Gray Matter (GM), as well as the Cerebrospinal Fluid (CSF), as shown in Fig. 5. Segmentation means a classification, whereas a classifier simply is for segmentation of input imagery. This means that problems in segmentation and classification are interconnected.

The results obtained after image segmentation have further applications in surgical planning, visualization, studying anatomical structures, and pathological regions. A large portion of research in segmentation is focused on 2D images, though it can be performed on a sequence of two-dimensional images, or three-dimensional volumetric imagery as well. The data in 3D is sourced from a bunch of MRI acquisitions in series, where every single image is subjected to segmentation separate and slicewise. This type of 3D segmenting involves postprocessing of 2D segment connection resulting in a continuous 3D volume. However, the resulting volume can be inconsistent and a non-smooth surface due to information loss in 3D space. Therefore, a need for 3D segmentation algorithms arises for accuracy in the segmentation of 3D brain MR images. Difference between 2D and 3D segmentation lies inside the concept of the pixels and voxels, the methods to process information in both of them and their neighborhoods in different dimensions over which the image features are defined. Often 2D segmentation methods can be converted to 3D spatial segmentation but this amounts to huge amount of data, and demand for huge computation power. This comes in the category of processing of big data which has its own challenges [24]. There are various techniques to process big data, and is often done on dedicated machines [25–28].

2.3   Mathematical Modeling of Pixel or Voxel Neighborhood

The mathematical modeling holds huge significance in segmentation of brain MRI. Provided that the input is not just randomly plotted noise, the pixel/voxel intensity highly depends statistically on gray intensity values of the adjacent (local) pixel(s) or voxel(s). The theory of Markov Random Field (MRF) efficiently makes available a firm base suiting the modeling of local-level features of the input picture, where the local scale patterns and trends are governing the global-scale patterns, or the "trends" as the term goes by, i.e., the local properties in the image. Lately, MRF models have previously been quite successfully integrated into existing brain MRI segmentation methods in an attempt to decrease the misclassification errors to some extent due to image noise. And regardless to say, the method has proved its mettle with the underlying algorithms proving to be effective in the application concerned.

As shown in Sect. 2.1, every pixel/voxel is capable of being represented in the lattice with a single node $\mathcal{P}$. Let x_i represent the intensity measurement of any singular voxel/pixel in the entire lattice with a position i in the image $\bar{x} = (x_1, ..., x_m)$, defined across a lattice, $\mathcal{P}$ finite in size.

Here, E stands for net count of image elements ($E = MN$ for any 2D image and $E = MND$ for any image in 3D).

Assume that, $\mathcal{N} = \{\mathcal{N}_i \mid \forall i \in \mathcal{P}\}$ denotes the neighborhood system of $\mathcal{P}$, where $\mathcal{N}$ stands for a significantly smaller neighborhood around i, (note—not including x_i). Now the nodes, which can be anything from pixels or voxels in a lattice $\mathcal{P}$, stand related to each other by a neighborhood system that can easily be defined or represented as

$$\mathcal{N} = \{\mathcal{N}_i \mid \forall i \in \mathcal{P}\} \tag{1}$$

The neighboring relation of the pixels/voxels has the properties as follows:

- Any node i is not belonging to own neighborhood as they both are mutually exclusive by definition.
- The neighboring relationship is mutual [29].

$$i \in \mathcal{N}_i \Leftrightarrow i \in \mathcal{N}_i \tag{2}$$

Hence, from Eqs. 1 and 2, the first-order and second-order neighborhoods in mathematical form are the most commonly used neighborhoods in the image segmentation.

MRF models can be represented with just a graph where $\mathcal{N}$ determines the links and $\mathcal{P}$ represents the nodes, which takes the nodes as per their neighborhood relations and connects them with the surrounding nodes. This relation derivation can prove to be helpful to a significant extent such that it can resolve the challenge of wrongly segmented portions being generated due to the presence of image noise, commonly called random noise. A graph structure in the form of nodes and edges will hence correspond to an image, wherein every lattice node corresponds to the pixels or even voxels, along with links existing between the nodes representing the mutual dependency on the basis of context, among the (any group of) voxel/pixel in the surroundings.

2.4   Analysis of Intensity Distribution in Brain MRI

Resultant intensity data of brain tissues in generated imagery is vital in segmentation; however, whenever the intensity values are corrupted or they seem to be affected by the MRI artifacts like the image noise or bias field effect or effects of partial volume, there the intensity-based methods, no matter what they are, will always lead to wrong results as they are affected by the visual garbage information in an already grayscale picture. Often the data is preprocessed for the algorithms to work on it so as to improve the resultant output quality. In the case when the extra skull structures, bias field, and background voxels are removed, the resulting histogram of an adult brain will effectively have three prime peaks in the intensity-based graph as shown in Fig. 6, which also confirms the fact that the three major types of brain components are having different intensity signatures in the MRI segmentation.

The peaks visible in Fig. 6 in order are for the following brain MRI components:

- Cerebrospinal Fluid (CSF),
- Gray Matter (GM), and
- White Matter (WM).

Sometimes, intensities of the brain may be in a form that the techniques can easily identify the parts. However, on a general note, the intensity of the brain can be taken as a piecewise constant, corrupted only by the noise, along with Partial Volume Effect (PVE) as talked about earlier. The PVE talks about the information lost due to the limitation of resolution on magnetic resonance imagery apparatus and also the limited time available to capture the scan iterations. The issue is highlighted further with relatively smaller neonatal brains, which are relatively more crowded and are hence harder to study by techniques like MRI, which are taking slices of a few mm(s) of resolution, thereby potentially missing chunks of information.

Fig. 6 Distribution of intensities in an adult brain MRI. The three peaks visible show the amount of CSF, WM, and GM, respectively, left to right, counted by number of pixels covering a certain material

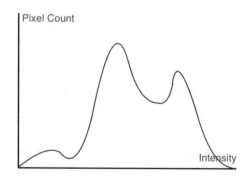

2.5   Segmented Features of MRI Image

The term "Image feature" is the collective term for the distinctive characteristics of the input picture to be subjected to segmentation. Resultant features or collections of such features are highly dependent on the underlying numerical measurements and calculations from the algorithms pre-fed to the system, and will also include the visual features and the specific shape descriptors used in the computation of segmentation portion in the said image.

All these components help the system, and hence technicians to distinguish between the background and the structures of interest. The result of image segmentation depends significantly on appropriate feature selection and accurate extraction of features. Usually, the approach that is used in the extraction of features and the MRI image classification is highly statistical, where a texture or pattern is a few features deduced from statistical math and represented in the space in vector form. These statistical features are based on gray intensity. They depend upon the first and second order of intensity statistics. Now talking about the origin/source of these statistics, the first-order statistics are derived from the gray value of the image in histogram form and involves the median, mean, intensity, and the standard deviation of the given pixel value. The image segmentation performance of the algorithms can enhance the probabilistic shape models. They are used frequently in segmentation of imagery in the medical field. These prior shape models specifically list the average shape and form variation of the object (tumor here), and are often approximated from a bunch of co-aligned pictures covering the object slices from different heights.

A prominent feature among all others in the segmentation for the identification of the tumor is the edge detection of the tumor in the brain scan, typically an MRI image, which is a ready to go algorithm and can easily be identified in relatively shorter time duration, on almost any machine. It may be a normal computer with a researcher or a professional analyst in a specialized institution. These edges are traditionally calculated out by threshold concept onto the first- and second-order spatial derivative of the pixel intensity in the taken picture. However, it must be kept in mind that the edges detected by this procedure are way more sensitive to the image noise that may creep in, and hence possibly hampers the result of the derivatives in a way or other, and hence these images often require preprocessing in the form of smoothening of the image as a significant step. In the conclusion of the algorithms talked about over here, they can be majorly assessed on their robustness, i.e., how much disturbance they can withstand and still be giving a significantly acceptable result.

Talking about robustness, another even robust technique for the detection methods oriented on edges happens to be the phase congruency method. This actually is a method constructed at the concept of frequency for detection of features. This method derives its inspiration from the methods mammals employ to identify an edge plausibly, along with studying it using the concept of local phase and energy. This method successfully explains how humans build up a psychophysical response to the edges and sharp lines in the visual feed. Mathematically, lines and edges are the spots in the image where the Fourier component is in the same maximal phase, also termed as "InPhase arrangement". Also, on a mathematical front, this is observed that the Rician distribution is the appropriate entity that governs the image noise contained within, with the proposal of this being formed from the fact that imaginary channel along with noise is following Gauss laws [30].

The probability density function is defined as

$$f_{\text{Rician}}(x) = \left(x/\left(\sigma^2\right)\right)\exp\left(-\left(\left(x^2\right) + \left(v^2\right)\right)/2\sigma^2\right) * I * \left(xv/\sigma^2\right),$$

Here, x is measured voxel/pixel density, and v stands for image intensity without noise, sigma is the standard deviation of the intermittent Gaussian noise within imaginary and real images, and also I_o stands for the zero-order, first kind of modified Bessel's function.

3 Image Preprocessing

The computer-based analysis of the MR images currently poses a challenging situation because of inconsistent intensity, changes in the range of intensity and contrast, and the noise. Hence, before proceeding for automated analysis, a few standard steps for the preparation of the data are needed to modify the imagery so as to look similar, also usually this is what is known as preprocessing, or sometimes as preparation steps. Usually, the steps taken for data preparation are in a sequence as discussed further.

3.1 Registering the Images in the System

Registration is space-based positioning of brain MRI images along with the "same axis" space, i.e., a plane with the same axis so as to align those images perfectly. Inter-patient image registration helps in making a standard and common notion of the images being registered and helps scientists and technicians in generalizing several attributes of the human body based on the common shape and design trends of the images on a general stereotaxic spatial arrangement. Oftentimes, Molecular Neuroimaging (MNI) or ICBM techniques are deployed for it. It is used to obtain complete information about the patients' health when using the images from different modes: (MRI, CT scans, PET, and SPEC techniques (or just SPECT)), and realignment is used in the process for the motion correction by the same subject (the patient) and the normalization process helps in inter-subject registration when several groups from the population are studied. Detection of transformation between the inputs is also involved in the process so that the corresponding features between two people or just two different models can be better understood. Transformation studied is usually one of either rigid or they can be affine. Rigid transformations are a Hexa-parameterized transformation consisting of translation of the models along with rotation too. If scaling and skewing to the parameters are allowed, the parameters change to 12 from 6. However, if the task aimed is the matching of the images belonging to the same subject but distinct stages of the brain or even other different subjects, a nonrigid registration is what is aimed for. But registration between two different brains is not possible when the brains include some disease or lesion as they cannot remain or maintain the original form due to the disease.

3.2   Skull Stripping/Extraction from the Image

Aimed at making the image free of skull in order to better concentrate on intracranial tissues, skull stripping takes out the skull elements from the image. Robex, BET, and SPM have been the usual methods to do this. The reason being the fact that the non-brain tissues like skull, fat, skin, or even the neck cavities have the intensities that overlap to that of the tissue in the human brain. Hence, the brain needs to be made free of such elements by some means so that only brain is processed and no extra matter is considered as the part of the brain erroneously. This step tags a voxel as either brain matter or not in a binary fashion. And the resultant can be an entirely separate picture with only voxels belonging to brain matter or binary format brain mask, which sets the value 1 for brain and a zero for the rest of the matter. Generally, the voxel of the brain is comprising the GM, WM, and CSF. The scalp, fat of the skull, skin over it, muscle for the movements, eyes, and even the bones are always classified as unwanted parts, and hence any part containing them (voxels containing) such parts are tagged as non-brain voxels. The common brain stripping uses the already available brain anatomy information beforehand so that the system can take references from the existent data, and hence decide efficiently about which voxel to ignore and which to not.

3.3   Bias Field Correction

This technique, bias field correction, also referred to as the image inhomogeneity correction is actually a low-frequency, space-varying MRI artifact, and is the rectification carried out in the image due to inconsistencies in the magnetic field. N4 approach stands out as the go to for this correction due to a solid record of performance in a large number of cases of removal of noise. The bias field arises from the space-based inhomogeneity of the magnetic field from the machine used in MRI process and also based on the sensitivity of the reception coil, sometimes on the human body to magnetic field interaction to some extent. The bias field is independent usually, but sometimes the output of the field depends significantly on the magnitude of the magnetic field when field applied is high. And in case of MRI machines as discussed here, the bias field does depend on the magnetic field when the MRI is taken at 0.5 T (unit = Tesla), the bias field is generally weak and can be neglected, but when the MRI is taken at 1.5 T or even 3 T or even higher, the bias field gets strong. In practice, trained medical experts can successfully perform analysis up to 30% of inhomogeneity. In practice, performance of the MRI analysis increases significantly in the presence of the bias field parameter because the algorithms assume the intensity homogeneity. In the literature of the MRI image segmentation technique, a number of methods with varying success and effectiveness rates have previously been proposed to provide a correction to the field bias. Earlier the manual labeling of tissue voxels was desired for this task. However, this must be kept in mind that the need of humans in the surface fitting segmentation process is also a drawback of the system as the elimination of human involvement in the entire process was the only aim of developing such algorithms in the first place. Another method having the potency to fulfill this objective is the low-pass method. This method, however, introduces unwanted entities in the picture by cutting out the original low-frequency components. Other methods include the image entropy minimization, histogram fitting of local neighbors to the global members, and the registered template methods. An alternative is the BET, called brain extraction tool, which targets for the center of gravity and expands spheres until the brain boundary is found. It works well in the T1 and T2 modes of data of good quality (in adults). But is inconsistent with neonatal brains.

3.4   Intensity Normalization

Intensity normalization is the arranging of all the involved images, intensity-wise to a specified range of 0–4095 or 0–255. Talking with respect to deep learning architectures, calculation of z-scores by subtracting the intensity mean from all the pixels is the main parameter which is desired to be optimized. MRI as everyone knows is noninvasive and hence that gives an excellent contrast between soft tissues without even the need of any minor incision, but a major drawback to be considered is the fact that here tissues do not have a singular value of intensity so it cannot use a constant value to be a specific intensity of the tissue, such as in computer tomography. Many techniques help in intensity normalization, a few are as follows:

- Histogram matching on generalized ball scale,
- STI, standardization of intensities,
- Gaussian method,
- Z-score method,
- Histogram matching on median, and
- MIMECS.

Some imaging techniques register different intensities for the same tissue even in the same subject due to the difference in orientation of the brain tissue, which is common to be in folds. And MRI may somehow detect these as different objects and not the same tissue in different orientations. These variations are machine dependent and cannot be corrected with just bias feed correction. The variations make segmentation process and hence image analysis very difficult. Hence, intensity normalization is an important preprocessing step. Now there are specific techniques based on deep learning that are used for image segmentation specifically. Section 4 talks about such techniques and elaborates their limitations, uses, and advantages.

4 Deep Learning Techniques Used in Image Segmentation

Classifying the dogs and cars using deep learning is quite easy but detecting and classifying tumors and lesions in the brain using deep learning is a challenging task. Locating the exact affected regions is a crucial step in planning the treatment and tracking the progression of various brain diseases. In this chapter, the case of brain tumors is considered. It is crucial to know where the tumor is located in order to decide whether or not to perform surgery. This section talks about some of the popular techniques used in medical image segmentation and enlists their effects, limits, and advantages. It should be noted that other techniques are also available [31].

4.1 Fully Convolutional Network (FCN)

Machine learning can learn various complex trends in data in ways that can range from traditional to abstract, but it can do various tasks [32–34]. Convolution networks are very effective in the visual mode of operation that yields feature hierarchies. In fact, they are so much preferred nowadays that the term FCS is almost omnipresent in the research field for the operations intended in the text. Convolutional networks train themselves pixel to pixel and create better results from their end-to-end training than other inferior segmentation techniques. The key focus in FCNs is the fact that these kinds of networks are the drivers and leaders of almost all the advances that are obtained in the field of recognition [35]. These networks very effectively rule out the limitation of hardcoded algorithms and they help in building better products that can recognize patterns and shapes better. These are not only improving image classification but also are making considerable progress in the logical tasks with a structured output, provided they are fed with the best data possible to train on. They account for the developments in the "Detection by Bounding Box" techniques and the part with a key-point prediction with local-level support of information better inferred from the data. Earlier approaches tend to use CNNs for the semantic segmentation where every individual pixel is labeled from its enclosing region's class; however, along with all the problem areas that various texts in the research field address in these networks, every data layer is an extremely large array of size "$a * b * c$", where "a" and "b" are space-based dimensions, and c is feature of the data. The initial layer is input picture, with "$a * b$" being pixel size and "c" the color channels. Locations in upper logical layers relate to location data in image in a path-connected manner, called receptive fields. While an ordinary deep network calculates a normal nonlinear function, a conv. network with structure layers and formation calculates a nonlinear FILTER. Any FCN normally operates on simple numerical input and gives an output of corresponding space-based dimensions but the output is resampled by means of calculations carried out in hidden network layers.

4.2 ParseNet

By using ParseNets, the global context is added to the information and the accuracy is increased in the classification process and final form of the shape is always kept in mind whenever the classification is done, and this enables the network to classify only the correct classes, and not some other class, subclass even if the numerical representations of both look similar in hidden layers [36]. Hence, the algorithms with global context know what they are doing and are not really making decisions in the dark based on just numbers being calculated on filters in hidden layers.

ParseNet Module: At the lower path, at certain convolutional layer, normalization using l2 norm is performed for each channel. At the upper path of the module, at a certain conv. layer, the "global average pooling" as shown in Fig. 7 is performed on these features and then l2 type normalization is performed, followed by un-pooling. Un-pooling is the unpacking of older information, adding new facts and making an even larger knowledge pool including new data. However, it should be noted that ParseNet has lower performance than DeepLab, etc., but is still competitive and is used till date. The prime key point that gives ParseNets the advantage they enjoy is the information about the global context. Context is vital to enhance the performance of

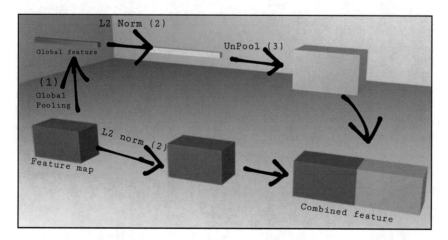

Fig. 7 The ParseNet model

detection, classification tasks, along with the use of deep learning in specific places illustrates how this can be applied to a number of different tasks. Talking about semantic segmentation, it is recommended that the system is provided the global context of the image it is working on so that it can work in an advance manner than just classifying each and every single pixel of the image as one class or the other. After this, the concepts of early and late fusion of the contexts come in, along with the normalization layers and their loops for iterative enhancement of the entire network.

4.3 Pyramid Scene Parsing Network (PSPNet)

PSPNet works on concept of scene parsing and is challenging the un-bound open vocabulary and scene with diversity [37, 38]. The goal of scene parsing and the PSPNets through it is to utilize the potential of the global context by distinct regions based on the aggregation of context of the picture by pyramid pooling. PSPNet allows a very effective architecture for pixel scale predictive capability in the picture and hence aides the classification. Objective of the scene parsing is providing every pixel of the picture a type-based tag as shown in Fig. 8. Scene parsing complete context understanding is possible. It provides prediction of the size shape and the location to the users of an element in the picture given, and hence is very similar to human visual approach. They actually locate and look at the object instead of just detecting the presence of something and calculating its coordinates.

Knowledge graphs also prove to be helpful since they infer knowledge based on previous scene information, i.e., previous layers of brain MRI. Most of the current techniques still cannot use the context of other neighboring scenes to generate output.

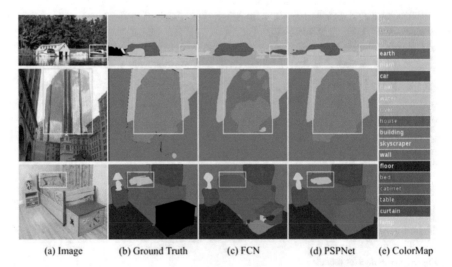

| (a) Image | (b) Ground Truth | (c) FCN | (d) PSPNet | (e) ColorMap |

Fig. 8 A comparison of PSPNet with other techniques

Initially, detection of global-level image features was done by space-based pyramid pooling with spatial statistics. But now spatial pyramid pooling networks strongly enhance the ability of image description. Here, in PSPNet, along with the usually employed FCNs, the pixel-level information is extended to explicitly engineered globally pooled attributes of the scene. This way their features enhance the ability of the algorithm to make even better predictions. Its worth noting that PSPNet(s) were the winner of the *ImageNet scene parsing challenge 2016,* and they claimed first spot in *PASCAL VOC 2012 semantic segmentation benchmark* along with winning in *urban scene Cityscapes data.* PSPNets give an extremely convincing direction for the pixel-level prediction [39–42]. Their code is available for testing and they are seen to be a way to help stereo matching on the basis of CNNs, estimating depth, etc. Hence, PSPNets are effective for interpreting sophisticated scenes.

4.4 DeepLab, DeepLabv3, and DeepLabv3+

Deep convolutional neural networks aided the vision system's performance graph to rise a significant step by pushing performance of the algorithms up by many scores, on a wide array of compatible problems of image classification, object detection, etc. The DCNNs have a very surprising edge of performance over other hand-coded algorithms in the said portion of computer vision applications. And to be specific about this success that they achieve, this is done by their inbuilt invariance to local image transformations which enables abstract data representations. This is accepted for classification operations but can be devastating to some extent for prediction tasks like semantic segmentation but spatial knowledge is desired.

This poses three main problems:

- Reduced feature resolution,
- Existence of objects at different scales, and
- Reduced localization accuracy to DCNN invariance.

This is the challenge(s) that the DeepLab overcomes or approaches to overcome. DeepLab is a very effective semantic segmentation model, entirely designed and later on open-sourced by Google in the year 2016 [43]. Multiple improvements have since been made in the model. Revised versions include the DeepLab, DeepLabV3, and DeepLabV3+. The DeepLab system once again proposes the networks that have been trained on picture classification technique, directly on the operation by applying the "atrous convolution" along with the filters that are up-sampled for dense extraction of the features. It further extends the tasks by spatial pyramid pooling, which encodes the image objects and encodes the image content at multiple scales as well. Now to produce detailed segmentation maps along with semantically accurate predictions along the boundaries of object, the ideas from the deep convolution neural networks and the fully connected random fields are combined too. As can be seen in Fig. 9, DeepLab uses Atrous convolution and fully connected conditional random field,

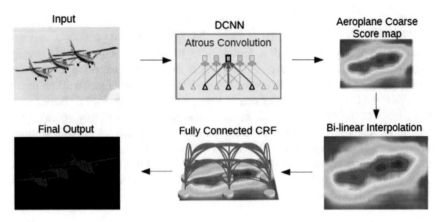

Fig. 9 DeepLab model

while the Atrous spatial pyramid pooling brings the additional piece of tech at our disposal with the next DeepLab version.

Talking of all three versions of the DeepLab, following must be discussed first:

- **Atrous convolution**: The term Atrous particularly comes from the French word "à trous", apparently meaning a hole. Hence, it is also called the hole algorithm, or going by the French naming: "algorithme à trous". Commonly used in wavelet transform because of mathematical potential and power enough to provide sufficient metrics or transformations to analyze the waves, nowadays this is applied in convolutions for deep learning. The following equation is used in Atrous convolution:

$$y[i] = \sum (k = 1, k)x[i + r \cdot k]w[k]; \quad r > 1 \text{ for atrous convolution.}$$

Atrous convolution facilitates the users in enlarging the view of the filter so as to incorporate the comparatively larger context than existing standard algorithms. In DeepLab, the LastPooling or Convolution5_1 is set to 1 to avoid the signal from being affected negatively too much. The output in this layer is much larger than the usual algorithms due to the enhancement in the size of the field of view.
- **Atrous Spatial Pyramid Pooling (ASPP)**: ASPP actually is an Atrous version of SPP, and this concept has been used in the SPPNet [44]. As the object of the same class can have different scales in the image, ASPP helps to account for objects of different scales (sizes) and this can help to improve the accuracy of the underlying algorithm by enhancing the output being thereby making ground truth look similar to it.
- **Fully Connected Conditional Random Field (FCCRF)**: This FCCRF is applied at the resultant output of network after the bilinear interpolation as shown below [45]:

$$E(x) = \sum (\theta_i)(x_i) + \sum \theta_{ij} x_i, x_j$$

where

(1) $\theta_i(x_i) = -\log P(x_i)$
(2) $\theta_{ij}(x_i, x_j) = \mu(x_i, x_j)[\omega_1 \exp\{-((p_i - p_j)_2)/2\sigma_{\alpha2}) - ((I_i - I)_2)/2\sigma_{\beta2}\} + \omega_2 \exp\{(p_i - p_j)_2)/2\sigma_2\}]$

In the formulae above, the first term θ_i is the log probability. The second term, θ_{ij} is a filter term. In the brackets of the filter, it is the weighted use of the two kernels. The first kernel depends on bilateral filter made from difference of the pixel value and the difference of pixel position. Bilateral filter has the edge-preserving property, i.e., it can preserve edges in the calculations. The second kernel only depends on the pixel position difference, which actually is a Gaussian filter. The σ and w are calculated by cross validation of both the equations. However, the CRF is a postprocessing step which makes DeepLab version 1 and DeepLab version 2 become not an end-to-end learning product. It is not found in further versions of Deeplab.

4.5   UNET

Olaf Ronneberger developed the UNET for medical image segmentation. CNN has a significant reputation when it comes to image segmentation because it generates considerably good results in simpler image segmentation problems. It does not work so well when it comes to the intricate problems of image segmentation. Here comes UNET in the picture of image segmentation. It was initially developed for image segmentation in the medical field. Eventually, the good results shown by UNET made it useful in many other fields too.

4.5.1   Concept Backing UNET

Learning the mapping of features of the image and using it to create increasingly refined feature mapping are the prime ideas that revolve around CNN. For further classification, the image gets converted to vector. This makes it work well in classification problems. Now coming to image segmentation, an image also needs to be reconstructed from its vector, along with the conversion of the feature map to vector form. Since the conversion of a vector to an image is more tedious than converting an image to a vector, this corresponds to a gigantic task. This is the problem around which the whole idea of UNET revolves around. The feature mapping of an image is used during the conversion of an image into a vector, and the very same mapping can be used to convert it back to an image. This is the main concept that backs UNET. To convert vector to image (after segmentation), the feature maps that were used for contraction are used again. This process would protect the constitutional stability of the image, which in turn would enormously reduce the distortion factor.

4.5.2   UNET Architecture

The name UNET is very well justified by its "U"-shaped architecture [46]. There are three sections in the UNET architecture:

- The contraction section,
- The bottleneck section, and
- The expansion section.

Contraction blocks are the building blocks of the contraction section. Each one of these blocks takes up an input on which double 3 * 3 convolution layers followed by a 2 * 2 max pooling are applied. For the architecture to learn the complex structures successfully, the number of kernels or feature maps gets doubled after each block. Between the contraction layer and the expansion layer arbitrates the lower most layer which uses double 3 * 3 CNN layers which are then followed by a 2 * 2 up-convolution layer. Expansion section is the core of this architecture.

It also consists of a number of expansion blocks just like the contraction layer. The input is then passed by each block to dual 3 * 3 CNN layers which are then followed by 2 * 2 up-sampling layers. In order to maintain symmetry, the number of kernels used by convolution layer gets halved after each block. The feature maps of the corresponding contraction layer get appended to the input. This ensures that for the reconstruction of the image, the features used would be the same as the features learned when contracting image. The number of contraction blocks and expansion blocks is same. Thereafter, the resultant mapping passes via one more 3 * 3 CNN layer having number of kernels equal to the number of segments required. The architecture of the same can be seen in Fig. 10.

All these techniques discussed above are actively deployed to process the images containing brain tumors. Some methods work better than others, and sometimes the input image defines the quality of results. But all these techniques have proved to be effective to at least one type of image and can segment the input effectively [47]. The effectiveness of any technique discussed depends on several other factors like the structure of the entire model used for segmentation, the classifiers used, the structure of input data and amount, consistency, and the quality of the data available. Apart from the techniques discussed, preprocessing the way the data is arranged in an ordered form is required, and after that it is fed to a segmentation model using one of the techniques discussed above. Only then can the segmented output be generated. However, more postprocessing may be involved to refine the outputs to suit a specific requirement.

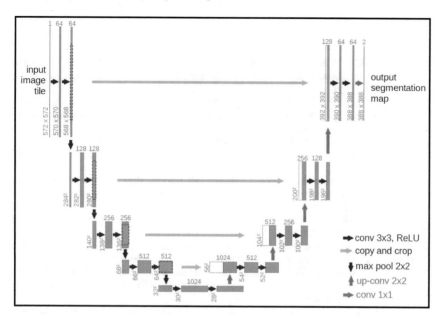

Fig. 10 UNET architecture

5   Conclusion

Image segmentation is the most difficult step in image processing and has been a vital and active research field since the past few years. It holds utmost importance in many medical applications like computer-aided diagnosis, image registration, and relevant fields. Its application also includes 3D visualizations. For brain MRI segmentation, there prevails a spread of state-of-the-art techniques and smart previous information. Still, it may be a difficult task and there is a necessity to boost the accuracy, precision, and speed of segmentation techniques for future analysis. Initially, in this chapter, the important concepts of image segmentation which are necessary for medical analysis have been discussed, including 2D and 3D image definition, modeling of neighborhood information, image features, and intensity distribution. After this, the image preprocessing steps which are necessary for preparing the data have been elaborated, including bias field correction, image registration, skull stripping, and intensity normalization. Lastly, the significance of deep learning and its various techniques used in image segmentation like FCN, ParseNet, PSPNet, DeepLab, DeepLabv3, DeepLabv3+, and UNET have been elaborated.

References

1. Mittal M, Verma A, Kaur I, Kaur B, Sharma M, Goyal LM, Roy S, Kim T (2019) An efficient edge detection approach to provide better edge connectivity for image analysis. IEEE Access 7(1):33240–33255
2. Kaur S, Bansal RK, Mittal M, Goyal LM, Kaur I, Verma A, Son LH (2019) Mixed pixel decomposition based on extended fuzzy clustering for single spectral value remote sensing images. J Indian Soc Remote Sens 1–11
3. Avendi MR, Kheradvar A, Jafarkhani H (2016) A combined deep-learning and deformable-model approach to fully automatic segmentation of the left ventricle in cardiac MRI. https://www.doi.org/10.1016/j.media.2016.01.005
4. Akkus Z, Galimzianova A, Hoogi A, Roobin DL, Erickson BJ (2017) Deep learning for brain MRI segmentation: state of the art and future directions. https://link.springer.com/article/10.1007/s10278-017-9983-4
5. Saxena A, Mittal M, Goyal LM (2015) Comparative analysis of clustering methods. Int J Comput Appl 118(21):30–35
6. Image segmentation in deep learning: methods and applications. https://missinglink.ai/guides/neural-network-concepts/image-segmentation-deep-learning-methods-applications/
7. Mittal M, Sharma RK, Singh VP (2011) Random automatic detection of clusters. In: IEEE international conference on image information processing, ICIIP-2011, JUIT Solan, 3–5 Nov 2011, proceedings of IEEE Delhi section, pp 91
8. Mittal M, Sharma RK, Singh VP, Goyal LM (2016) Modified single pass clustering algorithm based on median as a threshold similarity value. Collaborative filtering using data mining and analysis. IGI Global, pp 24–48
9. Despotovic I, Goossens B, Philips W (2015) MRI segmentation of the human brain: challenges, methods, and applications. Comput Math Methods Med 2015, Article ID 450341
10. Mittal M, Goyal LM, Hemanth DJ, Sethi JK (2019) Clustering approaches for high-dimensional databases: a review. WIREs Data Min Knowl Discov (Wiley) 1–14. https://doi.org/10.1002/widm.1300
11. Mittal M, Sharma RK, Singh VP (2019) Performance evaluation of threshold-based and k-means clustering algorithms using iris dataset. Recent Pat Eng 13(2)
12. Goyal LM, Mittal M, Sethi JK (2016) Fuzzy model generation using subtractive and fuzzy C-means clustering. CSI Trans ICT (Springer) 129–133
13. Mittal M, Sharma RK, Singh VP (2015) Modified single pass clustering with variable threshold approach. Int J Innov Comput Inf Control 11(1)
14. Mittal M, Sharma RK, Singh VP (2014) Validation of k-means and threshold based clustering method. Int J Adv Technol 5(2)
15. Early detection of cancer. https://www.who.int/cancer/detection/en/
16. Gliomas DA (2009) Recent results in cancer research, vol 171. Springer, Berlin
17. Cancer facts and figures 2015, American Cancer Society. https://www.cancer.org/content/dam/cancer-org/research/cancer-facts-and-statistics/annual-cancer-facts-and-figures/2015/cancer-facts-and-figures-2015.pdf
18. Norouzi A, Shafry M, Rahim M, Altameem A, Saba T, Ehsani RA, Rehman A, Uddin M (2014) Medical image segmentation methods, algorithms, and applications. IETE Tech Rev 31(3):199–213
19. Chen LC, Papandreou G, Kokkinos I, Murphy K, Yuille AL (2015) Semantic image segmentation with deep convolutional nets and fully connected CRFs. In: ICLR. https://arxiv.org/abs/1412.7062
20. Alberto G, Victor V et al (2018) A survey on deep learning techniques for image and video semantic segmentation. Appl Soft Comput
21. Geiger D, Yuille A (1991) A common framework for image segmentation. IJCV 6(3):227–243
22. Akkus Z, Galimzianova A, Hoogi A et al (2017) Deep learning for brain MRI segmentation: state of the art and future directions. J Digit Imaging 30:449. https://doi.org/10.1007/s10278-017-9983-4

23. Sharma S, Singh P, Mittal M (2017) S-ARRAY: highly scalable parallel sorting algorithm. In: Data intensive computing applications for big data. IOS Press Netherland
24. Bhatia M, Mittal M (2017) Big data & deep data: minding the challenges. In: Deep learning for image processing applications. IOS Press, Netherland, pp 177–193
25. Singh A, Mittal M, Kapoor N (2018) Data processing framework using apache and spark technologies in big data. In: Big data processing using spark in cloud. Studies in big data, vol 43. Springer, pp 107–122
26. Mittal M, Balas VE, Goyal LM, Kumar R (2018) Big data processing using spark in cloud, vol 43. Springer Nature Pte Ltd., Singapore
27. Mittal M, Hemanth DJ (2018) Big data for parallel computing. In: Balas VE, Kumar R (eds) Advances in parallel computing series. IOS Press
28. Kaur P, Sharma M, Mittal M (2018) Big data and machine learning based secure healthcare framework. Procedia Comput Sci (Elsevier) 132:1049–1059
29. Kaur B, Sharma M, Mittal M, Verma A, Goyal LM, Hemanth DJ (2018) An improved salient object detection algorithm combining background and foreground connectivity for brain image analysis. Comput Electr Eng 71:692–703
30. Glasbey CA, Horgan GW (1995) Image analysis for the biological sciences. Wiley, New York. ISBN: 0-471-93726-6
31. Alqazzaz S, Sun X, Yang X et al (2019) Automated brain tumor segmentation on multi-modal MR image using SegNet. Comput Vis Media. https://doi.org/10.1007/s41095-019-0139-y
32. Mittal M, Goyal LM, Sethi JK, Hemanth DJ (2018) Monitoring the impact of economic crisis on crime in India using machine learning. Comput Econ (Springer) 1–19
33. Shastri M, Roy S, Mittal M (2019) Stock price prediction using artificial neural model: an application of big data. SIS, EAI. https://doi.org/10.4108/eai.19-12-2018.156085
34. Bell S, Upchurch P, Snavely N, Bala K (2014) Material recognition in the wild with the materials in context database. arXiv:1412.0623
35. Long J, Evan S, Trevor D (2015) Fully convolutional networks for semantic segmentation. In: 2015 IEEE conference on computer vision and pattern recognition (CVPR). arXiv:1605.06211v1 [cs.CV], 20 May 2016
36. Liu W, Rabinovich A, Berg AC (2016) ParseNet: looking wider to see better. In: ILCR. https://www.cs.unc.edu/~wliu/papers/parsenet.pdf
37. Zhao H, Shi J, Qi X, Wang X, Jia J (2016) Pyramid scene parsing network. arXiv:1612.01105
38. Pyramid scene parsing network, CVPR2017. https://github.com/hszhao/PSPNet
39. Zhou B, Zhao H, Puig X, Fidler S, Barriuso A, Torralba A (2016) Semantic understanding of scenes through the ADE20K dataset. arXiv:1608.05442
40. Everingham M, Gool LJV, Williams CKI, Winn JM, Zisserman A (2010) The pascal visual object classes VOC challenge. In: IJCV
41. Cordts M, Omran M, Ramos S, Rehfeld T, Enzweiler M, Benenson R, Franke U, Roth S, Schiele B (2016) The cityscapes dataset for semantic urban scene understanding. In: CVPR
42. https://towardsdatascience.com/review-pspnet-winner-in-ilsvrc-2016-semantic-segmentation-scene-parsing-e089e5df177d
43. Chen J, Papandreou G, Kokkinos I, Murphy K, Yuille AL (2016) Deeplab: semantic image segmentation with deep convolutional nets, atrous convolution, and fully connected CRFs. arXiv:1606.00915
44. He K, Zhang X, Ren X, Sun J (2014) Spatial pyramid pooling in deep convolutional networks for visual recognition. In: ECCV
45. Chen LC, Papanderou G, Kokkinos I, Murphy K, Yuille A (2014) Semantic image segmentation with deep convolutional nets and fully connected CRFs. arXiv:1505.04597v1
46. Ronneberger O, Fischer F, Brox F (2015) U-Net: convolutional networks for biomedical image segmentation. arXiv:1505.04597v1 [cs.CV], 18 May 2015
47. Isin A, Direkoglu C, Sah M (2016) Review of MRI-based brain tumor image segmentation using deep learning methods. In: 12th international conference on application of fuzzy systems and soft computing, ICAFS 2016, 29–30 August 2016, Vienna, Austria

Chapter 4
Application of Machine Intelligence in Digital Pathology: Identification of Falciparum Malaria in Thin Blood Smear Image

Sanjay Nag, Nabanita Basu and Samir Kumar Bandyopadhyay

1 Introduction

Pathology is a branch of medicine that combines the science of disease, their cause, effect and diagnosis. A pathologist determines the cause of a particular disease conditions based on certain prescribed tests (chemical/clinical/microscopy), for accurate diagnosis and provide relief to the suffering patient. Most of these test are conducted by automated equipment using body fluids/tissue sample extracted from the patient. Microscopy is utilized for disease determination in cases where parasitic invasion within tissues occurs and to locate abnormality in histological/cytological body samples.

Microscopic examination of cellular and histological samples is widely used as a basis for disease detection. However, with the introduction of advanced digital microscope and high-resolution scanners, the approach towards pathology had a paradigm shift towards 'virtual microscopy' as an innovation in diagnostic workflow. Handling of glass slides across the labs is cumbersome and susceptible to loss of the slide or decreased quality of the specimen. The associated turnaround time (from the sample collection to report generation) is time-consuming. With the increase in reliability of digital equipment like digital imaging technologies, computer hardware

S. Nag (✉)
A. K. Choudhury School of Information Technology, University of Calcutta, JD-2, J D Block, Sector III, Salt Lake, Kolkata 700106, West Bengal, India
e-mail: sanjaynag75@gmail.com

N. Basu · S. K. Bandyopadhyay
Department of Computer Science and Engineering, University of Calcutta, JD-2, J D Block, Sector III, Salt Lake, Kolkata 700106, West Bengal, India
e-mail: nabs.basu@gmail.com

S. K. Bandyopadhyay
e-mail: 1954samir@gmail.com

© Springer Nature Singapore Pte Ltd. 2020
O. P. Verma et al. (eds.), *Advancement of Machine Intelligence in Interactive Medical Image Analysis*, Algorithms for Intelligent Systems,
https://doi.org/10.1007/978-981-15-1100-4_4

and software, there has been persistent acceptance of digital pathology in the medical community [1].

Conversion of a biological specimen in a glass slide to an image is referred to as virtual microscopy. This is the first step towards digital pathology. High-resolution Whole Slide Scanners (WSI) perform this task at resolutions of 40–60× magnification. A high-resolution image obtained of the whole slide can be extended to the size of a tennis court when projected/printed at 300 dpi resolution [2]. The information can be archived for training and technical education purpose, telemedicine applications, primary/secondary diagnosis or for a second opinion, review of consultation and for quality assurance mechanisms [3]. The consulting pathologist will view the slide images on high-resolution monitors that are specialized for medical purpose. The pathologist will be able to analyse remotely at the time of his preference without the sample being affected/destroyed or stained sample getting discoloured. The images can be distributed among consultants for double review or expert review and can result in faster workflow in pathological laboratories.

Digital pathology using WSI has been granted certification by Food and Drug Administration (FDA) [4] in 2017 for its application in primary diagnosis of disease. Medical images obtained from different equipment/vendors and of different modalities have been standardized to the DICOM (Digital Imaging and Communications in Medicine) standard. The DICOM images and Picture Archiving and Communication system or PACS system are being evolved to accommodate WSI for medical diagnosis. A working group, WG-26, was established by DICOM for this very purpose in 2005 [5]. Several studies indicate that the performance of WSI and glass slide is similar [6].

The use of digital imaging has also opened a new technological dimension for pathology. The images can be processed with Artificial intelligence and machine learning algorithms in computer-aided diagnosis systems or CAD system. Such a system can be used to identify abnormalities independent of human intervention and is referred as the third revolution in pathology [7].

1.1 Malaria

The disease malaria has been recorded as the oldest reported disease and accumulated the largest number of mortality over the ages to be considered as the deadliest of the human infectious diseases and is a primary cause of child mortality. The disease is predominantly widespread in tropical climatic regions that are backward and underdeveloped. Female Anopheles mosquito is the sole vector for the protozoan infectious disease that affects human population. The disease attains epidemic proportions in remote rural areas within a very short time. Prevention of the disease by curtailing the breeding grounds of the vector has proved futile in most areas of the world. The disease can only be managed with early detection, confirmation of species type, stage and density of parasite within the human blood.

Malaria is often designated as the 'King of Diseases' [8, 9] due to its predominance in the world of infectious diseases causing human mortality. The disease was a cause of dread for ages and still continues to be a threat to humankind for being the fifth deadliest infectious disease [8]. The disease was known to the medical practitioners for more than 50,000 years [10]; however, its effect on humans was identified at a later stage. The disease was known to almost all of the ancient civilizations of the world including Chinese, Indian, Greek, Babylonians and Romans. Various historical documents, scriptures and medical text finds the mention of a disease that is seasonal, causes intermittent fever, originates from marshy unhygienic areas, vector-borne and is responsible for mortalities of epidemic nature [9]. The disease finds its ways in the medical writings of Hippocrates, the Greek physician and also in the vedic texts, 'Sushruta Samhita', the Canon of Medicine and the ancient Chinese 'Nei Ching' [9]. The origin of the disease is pre-historic and its transmission to humans happened millions of years ago [11]. In the book of Sallares [12], the origin of the name for the disease and its impact on the ancient Italian society was discussed. The etymology of malaria originated from Italian 'mal aere', that has the meaning of 'bad air' found in the book entitled 'Scitture della laguna' of Marco Cornaro [13], published in 1440 in the city of Venice. However, the term malaria got its introduction in English literature from the letters of Horace Walpole to his cousin [13]. The word got associated to the specific disease in the publication of a book by Guido Baccelli called 'La Malaria di Roma' in 1878 [13].

1.2 Biological Aspect of Malaria Disease

The parasitic protozoan of genus Plasmodium is responsible for this deadly disease. There are several species that are known to infect human beings (five known to infect humans), the most serious type of malaria, called the 'Malignant Tertian Malaria' is affected by Plasmodium falciparum. The female mosquito (Anopheles) is key to the widespread dispersion of the malaria disease. Infected mosquitoes that had sucked the blood of an infected patient becomes the carrier of the disease without itself being affected by the disease. The parasites travel to the mosquito gut and multiply their numbers to create a pool of parasites that migrates to the saliva of the vector for further transmission to healthy human host. The blood-sucking event of such infected mosquito inoculates the parasite in the human vascular system. The plasmodium parasite finds the new host to propagate its progeny. The human host often becomes a reservoir for the parasite where they live multiple generations and increase their numbers at an exponential rate at the cost of human red blood cells. The parasites after entering the human bloodstream, travel to the liver. The hepatic tissue is the first region of the body that gets infected. The symptoms of malaria are not observed at this point of time. During this gestational period the parasite multiplies itself for further invasion of the vascular tissue and further defeating the host immune system. The host especially children with malnutrition are susceptible due to poor immunity. After invading the red blood cells, the parasites start their asexual life cycle. The parasites mature, destroys the invaded blood cell and the cycle repeats every 48/72 h. The cell bursts to release a colony of newly formed parasites that is ready to carry on further invasion of new, healthy host red blood cells. The parasite exhibits polymorphic life forms within the human host indicating different life cycle stages within the host. The bursting of cells typically associated with severe shivering and cramps exhibited by the patient. Typical Malaria symptoms of intermittent high fever are manifested by the host that occur in cycles, repeating the process every 2–3 days. The parasite multiplies by destroying the host cells resulting in blood loss and anaemia-related issues. Progression of the disease, if not controlled, leads to severe blood loss conditions, organ failure and mortality. The most virulent species, *Plasmodium falciparum* among the five species of parasites known to infect humans; infection spreads to fine blood capillaries of brain, where they block the flow of blood, resulting in rapid loss of brain functioning and mortality. The other species causes severe illness and suffering but mortality is less reported. This species, *P. falciparum*, is mostly reported in sub-Saharan Africa and Oceania. However, latest reports indicate that in India this species has progressed to be the leading cause of malaria infection relegating *P. vivax* to the second place.

The US government [14] under the aegis of CDC, 'The Centre for Disease Control and Prevention', create awareness for this disease along with other infectious disease. A ring form trophozoite in 100× and 40× resolutions along with a mature trophozoite depicting an amoeboid form [15] is shown in Fig. 1. Figure 2a shows the crescent-shaped gametocyte that typically distinguishes the falciparum species from the rest of the species of Plasmodium genus. These mature sexual forms initiate the sexual life cycle that happens within the female mosquito gut after a bite and sucking blood from a malaria infected person. Figure 2b, c shows the mature schizont that forms after repeated mitotic divisions of the parasite within the host cell. The bag full of spores contains 8–16 merozoites, which are released in the bloodstream to transmit the parasite to other healthy cells all over the vascular tissue.

(a)   **(b)**   **(c)**

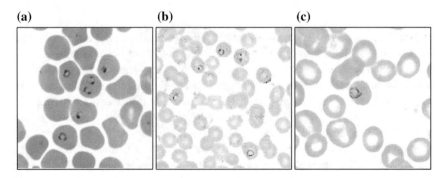

Fig. 1 Image of *P. falciparum* showing the presence of ring form trophozoite in thin smear slide images. **a** Multiple rings within RBC cell at 100× magnification, **b** rings of the parasite in 40× magnification and **c** mature form of trophozoite. Image obtained from DPDx image library of CDC [15]

(a)   **(b)**   **(c)**

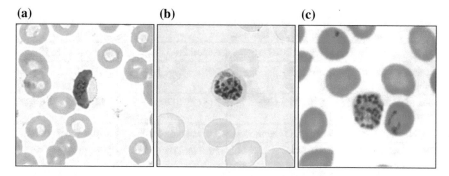

Fig. 2 Image of *P. falciparum* showing the presence of gametocyte and mature schizont in thin smear slide images. **a** Unique crescent-shaped sexual gametocyte that are found in peripheral bloodstream. **b, c** Mature form of asexual form containing a bag full of sporozoites waiting to burst from the host cell and disperse to infect other healthy cells in the vicinity. Image obtained from DPDx image library of CDC [15]

1.3 Malaria Statistics

The WHO, 'World Health Organization', under the aegis of United Nations collects malaria-related data from all over the world. They are the pioneers of malaria prevention and eradication programs in both poor and underdeveloped nations in the world that are worst affected by the disease. Malaria-related data is obtained from all member nations. The data is obtained by WHO, working closely with government hospitals and care centres from urban to rural and even from remote and inaccessible locations. The data is compiled and projected on the official website. The reports released each year reflect the efforts made by the member states towards malaria eradication, prevention and awareness. Table 1 summarizes the statistical details for last 3 years.

1.4 Malaria Diagnosis

The determination of the presence of malaria parasite involves analysis of patient blood to identify the presence of antigens or by-products in the blood sample. The diagnostic efficacy is dependent on several factor related to the parasite. Some important factors affecting diagnosis are the presence of different species and forms of the parasite, the pathogenicity exhibited by each of the different species effects diagnosis. Similarly, other factors like the amount of parasite transmission, movement of population as disease carriers, degree of immunity and reaction of drug and resistance towards known medications affect the volume of infection or parasitaemia. Factors that further contribute are 'recurrent malaria', 'persisting viable or non-viable' parasitaemia, and even the degree of penetration by the parasite into the host tissues, the use of preventive medicines and often use of presumptive medication based on

Table 1 A comparative analysis of malaria statistics for years 2016–2018 documented and distributed annually by World Health Organization

	Report 2018 [16]	Report 2017 [17]	Report 2016 [18]
Data representation (year)	2017	2016	2015
No. of cases globally in millions	219 (95% CI)	216 (95% CI)	212
Highest no. of cases	African Region 92%	African region 90%	African region 90%
Percentage of case in India (rank)	4% (4th)	7%< (NA)	7%< (NA)
Mortalities globally	435,000	445,000	429,000
Percentage of fatalities (5<)	61%	Not reported	70%
Percentage mortalities in India	4%	6%<	6%<
Complete elimination	Paraguay	Kyrgyzstan, Sri Lanka	10 countries reported no fatalities

clinical correlation can all influence the identification and parasitaemia calculation in a diagnostic test.

The laboratory techniques involve collection of blood sample and using scientific equipment or chemicals to identify the presence of malaria parasite. Use of conventional light microscopy is the primary method on blood smear slides that are often stained with Giemsa/Wright's/Field's stains. Staining of specimen slides for differential highlighting of parasite regions was formulated for the first time by Romanowsky in 1891 to achieve better diagnosis. The use of peripheral blood smear on glass slide (both thick and thin smear) and processing these slides with Giemsa-staining; initial screening for presence of parasite determined using the stained thick smear slide and using the thin smear slide for species identification. This process remains the 'gold standard' for laboratory technique for detection of malaria [19]. Though both the thick smear that has a better abundance of parasites as well as the more well define non-overlapping and single cell layered thin smear slide can be used for parasitaemia calculation [20]. The slides prepared by adhering to the best practice defined by WHO can be used for examining under the light microscope by a trained microscopist or pathologists [20]. The WHO has documented the guidelines that are needed to be followed in field microscopy for malaria detection and also presented a detailed list of the procedures to be carried out by laboratory technicians [20].

This simplest of the laboratory procedure has several advantages. The process workflow does not require extensive training and hence manageable by limited training of technicians. The simplicity of the process that does not involve costly equipment with high maintenance cost of such system, thus making it economically viable. Visual differentiation of normal and parasite-infected RBC by a trained microscopist/pathologist makes use of least of technological overheads. The preparation of slides are however laborious, time taking and requires skill by laboratory technicians. The process of detection and quantification is complex and challenging for the pathologists when there is a low level of infection (low parasitaemia in case of early detection). The primary disadvantage of such system is low sensitivity at detection level and species identification at low parasitaemia.

Other diagnostic methods that are rarely used are quantitative buffy coat test that prepares a concentrate of parasite region from collected blood. Serological test using fluorescent microscopy tries to find the presence of malaria antigen in blood. Similarly, molecular methods like Polymerase Chain Reaction (PCA) technique to trace malaria DNA are extremely accurate but are expensive systems. Flow Cytometry Methods (FCA) tries to trace the presence of Haemozoin that is formed on destruction of RBC. Similarly, DNA Microarrays and Mass Spectrophotometry are methods that are rarely used because of the exorbitant cost of equipment and maintenance. Rapid diagnostic test kits have become very popular due to their low cost and easy use in detection process. However, they can perform detection only and cannot contribute to stage differentiation, parasitaemia calculation and show low sensitive to species differentiation.

The extension of PBS technique with digital microscope/whole slide scanners to obtain a set of digital images is transforming pathology to the era of digital pathology. Combining the new method with CAD systems provides a complete system for diagnosis as an alternative to manual systems. Such a system can be used redundantly with manual method for additional accuracy in detection process. A digital pathology case study of *P. falciparum* detection is presented in this chapter to support the usefulness of this new technique.

2   Literature Review

Research work related to the medical/clinical methods for malaria parasite detection is dated to pre-historic times. However, research work on malaria in context to medical image analysis is a relatively new concept. As already discussed with the advent of digital imaging the focus has shifted to image analysis as a method for detection of parasite within the thin blood smear images. Quite a number of research works were focused on identifying malaria parasites as well as the type of species and life cycle stages from image/s. Most of the research work conducted during the early years of digital pathology was to identify simple graphical software tools for better understanding of the images. Subsequently, the use of image processing algorithms for pre-processing and segmentation, as a part of image analysis, was also utilized for parasite identification. Most of these systems were based on rule-based image processing techniques. Image segmentation techniques like image binarization, edge detection and colour-thresholding was widely used by authors for parasite segmentation. Morphological operators and Morphometry has also been effectively implemented for segmentation and parasite detection. However, recent works have shown a shift towards machine learning approach with supervised and unsupervised learning approach in both image segmentation and parasite identification. Authors have derived host of features including colour channels of different colour model, textural and morphological features from the images. Using these features they have implemented supervised/unsupervised machine learning algorithms to segment/classify parasite and cellular components. Some of the research works are focused on identifying the infection, stage and/or species directly from the slide images without segregating the infected region as region of interest. For development of an effective decision support system for the pathologists, the CAD system should be able to screen for infection, extract region of interest, segregate from normal cells, extract the parasite region, differentiate between species and stage, and quantify the infection regions and calculate parasitaemia. Some recent contributions in this domain of research is discussed in the following paragraphs in brief.

The images taken or prepared by the various authors show illumination and colour density variations that will vary both intra- and inter-dataset. The execution of a set of algorithm will require similar image conditions for the effectivity of the algorithm. Hence, certain correction are required essentially to maintain parity for all the images for each dataset and among images within a particular dataset. Illumination correction has been an integral part in many research work like authors Tek et al. [21, 22], used a reference image for subtraction of the test image from the known reference image. Similarly, Das et al. [23] employed the 'Gray World assumption' for illumination correction.

Image noise elimination is a major task in pre-processing as the presence of noise hinders accurate detection process. Median filtering, Gaussian filter and histogram equalizations are some of the common techniques available to reduce the presence of noise in images. Ruberto et al. [24], Ross et al. [25], Anggraini et al. [26], Das et al. [23], Rosado et al. [27], Predanan et al. [28], Bahendwar et al. [29] and Nugroho et al. [30] have used Median filters. While Dave et al. [31] and Savkare et al. [32] used a combination of median filtering with Laplacian filter for suppressing noise in the images and enhancement of the edge region for better segmentation outcome. Histogram equalization method by using a combination of adaptive and local histogram equalizations was proposed by Sio et al. [33], whereas, Gaussian filter was employed by Arco et al. [34]. Ahirwar et al. [35] in his research method used SUSAN filter algorithm for noise reduction. In the research citation by Reni et al. [36], the authors have performed contrast enhancement while converting to grayscale image by calculating optimum weights for R, G and B channels.

Research work has emphasized on enumeration post segmentation. Cell clump removal is essential for getting accurate enumeration results. The red blood cell are expected to be non-overlapping for the purpose of counting but often they remain in a group overlapping each other so the process of de-clumping is required to be performed. A rule-based binary splitting algorithm as proposed by Kumar et al. [37] and Sio et al. [33] whereas, Diaz et al. [38] performed template matching by sliding a template over the clump region to separate each of the cells present within the clump. Watershed transform method for clump splitting was proposed by Preedanan et al. [28], while Bairagi et al. [39] performed watershed transform with Euclidean distance transform.

The accuracy of parasite detection depends upon good segmentation of the images. This helps to differentiate between parasitic/non-parasitic regions within the image. Mathematical morphology and/or granulometry is a common segmentation procedure when the size or shape of a particular object is known. Since the cellular material can be distinguished by their characteristic shape and size, Ruberto et al. [24], Tek et al. [21], Das et al. [23], Ahirwar et al. [35], Prasad et al. [40], Dave et al. [31] and Savkare et al. [32] have utilized this information with morphological operators for red blood cell segmentation. This was also used for obtaining the foreground of the image for further consideration. Reni et al. [36] preserved maximum foreground information and performed artefact removal using a modified form of 'Morphological Closing' operation. Image binarization reduces the colour/grayscale parameters to binary colour and can be achieved by using Otsu thresholding. Very often the popular Otsu method is found effective in performing image segmentation. This idea was extended to digital pathology by Das et al. [23], Ahirwar et al. [35], Anggraini et al. [26], Mehrjou [41], Rosado et al. [27] and Savkare et al. [32]. Similarly, Dave et al. [31] have used grayscale histogram along with Kurtosis for determination of whether the histogram depicts a uni/bimodal histogram before performing Otsu thresholding. Bairagi et al. [39] employed Otsu thresholding on each individual colour channels of images in RGB and HSV colour model. Preedanan et al. [28] adopted an adaptive histogram thresholding technique for segmentation of the foreground image. Use of watershed transform with distance transform is found to be effective in segmentation outcome and have been used by Savkare et al. [32] and Mehrjou [41] for segregation of red blood cells from the entire image. A variation to classical watershed transform called marker controlled watershed transform was proposed by Das et al. [42] and Khan et al. [43]. Another image binarization technique called Zack thresholding method was implemented by Damahe et al. [44]. The author considered the 'V' or 'value' component of HSV image for thresholding to achieve segmentation. Purwar et al. [45] have used active contour model for segmentation of foreground image.

Parasite region identification has been performed by a group of authors involving rule-based image processing methods. Ruberto et al. [24] proposed two distinct methods, using morphological thinning operation for obtaining a skeleton of the ring form of the parasite and subsequently implementing colour histogram similarity measure to determine the infected parasite region. A statistical Variance based approach and use of texture analysis by colour-based co-occurrence matrix for matching of parasite region was implemented by Halim et al. [46]. To locate the parasite region within the image, Tek et al. [22] used RGB histogram with probability density function for detection. Toha and Ngah [47], estimated a threshold value to distinguish between normal and parasite region and subsequently calculated the Euclidean distance between the regions to separate each parasite cluster. Makkapati et al. [48] also employed Otsu threshold method with HSV colour channel information to compute distance between red blood cell region and parasitic dense chromatin regions. The method was used by the author to differentiate the nucleus of white blood cells that might be falsely considered as a parasite region. Damahe et al. [44] proposed a thresholding scheme using the 'S component histogram' for HSV images, and Dave et al. [31] performed thresholding on the 'Hue channel' of images in HSV colour model to obtain parasite regions. Ross et al. [25] utilized the green channel histogram to determine two distinct threshold values from the image for segmenting the parasite region from normal red blood cell regions. Ghosh et al. [49] identified parasite regions by using a computed threshold value with gradient operator. Fang et al. [50] used Quaternion Fourier Transform (QFT) and the inverse QFT to locate parasite region. Elter et al. [51] obtained threshold values from green and blue histogram of image and then implemented morphological top-hat to determine parasite region.

Use of learning models for segmentation and parasite detection has also been widely used by various authors to prove the efficacy of these algorithms in this particular research domain. Tek et al. [22] used K-nearest neighbour classifier and considered 20 classes. Each class attributed to the four species and four stages for each species including normal cases. The authors obtained colour information and textural information as features to assign the images to a specific class. Khan et al. [52] performed unsupervised clustering on the image in lab colour space for parasite detection. Ross et al. [25] extracted geometric features along with textural features and implemented backpropagation neural network for classification of parasite. The use of Artificial Neural Networks (ANN) for segmentation and parasite classification was used by several authors with a different set of features in each case. Bahendwar et al. [29] considered RGB and HIS histogram information as features while Nugroho et al. [30] considered textural features. Diaz et al. [38] used colour histogram features in combination with Support Vector Machine (SVM) and multilayer perceptron as classifiers. Anggraini et al. [26] proposed a multilayer perceptron model for classification of parasite region. Feed-forward backpropagation neural network was implemented by Khan et al. [43] and the author also used Grey Level Co-occurrence Matrix (GLCM). Statistical-based features were extracted for classification. Das et al. [23] and Ghosh et al. [49] implemented SVM and Bayesian classifier while Das et al. [42] proposed the use of texture-based features with multivariate logistical regression. SVM classifier along with colour, texture and shape features was proposed by Savkare et al. [32] for parasite detection. Similarly, statistical features from grayscale image with SVM classifier using RBF kernel was implemented by Preedanan et al. [28] for parasite classification. A combination of statistical and texture based features with SVM classifier was proposed by Bairagi et al. [39] while texture and colour channel histogram features with SVM classifier was used by Rosado et al. [27]. Annaldas et al. [53] obtained the energy feature from Grey Level Co-occurrence Matrices (GLCM), statistical features and phase of image features and then implemented SVM and ANN classifier for parasite segregation.

There are other notable works that include Somasekar et al. [54], Ghate et al. [55], Suryawanshi et al. [56], Nasir et al. [57], Chayadevi et al. [58] that have used a combination of aforementioned works to obtain good detection accuracy. Similarly, some recently published works of Rajaraman et al. [59], Gopakumar et al. [60], Rosado et al. [61], Bibin et al. [62], Devi et al. [63], Park et al. [64] and Widodo et al. [65] have used a combination of features and classifiers and have made significant contribution to the domain of knowledge.

3 Dataset

The most vital component of any research work is the development of dataset for implementation of the algorithm. Most authors have relied upon gathering smear slides from hospitals and digitizing them. Moreover, such customized dataset need a pathologist to prepare the ground truth before the dataset can be practically useful. Publicly available and benchmarked dataset for peripheral blood smear digitized images are rare. Some websites do have images for depiction of the disease but such images are not suitable for research purpose. One notable website, the Centre for Disease Control (CDC) provides a dedicated digital library for malaria [15]. This American government website basically creates public awareness on different types of diseases and also to serve for educational and training purposes. Some authors have used images available in this digital library for the purpose of research. However, the library does not contain adequate number of images for practical implementation of an algorithm or proof of concept. The images from this library were used in this research for the purpose of representation only. The datasets used in this research work is a publicly available dataset from MaMic [66].

The glass slide of a thin smear of blood contains a spread of vascular tissue of an individual probably containing malaria infection. For observation and analysing a digitized image requires the obtained image to be captured at 600× to 1000× magnification. To identify species and different life cycle stages of the parasite and classification, 1000× magnification is the standard. For each patient, a single slide is prepared; however as per WHO standards, more than 100 non-overlapping thin smear images are required to be observed to report the absence of infection. Similarly, the presence of infection can be established after observing at least 300 number of red blood cells (normal and infected) to calculate and report parasitaemia. Figure 3 shows sample images of the dataset on which the proposed algorithms were implemented. The detailed image specifications and dataset details are shown in Table 2.

The database that was acquired from MaMic [66] (which is a publicly available database) pertains to snapshots taken from a whole thin blood smear slide scanned at 100× resolution of *P. falciparum* infection. Some of the infected slide scan areas were devoid of infection. This is due to low parasitaemia for those slide scans. Some of these slides were taken into consideration to test the robustness of the proposed system.

4 Methodology

The process of *P. falciparum* malaria detection requires that the images are properly segmented. In contrast to other forms of the disease, which is caused by other four species, the parasite is mostly confined within the erythrocyte walls. This infection does not result in the enlargement of the RBC. This makes the process of identifying the parasite easier as the search will be confined within the red blood cells. For this

(a)   **(b)**

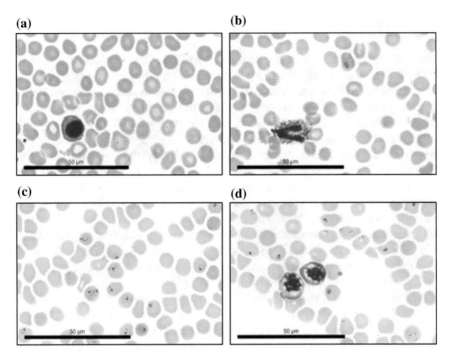

(c)   **(d)**

Fig. 3 Sample dataset from MaMic. **a** A normal image with a white blood cell (leucocyte), **b** infection with trophozoite and another white blood cell (neutrophil), and **c** multiple cells infected with *P. falciparum*, early ring stage. Some of the cells show multiple infection. **d** Several infected cells with early ring form and mature trophozoite form with two white blood cells (neutrophils)

Table 2 Parameter followed for dataset development

Specification	Value
Total number of scanned slides	54 slides
Sample size based on Cochrane's sample size selection for small datasets	47 slides (23 normal and 24 infected)
Number of non-overlapping blocks from each slide	30 [convenience based sampling] (understanding power and rules of thumb for determining sample sizes) (each of size 5.08 $\times$ 3.39 cm^2 [600 $\times$ 400 pixel^2])
Image resolution	300
Magnification used for each digital image	100$\times$
Total number of images used for the study	30 $\times$ 47 = 1410 images
Normal images	743 (some slide with infection show normal images due to low parasitaemia)
Images with infection	667

purpose, segmentation of the foreground image containing the cells is essentially the first step. However, illumination and noise can affect the segmentation outcome. Hence, an initial pre-processing of images is performed before the process of segmentation. The segmentation of different cellular components to obtain only the red blood cells is the next logical step towards parasite identification. Often the red blood cells are clumped together. The clumps are formed during the smear process. Though an experienced technician/phlebologist can prepare a single layer cellular spread but certain degree of overlapping cells are still observed in a thin Peripheral Blood Smear (PBS) slide. Removal of clumped cells is necessary for enumeration of cells and to calculate parasitaemia. Once the individual cells are obtained, the process of locating any infected region is required to be performed. At this stage, it is possible to segregate infected and normal images. Any image that does not exhibit any infected red blood cell region is designated as normal and is eliminated from further investigation. Multiple and non-overlapping images of a slide is considered for investigation. The system will report a normal case only when all the images from a particular slide show absence of infection. The infected images will be further investigated to determine the nature of infection and the stage of the parasite. This process requires that the infected region to be extracted and matched with the known form of the parasite. Identification of the stage and the knowledge of parasitaemia level exhibited by the image/s will provide an adequate observation to the pathologist for disease prognosis and treatment.

The method should be able to recognize the cellular components and differentiate the RBC from the WBC and platelets. The proposed system should be able to detect the presence of abnormality within a red blood cell. Finally, if an abnormal region is located, it has to be extracted and the stage of parasite to be identified. Figure 4 shows the stages of parasite as well as the cellular components the algorithm needs to encounter with.

4.1 Illumination Correction

Illumination of images plays a significant role in segmentation process. It has been observed statistically that images within the database show significant variations. For successful execution of segmentation algorithm, it is important to standardize the illumination across all the images. One of the images was identified to form the reference image in the illumination correction process. Equation 1 is used to correct the images.

$$\text{Correction} = (\sigma_{test} - \sigma_{ref})/\sigma_{ref} \qquad (1)$$

Figure 5 shows the outcome of illumination correction process with Fig. 5b is used as a reference image that was empirically defined.

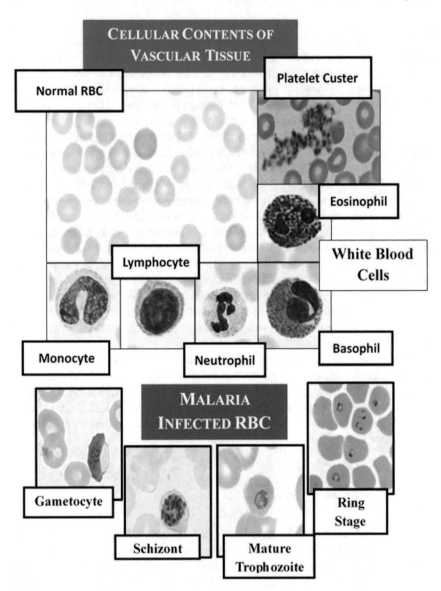

Fig. 4 The graphical representation of the vascular tissue showing different cellular composition, including red blood cells, white blood cell (including types) and the *P. falciparum* stages that the algorithm needs to consider for detection and analysis

(a)   (b)   (c)

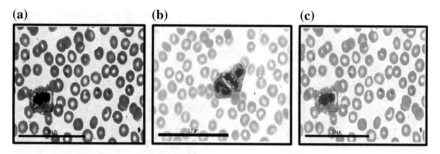

Fig. 5 Illumination correction of images: **a** the image to be corrected, **b** the reference image for the dataset and **c** the illumination corrected image

(a)   (b)

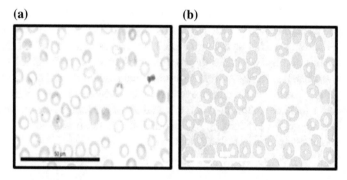

Fig. 6 **a** The image with scale artefact and noise; **b** the image after elimination of noise and artefact removal

4.2 Artefact Removal and Noise Reduction

Noise reduction in images is required because some salt and pepper noise may be introduced during scanning process. This noise will interfere with proper execution of automated algorithms. Such noise is eliminated using Median filtering with a 3 × 3 kernel window. The images exhibit label artefacts that are needed to be suppressed as they are not a part of the image. Simple pixel replacement strategy is employed for removal of label artefacts. Figure 6 shows an artefact and noise removed image.

4.3 Background Suppression

The foreground image contains the cellular matter whereas the background represents the non-uniform colour variations. Since the foreground cellular matter is the region of interest, the non-uniform background is irrelevant. Due to the colour variation, this region poses difficulty for subsequent algorithm execution. The region is removed

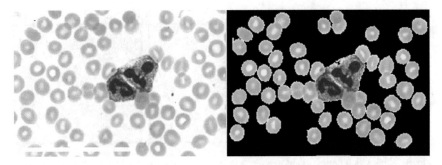

Fig. 7 Image depicts the outcome of background suppression where **a** input image after artefact removal and **b** background separated image with incomplete border cells removed

and replaced by black colour for further processing of the image. The process involves converting the image to La*b* colour space. Then unsupervised K-means clustering is performed on the image where three number of clusters are obtained ($K = 3$). The first two clusters represent the foreground image and the third cluster contains the background pixels that are eliminated and replaced by black pixel. Similarly, the cellular matter that gets intersected at the boundary of the images is removed as per the practices suggested by WHO [20]. Figure 7 shows the background suppressed image that is considered for further investigation.

4.4 Suspect Region Identification

The presence of any suspect region within the cells will be depicted by the presence of chromatin matter of the parasite. Since the red blood cells are non-nucleated, hence the presence of nucleated matter will indicate presence of parasite. Such nucleated matters are highlighted by the staining process (Giemsa stain). The only impediment to this hypothesis is the presence of nucleated white blood cells and platelets that are also marked as suspect region. The algorithm at a later level will differentiate between them. The RGB colour image will be non-uniform if it contains abnormalities. This is the basic assumption for the algorithm.

Colour Down-Sampling

However, a large number of colour variations in RGB colour space are difficult for the algorithm to accommodate. For this purpose, colour down-sampling is done by converting the image into YC_bC_r colour space. This suppresses the colour variations. The image is re-coloured to RGB colour space by considering the largest of the magnitude values of the three components. Figure 8 shows the down-sampling result.

Abnormal Region Identification

For establishing the presence of suspect region a colour-matching scheme is implemented. The presence of parasite will show more colour variation within a cell and

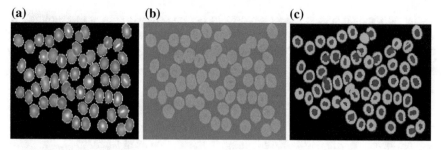

Fig. 8 The down-sampling outcome with **a** the background suppressed image, **b** the image after conversion to YC_bC_r colour space and **c** RGB colour down-sampled image

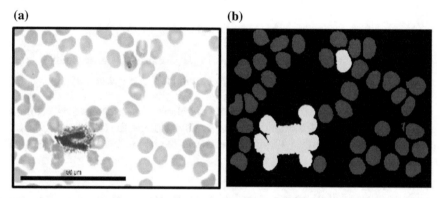

Fig. 9 The **a** original image containing an infection and a WBC element; **b** the regions marked as suspect. The WBC is clumped with the surrounding RBC, and the entire connected component is considered a suspect region

hence more number of colours within the cell in comparison to the surrounding normal cells. This assumption has been used in which a majority alignment rule is devised to compare the number of colours representing a particular closed component with the number of colours depicted by 90% of the surrounding normal cells. A deviation in the number indicates the presence of abnormalities and the closed component is marked as a suspect region for further investigation. Figure 9 shows the marked suspect region in the image.

4.5 Cell Clump Removal

While performing suspect region identification it was observed that there was a white blood cell and red blood cell clumps. Similar clumps are formed among red blood cells. Some authors have used clump splitting using the entire image; however, this approach may result in over segmentation of the clump region. So it is necessary to devise mechanisms to detect clumps. In the present work, area thresholding of connected components have been devised for candidate selection. After identifying clumps, the candidate closed components will be de-clumped and identified whether they are RBC or WBC.

Candidate Selection
Two basic approaches that were implemented for clump selection were, namely, third quartile and Tukey's hinge.

- Area-driven third quartile-based clump identification as in Eq. 2.

$$\text{Clump area} >= \text{Mean/Median\_Area\_of\_Connected\_Components} \times (3/2) \tag{2}$$

- Tukey's hinge-based identification of clump area

$$\text{Clump area} >= \text{Mean/Median\_Area\_of\_Connected\_Components} +$$
$$2 \times \sigma(Area\_of\_Connected\_Components) \tag{3}$$

While Tukey's upper bound for marking out clump area excludes small clumps and takes only into consideration significantly large clumps, the third quartile-based threshold function identifies all of the clusters in the image and is used for clump candidate selection.

De-Clumping of closed components
For the purpose of de-clumping watershed algorithm is used as proposed by Kumar et al. [19]. Watershed algorithm is applied to only those closed components that are identified to contain clump of cells. The cell partitions defined by the watershed algorithm are implemented to partition the clumps into individual cells. Figure 10 shows the process of de-clumping of cells in an image.

Differentiating Red Blood cells and White Blood cells
The de-clumping algorithm is not only constricted to de-clumping connected RBC, the same algorithm is also utilized to de-clump the cluster formed between WBC and RBC. The algorithm calculates the area of each of the components in the de-clumped image. The area value of the connected component which is greater than the Tukey's upper bound (>(median of area + 2 × standard deviation of area) area value for connected components in the image is marked as white blood cell while the rest are marked as red blood cells. Figure 11 shows the differentiation of RBC from the WBC.

4.6 Enumeration

After the process of de-clumping, each RBC are marked for enumerations. The WBC are suppressed in the images. The cells that show suspect region is re-coloured to

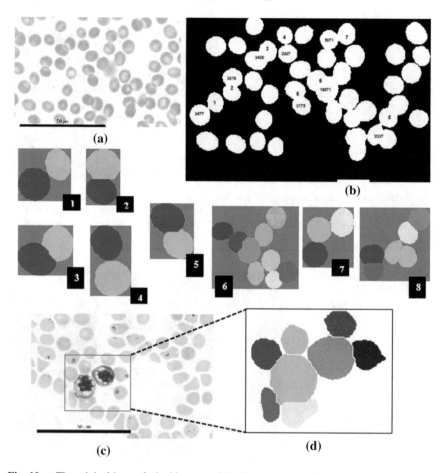

Fig. 10 a The original image **b** the binary mask in (**a**) is represented along with the area of the clumped cells. The marked out clumps in the image are numbered in order to match the same with the corresponding de-clumped versions. **c** Image from the dataset consisting of a mixed cluster of RBC and white blood cell/s. **d** Image in (**b**) consists of only one cluster. De-clump of the mixed white and red blood cell/s

depict infected cells. The ratio of infected cells against normal cells is calculated for parasitaemia estimation. Figure 12 shows the image of enumerated red blood cells along with the infected cells.

4.7 Feature Selection

The WBC cluster formed as a part of the aforementioned segmentation (second cluster of the 3-means clustering in La*b* colour space) process consists of normal WBC

(a)   **(b)**

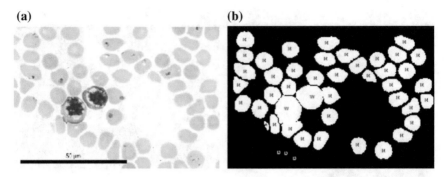

Fig. 11 a A sample image containing infected RBC and two WBC clumped with other RBC which are normal/infected. **b** The binary mask of the de-clumped and reconstructed image with the white blood cells marked with 'W', the red blood corpuscles marked with 'R' and the elements to be discarded marked with 'D'

Fig. 12 The image shows the red blood cells enumerated as well as the suspect region identified and marked for further investigation by subsequent algorithm

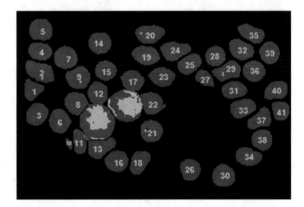

cell, infection, artefacts and certain RBC cell outliers that are particularly bigger in size (based on Tukey's Hinges) that RBC and have taken a stain colour similar to the WBC cell cluster. Features were calculated for each connected component in the predominantly white blood cell cluster. The features used to segregate a white blood cell from infection and an outlier red blood cell are morphological and textural features. Table 3 represents the feature list that was calculated for segregation of WBC from other connected components in the cluster under consideration. The features were normalized with mean zero ('0') and standard deviation one ('1'). Three-nearest neighbour classification was performed based on the feature set. The performance of the algorithm was evaluated based on sensitivity, specificity and accuracy values.

The connected components that were marked out by the algorithm as normal RBC were incorporated in the RBC cluster. Only the infected cell components in the WBC cluster were considered for further investigation.

So for stage classification, two images have been used, infected WBC cluster and the infected RBC cluster. For the RBC infected cluster semi-supervised lab

Table 3 Feature list used for segregation of WBC Cluster from other connected components

Cell level features	
Feature name	Number of features
Nuclear mass presence	Number of colour coded nuclear mass in a coloured component; present/absent feature (2 features)
Difference of area from median area of complete uninfected RBC in a radius of 2 × major axis length of the component area under consideration	Numerical feature, a negative value indicates the size of the area under consideration is smaller than the area of the average uninfected RBC in the given area and likewise for positive value (1 feature)
Proportion of marked out infection to normal cell area (ratio of colour pixel values)	Numerical float value(1 feature)
Textural features	
Texture features—Tamura features	3 features for each grayscale cell component
Texture features—GLCM	88 features for each grayscale cell component
Total features	95 features

colour-based clustering was performed for pronounced ring identification within the RBC cell (refer Fig. 13). Table 4 represents the features that have been used for stage classification of RBC cluster-based infection. For the RBC infected cluster, the stages are *P. falciparum* trophozoite, ring, schizont and gametocyte (male, female).

4.8 Training and Cross-Validation

For either of the two infection clusters, the feature space dimensionality is reduced to a set of 50 features using Conditional Mutual Information Maximization (CMIM) algorithm [67]. The top 50 features from both the feature list are utilized for classification. Training of dataset is performed for supervised learning by the system so that it can predict the class of a test data case accurately. Now the 50 feature strong dataset of 1410 images were all subjected to tenfold cross-validation to prevent overfitting of the data model developed.

4.9 Classification

Classification for either infected cluster is performed in different ways to test the performance among singular classifiers and also compare their performance to ensemble classifiers. Three classifiers that are used are k-NN where $k = 3$, Naïve Bayes and SVM classifier with RBF kernel function is used. The classification scheme is presented in Fig. 14.

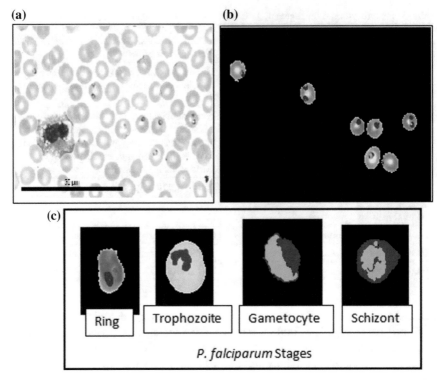

Fig. 13 **a** The original image from dataset, **b** region extraction for parasite region after semi-supervised k-means clustering in lab colour space and **c** the four different life cycle stages of the parasite region that is extracted as feature

5 Platform for Development and Testing

The work has been conducted in MATLAB R2017a. It is a popular tool for simulation of the proof of concept. It has been popularly used by other researchers in the image processing domain for proof of concept and simulation. Hence, this tool was used to develop and implement the proposed algorithms. The machine used is HP Pavilion 15 Notebook PC with Processor Intel(R) Core(TM) i7-4510U CPU @2.00 GHz, 8 Gb RAM and Windows10 OS.

6 Result Analysis

Based on the annotations provided by a registered medical practitioner the performance of the hybrid algorithm for detection of infected RBC was evaluated. In accordance with the proposed methodology, the algorithm can particularly be classified into four essential blocks namely, image pre-processing along with ROI extraction,

Table 4 Feature list used for species and stage classification from RBC cluster

Morphological features	
Feature name	Number of features
Infection morphology	Area, perimeter, eccentricity (three features)
Cell morphology	Area, perimeter, eccentricity (three features)
Number of infection instances in a cell	1 feature
Cell level features	
Difference of area from median area of complete uninfected RBC in a radius of 2 × major axis length of the component area under consideration	Numerical feature, a negative value indicates the size of the area under consideration is smaller than the area of the average uninfected RBC in the given area and likewise for positive value (one feature)
Proportion of marked out infection to normal cell area (ratio of colour pixel values)	Numerical float value (one feature)
Textural features	
Texture features—Tamura features	3 features for each grayscale cell component
Texture features—GLCM	88 features for each grayscale cell component
Total features	100 features

De-clumping of cluster (RBC cluster, mixed RBC with WBC clusters) along with red blood cell enumeration, malaria parasite detection and parasite stage classification.

For the first section, the foreground identification accuracy achieved by 3-means clustering was found to be 92.19% across all images Again, these values were obtained on images for which illumination was not corrected (refer to Table 5). After performing illumination correction, there was a significant improvement in the accuracy of the segmentation process. The 3-means clustering achieved average Accuracy of 98.96%.

The second integral section can further be divided into two serially dependent blocks, namely, de-clumping and red blood cell enumeration. Modified watershed-based methodology was used for automatic de-clumping of red blood cell clusters as also mixed red blood cell–white blood cell clusters. In coherence with the proposed methodology, it was integral to evaluate the performance of different statistical metrics to clearly mark out the clumps (i.e. RBC clumps or mixed white and red blood cell clumps) from the individual cells in the digitized thin blood smear image. Table 6 provides the performance of the different statistical metrics used for differentiating clumps from solitary cell particles (for MaMic database).

After two-stage de-clumping red blood cells were segregated from white using a threshold value. Tukey's upper hinge was used as a threshold to segregate red from white blood cells, namely, Table 7 represents the performance metric for red blood cell segregation from white blood cells. While the first section forms the basis of the algorithm as a whole, the second block is precursor to parasitaemia estimation that is worked on in the third block of the research design. The third block is aimed to detect malaria parasite in a thin blood smear image.

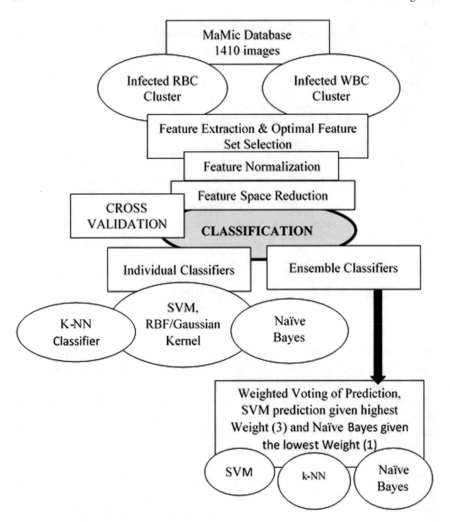

Fig. 14 Data model for *P. falciparum* detection and stage classification

Table 5 Comparative account of the accuracy achieved by 3-means clustering used for image background separation before and after illumination correction (for MaMic dataset)

3-means clustering method used for background separation	Accuracy achieved (%)
Algorithm performance without illumination correction of the images	92.19
Algorithm performance after image illumination correction	98.96

Table 6 Performance statistics for de-clump algorithm

Statistical metric used/proposed	Accuracy	Sensitivity	Specificity
Use of third quartile of area as a threshold	0.9875	0.9825	0.9877
Use of Tukey's upper hinge as a threshold	0.9782	0.3829	0.9982

Table 7 Red blood cell segregation from white blood cell in the digitized thin blood smear for the dataset under consideration

Dataset	Accuracy	Sensitivity	Specificity
MaMic database	0.9925	0.9875	0.9942

Table 8 Performance statistics for identification of infected RBC from normal RBC Cells and for identification of WBC cell cluster using one versus all strategy

Cluster information	Accuracy	Sensitivity	Specificity
RBC cluster	0.9946	0.9963	0.9949
WBC cluster	0.9623	0.9609	0.9726

For parasitaemia estimation the algorithm identifies a WBC based on the presence of nucleus (represented in green) and area value along with other texture features as highlighted in Table 3, is marked as normal, the other problem area/s present within a red blood cell or as standalone are retained for further investigation.

As per the proposed model, the performance of the hybrid algorithm towards detection of infected RBC at cellular level was recorded (accuracy 0.9946, sensitivity 0.9963, and specificity 0.9949). In case of the WBC Cluster, a 3–NN classifier was used to separate out white blood cell, infection, artefact (i.e. clustered platelets) and outlier red blood cell.

The evaluation of the algorithm for infection detection within RBC cell and for segregation of WBC from infection, artefact and outlier RBC using a one versus all strategy for the database under consideration has been represented in Table 8.

The final feature set of size 50 is selected using conditional mutual information maximization algorithm. Table 9 represents the final average classification accuracy across all infection classes with single and ensemble classifiers, respectively, at image level.

Table 9 Performance statistics for final average classification accuracy across all infection classes with single and ensemble classifiers (at image level)

Classifier (species and stage classification)	Accuracy (for 50 feature set selected by CMIM)	Sensitivity	Specificity
K-NN ($K = 3$)	0.9653	0.9472	0.9677
SVM (RBF Kernel) C-1, gamma-½	0.9759	0.9531	0.9753
Naïve Bayes classifier (Gaussian distribution assumed)	0.9245	0.9168	0.9344
Ensemble (SVM, 3-NN, Naïve Bayes)	0.9874	0.9925	0.9820

7   Discussion

Analysis of blood smear images of both thick/thin for detection of malaria parasite is a well-tread domain over the last two decades. However, the complex staining of the RBC, malaria parasite and WBC, non-uniform illumination makes it difficult to develop an automated algorithm for malaria parasite detection that works across all dataset and all staining practices followed. In an attempt to extend the works of Diaz et al. [38], Seo et al. [33], and Tek et al. [21, 22], a method has been proposed in this paper that combines unsupervised learning techniques with rule-based and supervised learning-based methods for optimal performance.

Performance comparison is performed with other notable works is represented in Table 10. The consulting doctor provided us with the ground truth for the selected 1410 images. The selected dataset have adequate representation of all classes of infection and all features necessary for the algorithm to provide good prediction result. The table was compiled as part of the overall detection and classification work.

Table 10   Comparative study of the present work with other notable contribution to the domain of digital pathology with perspective towards malaria parasite detection

Authors	Methods (segmentation, features, classifications)	Stage and species classification/parasitaemia	Overall performance
Nag et al. (proposed method)	Segmentation: using unsupervised clustering. Non-uniform background separation, colour down-sampling and matching for suspect region identification, candidate selection for de-clumping (area-based threshold) Features: Morphological (features introduced based on parasite and modified cell morphology) and textural features—95/100 feature set Classification: k-NN, SVM, Naïve Bayes, Ensemble (combination of three weighted)	Stage classification (ring, trophozoite, schizont and gametocyte—four stages) Species *P. falciparum* Prevalent in India RBC enumeration and parasitaemia calculation	Stage and species classification results Ensemble (SVM, 3-NN, Naïve Bayes) method Accuracy: 0.9874 Sensitivity: 0.9925 Specificity: 0.9820

(continued)

Table 10 (continued)

Authors	Methods (segmentation, features, classifications)	Stage and species classification/parasitaemia	Overall performance
Rajaraman et al. [59]	Segmentation: A level-set-based algorithm to detect and segment the RBC; Classification: Convolution neural network has three convolutional layers and two fully connected layers	Only detection performed	Cell/patient level Accuracy: 0.986, 0.959 Sensitivity: 0.981, 0.947 Specificity: 0.992, 0.972
Gopakumar et al. [60]	Segmentation: Otsu thresholding and marker controlled watershed algorithm Classification: Convolution neural network has two convolutional layers and two fully connected layers	Only detection performed	Method B (best focused image) Sensitivity: 0.9891 Specificity: 0.9939
Rosado et al. [61]	Segmentation: Adaptive thresholding for segmentation of respective region Features: Geometric, colour and texture Classification: Using SVM by altering the gamma and C values for each species and stage	Trophozoites, Schizonts and Gametocytes of *P. falciparum, P. Malariae and P. ovale*	Sensitivity: 73.9–96.2 Specificity: 92.6–99.3
Bibin et al. [62]	Segmentation: Level-set method Features: Colour histogram, coherence vector, GLCM and GLRLM and LBP for texture Classification: Deep belief network input layer 484 nodes, four layers of hidden nodes (600 each) and two output nodes	Only detection performed	Sensitivity: 97.6 Specificity: 95.92

(continued)

Table 10 (continued)

Authors	Methods (segmentation, features, classifications)	Stage and species classification/parasitaemia	Overall performance
Devi et al. [63]	Segmentation: Binarization using Otsu threshold, morphological operations, marker controlled watershed for clump splitting Features: 134 features including intensity-based and textural like GLCM, Gabor Classification: Hybrid classifiers; combination of k-NN, SVM, Naïve Bayes and ANN in every possible combinations are compared	Differentiation between normal, *P. falciparum* (ring, schizont), *P. vivax* (ring, ameboid ring, schizont and gametocyte)— seven class problem	Accuracy 96.54 with hybrid classifier
Park et al. [64]	Segmentation: Not performed Features: 23 Morphological features Classification: Fisher discriminant analysis, logistic regression and k-NN	Differentiation into falciparum ring, trophozoite and schizont	Sensitivity, specificity, accuracy of three classifiers for three classes are provided
Widodo et al. [65]	Segmentation: Active contour Features: First-order texture and second-order GLCM texture features Classification: SVM	Differentiation of stage performed (three-stage)	Individual accuracy for each stage ranging between 85 and 100%
Das et al. [23]	Segmentation: Foreground separation using Sobel and segmentation using marker controlled watershed Features: 80 texture features and 16 morphological features (96 features reduced to 94 features) Classification: Naïve Bayes and SVM	*P. vivax* and *P. falciparum* stage classifications performed. (6-Class) PV (ring, S, G) PF (ring, G) and normal	SVM classifier (9 features subset) Sensitivity— 96.62% Specificity— 88.51%

(continued)

Table 10 (continued)

Authors	Methods (segmentation, features, classifications)	Stage and species classifica-tion/parasitaemia	Overall performance
Purwar et al. [45]	Segmentation: Using Chan–Vese method for edge detection followed by Morphological operations Features: Intensity Classification: Probabilistic k-means clustering	Only detection performed	Comparative study between automated and manual detection (overall sensitivity 100%, Specificity 50–80%)
Tek et al. [22]	Segmentation: Background separation with area granulometry and intensity-based thresholding Features: 83 features computed based on morphology and colour information Classification: k-NN compared with Fisher discriminant analysis and BPNN	Detection, stage and species differentiation with parasitaemia calculation 20 class—including non-parasite, 16 class (4 stage × 4 species) and 4 class	Overall Detection results Accuracy: 93.3 Sensitivity: 72.4 Specificity: 97.6 Parasitaemia results not provided
Ruberto et al. [24]	Segmentation: Using morphological operation and area granulometry. Watershed transform for de-clumping Classification based on colour similarity	Detection and stage differentiation performed only	Comparative study present. Among detection result with manual observations

8 Conclusion

The contribution of the work is particularly twofold. In terms of application software, it stands as a tool to assist medical practitioners at effective detection of *P. falciparum* parasite and stage classification. As opposed to other toolkits having similar functionality, this particular tool investigates parasites at cellular level which is much preferred by medical practitioners with significantly lower computational overhead. In perspective of computer science, it has been a long-standing debate with regard to the predictive power of the classifiers. This paper adds on to the vast domain by putting forth a comparative study of single and ensemble classifiers using the same set of normalized filter features in perspective of the computer science domain.

As a result of tenfold cross-validation, the final or best performance that has been achieved by the system is accuracy of 0.9874, sensitivity of 0.9925 and specificity of 0.9820. There are similarly few drawbacks of the system. The model requires further extensive testing before it can be reliably used within the medical domain as an effective aid to the medical practitioner. Similarly, the system does not take into consideration other forms of malaria parasite infections caused by different species. This can be considered as a part of future work where the proposed system can be extended to detect all type of species and stages of malaria parasite, thus becoming a better and comprehensive tool for CAD application in digital pathology.

Acknowledgements This work was conducted under the guidance of Dr. Pradip Saha (Radiologist & Medical Practitioner) and Dr. Debashis Chakraborty (Pathologist) of SSKM Hospital. The doctors helped the research team with domain knowledge and provided the ground truth for the images. Their insights and approach towards disease detection helped the research team prepare the data model and were vital in achieving good results. University of Calcutta provided us all academic and infrastructural support for the research work.

References

1. García-Rojo M, Blobel B, Laurinavicius A (2012) Perspectives on digital pathology. IOS Press, Amsterdam
2. Treanor D, Williams B (2019) The leeds guide to digital pathology. The Leeds Teaching Hospitals NHS, University of Leeds. [Online]. https://www.virtualpathology.leeds.ac.uk/Research/clinical/. Accessed 15 May 2019
3. Bueno G, Fernández-Carrobles MM, Deniz O, García-Rojo M (2016) New trends of emerging technologies in digital pathology. Pathobiology 83(2–3):61–69
4. Pantanowitz L, Sharma A, Carter AB, Kur TM, Sussman A, Saltz JH (2018) Twenty years of digital pathology: an overview of the road travelled, what is on the horizon, and the emergence of vendor-neutral archives. J Pathol Inform
5. Singh R, Chubb LG, Pantanowitz L, Parwani AV (2011) Standardization in digital pathology: supplement 145 of the DICOM standards. J Pathol Inform
6. Bauer TW, Slaw RJ (2014) Validating whole-slide imaging for consultation diagnoses in surgical pathology. Arch Pathol Lab Med 138(11):1459–1465
7. Salto-Tellez M, Maxwell P, Hamilton P (2019) Artificial intelligence—the third revolution in pathology. Histopathology 74(3):372–376
8. Kakkilaya BS (2018) The challenge of Malaria. [Online]. https://www.Malariasite.com/challenge-of-Malaria/. Accessed 10 May 2019
9. Neghina R, Iacobiciu I, Neghina AM, Marincu I (2010) Malaria, a journey in time. In search of the lost myths and forgotten stories. Am J Med Sci 340(6):492–498
10. Otto TD et al (2018) Genomes of all known members of a Plasmodium subgenus reveal paths to virulent human Malaria. Nat Microbiol 3(6):687–697
11. Carter R, Mendis KN (2002) Evolutionary and historical aspects of the burden of Malaria. Clin Microbiol Rev 15(4):564–594
12. Sallares R (2002) Malaria and Rome: a history of Malaria in ancient Italy. Oxford University Press, Oxford
13. Ziegler M (2014) Early use of the term 'Malaria,'" Contagions. [Online]. https://contagions.wordpress.com/2014/08/07/early-use-of-the-term-Malaria/. Accessed 14 May 2019
14. D. of P. D. Global Health, "Parasites-Malaria," CDC Govt. of USA, 2019. [Online]. https://www.cdc.gov/parasites/Malaria/index.html. Accessed 15 May 2019
15. D. of P. D. and M. Global Health, "DPDx—laboratory identification of parasites of public health concern—Malaria," CDC Govt. of USA, 2017. [Online]. https://www.cdc.gov/dpdx/Malaria/index.html. Accessed 14 May 2019

16. World Health Organization, "WHO | This year's world Malaria report at a glance," WHO, 2019. [Online]. https://www.who.int/Malaria/media/world-Malaria-report-2018/en/# Global and regional Malaria burden, in numbers. Accessed 14 May 2019

17. WHO, "Malaria key points: world Malaria report 2017," World Health Organization, 2018. [Online]. https://www.who.int/Malaria/media/world-Malaria-report-2017/en/. Accessed 14 May 2019

18. WHO, "World Malaria report 2016. Switzerland," World Health Organization, 2016. [Online]. http://apps.who.int/iris/bitstream/10665/252038/1/9789241511711-eng.pdf?ua=1. Accessed 15 May 2019

19. Garcia LS (ed) (2007) Diagnostic medical parasitology, 5th ed. American Society of Microbiology

20. Storey J (2010) Basic Malaria microscopy—Part I: Learner's guide, 5th edn. World Health Organization, Geneva

21. Tek FB, Dempster AG, Kale I (2009) Computer vision for microscopy diagnosis of Malaria. Malar J 8(1):153

22. Tek FB, Dempster AG, Kale İ (2010) Parasite detection and identification for automated thin blood film Malaria diagnosis. Comput Vis Image Underst 114(1):21–32

23. Das DK, Ghosh M, Pal M, Maiti AK, Chakraborty C (2013) Machine learning approach for automated screening of Malaria parasite using light microscopic images. Micron 45:97–106

24. Di Ruberto C, Dempster A, Khan S, Jarra B (2002) Analysis of infected blood cell images using morphological operators. Image Vis Comput 20(2):133–146

25. Ross NE, Pritchard CJ, Rubin DM, Dusé AG (2006) Automated image processing method for the diagnosis and classification of Malaria on thin blood smears. Med Biol Eng Comput 44(5):427–436

26. Anggraini D, Nugroho AS, Pratama C, Rozi IE, Iskandar AA, Hartono RN (2011) Automated status identification of microscopic images obtained from Malaria thin blood smears. In: Proceedings of the 2011 international conference on electrical engineering and informatics, pp 1–6

27. Rosado L, da Costa JMC, Elias D, Cardoso JS (2016) Automated detection of Malaria parasites on thick blood smears via mobile devices. Procedia Comput Sci 90:138–144

28. Preedanan W, Phothisonothai M, Senavongse W, Tantisatirapong S (2016) Automated detection of plasmodium falciparum from Giemsa-stained thin blood films. In: 2016 8th international conference on knowledge and smart technology (KST), pp 215–218

29. Bahendwar YS, Chandra UK (2015) Detection of Malaria parasites through medical image segmentation using ANN algorithm. Int J Adv Res Comput Sci Softw Eng 5(7):1063–1067

30. Nugroho HA, Akbar SA, Murhandarwati EEH (2015) Feature extraction and classification for detection Malaria parasites in thin blood smear. In: 2015 2nd international conference on information technology, computer, and electrical engineering (ICITACEE), pp 197–201

31. Dave IR, Upla KP (2017) Computer aided diagnosis of Malaria disease for thin and thick blood smear microscopic images. In: 2017 4th international conference on signal processing and integrated networks (SPIN), pp 561–565

32. Savkare SS, Narote SP (2015) Automated system for Malaria parasite identification. In: 2015 international conference on communication, information & computing technology (ICCICT), pp 1–4

33. Sio SWS et al (2007) MalariaCount: an image analysis-based program for the accurate determination of parasitemia. J Microbiol Methods 68(1):11–18

34. Arco JE, Górriz JM, Ramírez J, Álvarez I, Puntonet CG (2015) Digital image analysis for automatic enumeration of Malaria parasites using morphological operations. Expert Syst Appl 42(6):3041–3047

35. Ahirwar N, Pattnaik S, Acharya B (2012) Advanced image analysis based system for automatic detection and classification Malarial parasite in blood images, vol 5

36. Reni SK, Kale I, Morling R (2015) Analysis of thin blood images for automated Malaria diagnosis. In: 2015 E-health and bioengineering conference (EHB), pp 1–4

37. Kumar S, Ong SH, Ranganath S, Ong TC, Chew FT (2006) A rule-based approach for robust clump splitting. Pattern Recogn 39(6):1088–1098
38. Díaz G, González FA, Romero E (2009) A semi-automatic method for quantification and classification of erythrocytes infected with Malaria parasites in microscopic images. J Biomed Inform 42(2):296–307
39. Bairagi VK, Charpe KC (2016) Comparison of texture features used for classification of life stages of Malaria parasite. Int J Biomed Imaging 2016:7214156
40. Prasad K, Winter J, Bhat UM, Acharya RV, Prabhu GK (2012) Image analysis approach for development of a decision support system for detection of Malaria parasites in thin blood smear images. J Digit Imaging 25(4):542–549
41. Mehrjou A, Abbasian T, Izadi M (2013) Automatic Malaria diagnosis system. In: 2013 first RSI/ISM international conference on robotics and mechatronics (ICRoM), pp 205–211
42. Das D, Ghosh M, Chakraborty C, Maiti AK, Pal M (2011) Probabilistic prediction of Malaria using morphological and textural information. In: 2011 international conference on image information processing, pp 1–6
43. Khan MI, Singh BK, Acharya B, Soni J (2011) Content based image retrieval approaches for detection of Malarial in blood images. Int J Biometrics Bioinform 5(2):97–110
44. Damahe L, Thakur N, Krishna RK, Janwe N (2011) Segmentation based approach to detect parasites and RBCs in blood cell images, vol 4
45. Purwar Y, Shah SL, Clarke G, Almugairi A, Muehlenbachs A (2011) Automated and unsupervised detection of Malarial parasites in microscopic images. Malar J 10(1):364
46. Halim S, Bretschneider TR, Li Y, Preiser PR, Kuss C (2006) Estimating Malaria Parasitaemia from blood smear images. In: 2006 9th international conference on control, automation, robotics and vision, pp 1–6
47. Toha SF, Ngah UK (2007) Computer aided medical diagnosis for the identification of Malaria parasites. In: 2007 international conference on signal processing, communications and networking, pp 521–522
48. Makkapati VV, Rao RM (2009) Segmentation of Malaria parasites in peripheral blood smear images. In: ICASSP, IEEE international conference on acoustics, speech and signal processing, pp 1361–1364
49. Ghosh P, Bhattacharjee D, Nasipuri M, Basu DK (2011) Medical aid for automatic detection of Malaria, vol 245 CCIS
50. Yuming Fang Y, Wei Xiong W, Weisi Lin W, Zhenzhong Chen Z (2011) Unsupervised Malaria parasite detection based on phase spectrum. In: 2011 annual international conference of the IEEE Engineering in Medicine and Biology Society, vol 2011, pp 7997–8000
51. Elter M, Hasslmeyer E, Zerfass T (2011) Detection of Malaria parasites in thick blood films. In: 2011 annual international conference of the IEEE Engineering in Medicine and Biology Society, vol 2011, pp 5140–5144
52. Khan NA, Pervaz H, Latif AK, Musharraf A, Saniya (2014) Unsupervised identification of Malaria parasites using computer vision. In: 2014 11th international joint conference on computer science and software engineering (JCSSE), pp 263–267
53. Annaldas S, Shirgan SS (2015) Automatic diagnosis of Malaria parasites using neural network and support vector machine. Int J Adv Found Res Comput 2:60–66
54. Somasekar J, Reddy BE, Reddy EK, Lai C-H (2011) An image processing approach for accurate determination of parasitemia in peripheral blood smear images. Asp Digit Imaging Appl (IJCA Spec Iss Nov) 1:23–28
55. Ghate AMD (2014) Automatic detection of Malaria parasite from blood images. Int J Adv Comput Technol 4(1):129–132
56. Suryawanshi S, Dixit VV (2013) Comparative study of Malaria parasite detection using euclidean distance classifier & SVM. Int J Adv Res Comput Eng Technol 2(11):2994–2997
57. Abdul Nasir AS, Mashor MY, Mohamed Z (2012) Segmentation based approach for detection of Malaria parasites using moving k-means clustering. In: 2012 IEEE-EMBS conference on biomedical engineering and sciences, pp 653–658

58. Chayadevi ML, Raju GT (2014) Usage of ART for automatic Malaria parasite identification based on fractal features. Int J Video Image Process Netw Secur IJVIPNS-IJENS 14(04):7–15

59. Rajaraman S et al (2018) Pre-trained convolutional neural networks as feature extractors toward improved Malaria parasite detection in thin blood smear images. PeerJ 6:e4568

60. Gopakumar GP, Swetha M, Sai Siva G, Sai Subrahmanyam GRK (2018) Convolutional neural network-based Malaria diagnosis from focus stack of blood smear images acquired using custom-built slide scanner. J Biophotonics 11(3):e201700003

61. Rosado L, da Costa J, Elias D, Cardoso J (2017) Mobile-based analysis of Malaria-infected thin blood smears: automated species and life cycle stage determination. Sensors 17(10):2167

62. Bibin D, Nair MS, Punitha P (2017) Malaria parasite detection from peripheral blood smear images using deep belief networks. IEEE Access 5:9099–9108

63. Devi SS, Laskar RH, Sheikh SA (2018) Hybrid classifier based life cycle stages analysis for Malaria-infected erythrocyte using thin blood smear images. Neural Comput Appl 29(8):217–235

64. Park HS, Rinehart MT, Walzer KA, Chi J-TA, Wax A (2016) Automated detection of P. falciparum using machine learning algorithms with quantitative phase images of unstained cells. PLoS ONE 11(9):e0163045

65. Widodo S, Widyaningsih P (2015) Software development for detecting Malaria tropika on blood smears image using support vector machine. Int J Eng Sci Res Technol 4(1):39–44

66. "The MaMic Image Database." [Online]. http://fimm.webmicroscope.net/Research/Momic/mamic. Accessed 07 Apr 2019

67. Fleuret F (2004) Fast binary feature selection with conditional mutual information. J Mach Learn Res 5:1531–1555

Chapter 5
Efficient ANN Algorithms for Sleep Apnea Detection Using Transform Methods

Jyoti Bali, Anilkumar Nandi and P. S. Hiremath

1 Introduction

Physiological signals, namely, ECG, EEG, EMG, blood oxygen, etc., depicting the health status of heart, brain, and other vital organs of human being are very important for study apart from the monitoring of some important activities like breathing effort, body position, etc. Sound sleep is an important parameter of health as it is directly related to the health of the heart as well of the brain and psychological health of the human being. Sleep disorder or sleep apnea (SA) hampers the quality of sleep, which drastically affects the individual's daytime work. SA occurs in three forms, namely, obstructive sleep apnea (OSA), central sleep apnea (CSA), and mixed sleep apnea (MSA). OSA is caused by the blocking of the upper airway gap because of relaxed throat muscles or falling of tongue into the airway or the enlarged tonsils and adenoids impeding the airflow. However, during apnea condition, the soft tissues around throat relax, causing partial closure of the airway. This is referred to as hypopnea condition, a condition of difficulty in breathing and thus leading to snoring. If the muscles around the throat relax too much, complete blocking of the air gap is resulted in causing apnea condition. The most prevalent of all types of sleep apnea is OSA and occurs in 84% of the apnea cases [1–4]. CSA is caused by the lack of respiratory effort initiated by the brain that causes the stoppage of breathing during sleep. Lastly, the MSA is resulted due to the occurrence of both OSA and CSA.

J. Bali (✉) · A. Nandi · P. S. Hiremath
KLE Technological University, BVB Campus, Hubballi 580031, Karnataka, India
e-mail: jyoti_bali@kletech.ac.in

A. Nandi
e-mail: anilnandi@kletech.ac.in

P. S. Hiremath
e-mail: pshiremath@kletech.ac.in

© Springer Nature Singapore Pte Ltd. 2020
O. P. Verma et al. (eds.), *Advancement of Machine Intelligence in Interactive Medical Image Analysis*, Algorithms for Intelligent Systems,
https://doi.org/10.1007/978-981-15-1100-4_5

Sleep apnea is caused by interruptions in breathing, causing the shortage of oxygen supply to the heart, in turn affecting the pumping action of the heart. Hence, the heart rate reduces, in turn, causing Bradycardia, i.e., slow heart condition and this condition of reduced oxygen level in the blood is sensed by the brain that constricts the arteries to pump blood at a faster rate to supply the deficient oxygen. As a result, the heart rate increases temporarily. Thus, the resulted episode of bradycardia followed by tachycardia is referred to be an Apnea event. The adverse conditions of repeated apnea events cause continuous episodes of tachycardia. An important indicator of sleep apnea is the variations in heart rate, which can be easily noted from the ECG signal study. The apnea events do occur in a period of about 10–20 s. Apnea–hypopnea index (AHI) is used as the measure of the count of apnea episodes per hour. OSA is classified into three types, namely, mild OSA, if $5 < AHI < 15$; moderate OSA, if $15 < AHI < 30$; and severe OSA, if $AHI > 30$. OSA is the most prevalent among the different types of OSA. Thus when AHI is greater than 5, then the disorder is considered as sleep apnea [1–6].

Respiration activity is disturbed in case of sleep apnea condition, which affects the normal functioning of the heart. ECG signal is a well-established diagnostic modality to track the function of the heart. ECG signal recorded using the surface electrodes gets modulated in amplitude and frequency due to the relative motion of electrodes concerning the heart. The direction of the mean cardiac electrical axis gets modulated as a result of expansion and contraction of heart, which in turn affects the beat morphology. This also causes the rotation of mean cardiac electrical axis, in turn, resulting in the frequency modulation of the ECG, thus resulting in the variations of ECG parameters. Heart rate variability is the result of variations in heart rate caused by the variation of ECG parameters. The respiration process is associated with respiratory sinus arrhythmia (RSA) responsible for the beat to beat fluctuations. Hence, the heart and respiration functions are interdependent [1–6].

1.1 Diagnostic Methods and Treatment

For cases of Sleep apnea, the affected person needs to consult either a cardiologist or pulmonologist or a neurologist. As per the clinical observations of the doctor, the affected person is recommended to undergo Polysomnography (PSG) testing in a sophisticated and expensive sleep lab set up under the supervision of expert personnel. The vital parameter values and the other information related to the functions of heart, lung, brain activity, breathing patterns, and arm and leg movement are gathered through a set of electrodes for ECG, a respiration sensor, EEG, abdomen electrodes, motion sensors, etc. PSG is a costly setup available only in tertiary facilities, and the test proves to be cumbersome and complex, conducted during the entire night and a set of sensor electrodes put on the subject for monitoring of different vital parameters continuously. Hence, the ECG signal can be used for gathering the parameters that can help in the study of sleep apnea. The detection and diagnosis can be error-prone if it is a manual operation by an expert as the decision can vary from person to person. Hence, the use of computer algorithms can simplify the task of detecting sleep apnea from ECG parameters. Automation of Sleep apnea detection using ECG-based systems can provide noninvasive, accurate, and cost-effective diagnosis methods [1–6]. The treatment for CSA is the use of specific drugs to stimulate breathing. Continuous positive air pressure (CPAP) is a technique used to keep the airway open with forced mild air pressure. The second method is the use of Bi-level airway pressure that supplies stable and continuous pressure in the upper airway, during breathing inhale and exhale operation, which is employed after the condition of no airflow is detected. The third method is the adaptive servo ventilation (ASV) system, which is an intelligent, trained system that uses the forcing of pressure to regulate the breathing pattern [4–6]. The focus of the proposed study is OSA.

1.2   ECG Signal Characteristics

ECG signal is an important diagnostic modality used for the study of the heart condition. ECG signal is of small amplitude, typically 0.5 mV in an offset environment of 300 mV and having the frequency range of 0.05–100 Hz. An ECG signal is formed by different wave components and variations in terms of amplitudes and time intervals as a result of electrical activity caused by atrial and ventricular depolarization (contraction) and repolarization (expansion) as captured from different leads. ECG signal is thus a recurrent occurrence of wave components, namely, P wave, QRS complex, and T wave in every heartbeat cycle, and its ideal form is graphically represented as shown in Fig. 1. The ECG signal can be analyzed to understand information based on heart rate, rhythm, and morphology. ECG parameters are represented in terms of

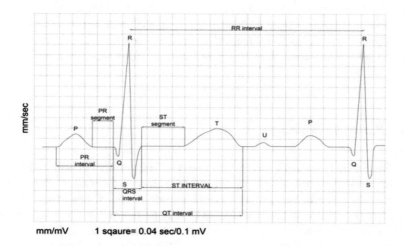

Fig. 1 Intervals and amplitudes for an ideal (normal) ECG waveform [12–17]

Table 1 Time values and amplitudes of ideal ECG waveform [6, 7]

ECG component	Time (s)	ECG wave	Amplitude (mV)	Frequency (Hz)
PR interval	0.12–0.20	P wave	0.25	0.5–10
QT interval	0.35–0.44	Q wave	0.4	QRS complex 3–40
ST segment	0.05–0.15	R wave	1.6	
P wave interval	0.11	S wave	<2	
QRS complex	0.09	T wave	0.5	0.5–10

its wave amplitudes, time intervals, namely, PQ, QRS, and TU intervals as well as PR segment, ST segment, and TP segments as indicated in Fig. 1.

In the sleep apnea condition, ECG signal gets changed in amplitude and frequency characteristics due to breathing cessations causing variations in its parameters by deviating from their ideal values shown in Table 1. The measure of the deviations is of interest for the proposed study of sleep apnea detection using benchmark databases. It is a well-established fact that electrical activity of the heart captured by ECG signal provides distinct and discriminatory characteristics that are suitable for analysis of various applications and diagnosis including the study of sleep disorders [6–15].

1.3 Standards for ECG Signal Processing and Analysis

There are scientific statements released from the AHA Electrocardiography and Arrhythmias Committee, Council on Clinical Cardiology; the American College of Cardiology Foundation and the Heart Rhythm Society on recommendations for the standardization and interpretation of the electrocardiogram as Part I and Part II for Electrocardiogram and its technology. Campbell et al. [5] lists the standard guidelines for recording ECG as applied to the clinical and hospital environment, the need for trained personnel in recording and explains the consequences of incorrect recording [4–6].

1.4 Basic Methodology

The objective of the present study is to propose low-power, high-performance methods for sleep apnea detection based on ECG signal analysis using artificial neural networks (ANN) trained with appropriate training algorithms. The proposed methodology as given in Fig. 2, comprises steps namely preprocessing for noise removal followed by QRS complex detection, extraction of features based on QRS complex reference and classification of features using ANN for sleep apnea detection [12–14].

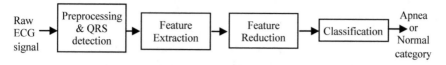

Fig. 2   Basic methodology of implementation

1.5   Preprocessing and QRS Complex Detection of ECG Signal

ECG signal processing for any application requires the basic steps as sampling and filtering, template formation, followed by feature extraction and measurement. QRS detection mechanism involves two important processes, namely, (i) preprocessing stage comprising linear filtering or Nonlinear filtering and (ii) decision stage having the peak detection logic and decision-making function. Preprocessing of ECG signal is carried out as a prerequisite before QRS detection to free the signal from undesired noise due to baseline wander, motion artifacts, and power-line interference by various means. A typical ECG signal is preprocessed for noise removal. The ECG signal is characterized by QRS complexes occurring in it. The QRS detection is implemented by employing methods, namely, Pan–Tompkins method, Hilbert transform, and wavelet transform methods [12–17].

1.6   Pan–Tompkins (PT) Algorithm

The algorithm uses the combination of digital filters, I derivative filter followed by squaring stage and moving window integrator stage to detect the QRS complex as shown in Fig. 3. The proposed work uses the Pan–Tompkins algorithm for the detection of peaks in the ECG signal, and finally, the decision-making on QRS complex detection is made using the adaptive threshold method [17–19].

It is a real-time algorithm that uses slope, amplitude, and width of the QRS complex for its detection. Its implementation is done in the following four stages, as shown in Fig. 4.

After MWI, the decision-making of the detection process uses the peak detection logic with an adaptive threshold technique.

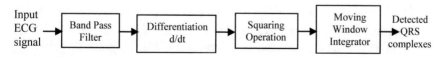

Fig. 3   PT algorithm to locate QRS complexes in ECG signal

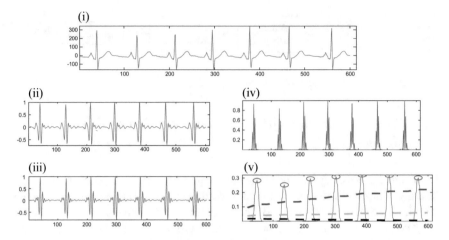

Fig. 4 Stages of Pan–Tompkins algorithm: (i) a raw ECG signal X05, (ii) BPF output, (iii) differentiator output, (iv) squared output, and (v) output of MWI

1.7 QRS Detection Methods in the Frequency Domain

The basic methodology of detection of QRS complex involves two steps, namely, QRS enhancement followed by QRS detection. The methods based on filter banks, Hilbert transform (HT), and wavelet transform (WT) are employed under frequency domain to ensure accurate QRS detection and robust noise performance [20–27].

1.7.1 Hilbert Transform-Based QRS Complex Detection Method

The Hilbert transform (HT) method is better suited to process long-term recordings of large ECG Apnea database. It is used to calculate instantaneous values of time series ECG signal with respect to its amplitude and frequency. In the time domain, it effectively detects the envelope of the ECG signal. Thus, the generated analytical signal contains real and imaginary parts and is represented as in Eq. (1).

$$Y(t) = y(t) + jh(t) \tag{1}$$

where $Y(t) = A(t) \exp[j\psi(t)]$ with $A(t) = \sqrt{y^2 + h^2}$ is the amplitude and $\psi = \mathrm{Tan}^{-1}\left(\frac{h}{y}\right)$ is the phase of the complex analytical signal.

The phase of the signal provides information about the local symmetry of the signal and helps in its R peak detection. Thus, around this maximum peak, the other parameters are determined for the input ECG recording for every 1 min [22, 23]. The result of HT-based QRS detection in a sample ECG signal record is shown in Fig. 5.

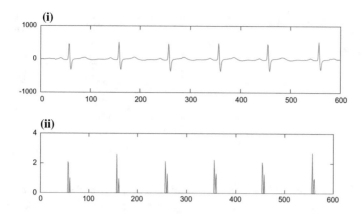

Fig. 5 (i) Preprocessed ECG signal and (ii) R peak detected signal using HT

1.7.2 Wavelet Transform-Based QRS Detection

In the proposed work, DWT is applied to the raw ECG signal in order to de-noise as well as to detect QRS complex. The stages involved in the application of WT on the ECG signal for de-noising as well the QRS detection are as shown in Fig. 6.

Hence in the proposed system, db4 wavelet as shown in Fig. 7 is convolved with ECG signal for de-noising and extraction of QRS complex accurately. Discrete wavelet transform (DWT) uses the two filters, an LPF and an HPF. The output coefficients of DWT got from high-pass filter are called as detail coefficients, and the output coefficients from LPF are called as approximate coefficients. This is the first level of decomposition. The approximation signal can be sent again to HPF and LPF to decompose it into detail and approximate coefficients at the second

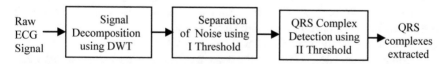

Fig. 6 Stages in the application of the wavelet transform of ECG signal

Fig. 7 Daubechies-Db4 wavelet

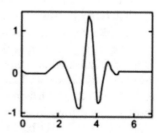

Table 2 Eight decomposition levels of ECG signal with their frequency ranges

DWT coefficients	Range frequencies (Hz)	DWT coefficients	Range frequencies (Hz)
$d1$	65–130	$d6$	2.031–4.06
$d2$	32.5–65	$d7$	1.015–2.03
$d3$	16.25–32.5	$d8$	0.507–1.01
$d4$	8.125–16.25	$a8$	0–0.50
$d5$	4.062–8.12		

decomposition level. This is continued until the desired number of decomposition levels is achieved to separate the signal components in different frequency bands as illustrated in Table 2.

The reconstruction of the signal can be easily achieved keeping the desired frequency bands and rejecting the undesired frequency bands. Here the de-noising operation uses the first threshold to separate the undesired noise frequency components from the signal to improve its signal-to-noise ratio (SNR). Using the wavelet decomposition process db04, the eight-level decomposition of the ECG signal is performed by dividing it into detail and approximate coefficients at each level [20, 21]. Table 2 provides the list of DWT, detail coefficients and the range of their frequencies. Daubechies-db4 wavelet is quite similar to the QRS complex of the ECG signal with its energy spectrum concentrated in the low-frequency range and its application to the ECG signal is shown in Fig. 7.

High-frequency noise due to power-line interference and muscle artifacts is removed by discarding the detail coefficients d1 and d2 after the first two decomposition levels to smoothen the ECG signal. Baseline wander noise is obtained separately from the last decomposition level and can be separated from the signal by subtracting d8 coefficient from the smoothened ECG signal got from the previous de-noising operation. Thus, the resulting ECG signal is free from all types of noise [12–15, 20, 21]. The selected significant frequency sub-bands of interest are recursively combined for the reconstruction of the output signal.

QRS Complex Detection of ECG Signal Using DWT

QRS complex extraction is done by the decomposition process using wavelet transform. The modulus maxima and zero crossings of WT help in locating the sharp changing QRS complex. Later the second threshold is applied to distinguish the QRS complex peak from other spurious peaks resulted due to P and T wave peaks of the signal (Fig. 8). The finer decomposition ability of WT ensures the accurate detection of QRS complex, in turn yielding to efficient feature extraction process along with the saving of computation time. The de-noised ECG signal is reconstructed keeping the desired decomposition levels having signal content and rejecting the decomposition levels with noise content. Hence, the choice of decomposition levels is important along with the choice of wavelet function [20, 21]. The feature extraction process of gathering 30 features belonging to both time and frequency domain is the same, followed for HT-based method. Feature reduction is implemented using PCA

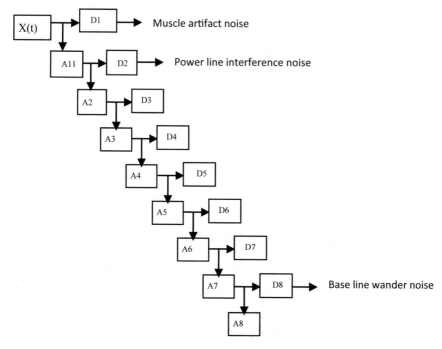

Fig. 8 ECG signal decomposition for de-noising and extraction of QRS complexes

to reduce the data redundancy, and the experimentation was conducted to decide on an optimum number of PCA components. The final classification of the ECG data as either belonging to the apnea category or normal category is done using ANN_LM and ANN_SCG classifiers (Fig. 9).

1.8   Performance Measures of QRS Detection

The results of QRS detection are compared with an expert decision and are recorded in the confusion matrix as number of true positives (TP), number of false negatives (FN), number of true negatives (TN), and number of false positives (FP). TP is the number of ORS complexes detected accurately as positive. The performance measures of QRS detection are computed in terms of accuracy, sensitivity (recall), specificity, precision in percentage values using TP, TN, FP, and FN as shown in Table 3 F-measure is the harmonic mean of Recall and Precision. Receiver operating characteristics (ROC) is the curve that is got by plotting false rejection rate (FRR) against false acceptance rate, where FRR = (1 − Sensitivity) and FPR = (1 − Specificity) [28, 29].

ROC curves are constructed by stepwise changes of the decision threshold to demonstrate the ability of detection to discriminate between classes as presence or

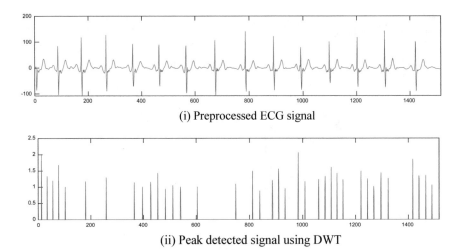

(i) Preprocessed ECG signal

(ii) Peak detected signal using DWT

Fig. 9 (i) Preprocessed ECG signal and (ii) peak detected ECG signal using DWT

Table 3 Performance measures of detection

$\text{Accuracy(Acc)} = \dfrac{\text{TP+TN}}{\text{Total No.of Records}}$	$\text{Sensitivity(Se)} = \dfrac{\text{TP}}{\text{TP+FN}} = \text{Recall}$
$\text{Specificity(Sp)} = \dfrac{\text{TN}}{\text{TN+FP}} = (1 - \text{FPR})$	$\text{Precision(Pr)} = \dfrac{\text{TP}}{\text{TP+FP}}$
$\text{F} - \text{measure} = \dfrac{2*\text{Pr}*\text{Se}}{\text{Pr+Se}}$	$\text{EER} = \text{Equal error rate} = 1 - \text{FRR} = 1 - \text{FPR}$

absence of QRS complex. The measure of EER provides the optimal operating point where the value of FRR is same as the value of FPR.

1.9 ECG Signal Feature Extraction in Time and Frequency Domain

The 30 typical ECG features are extracted keeping the QRS complex as a reference and are used for sleep apnea detection. ECG feature extraction. The features extracted from heart rate variability and RR interval variability are used for analysis in time and frequency domain, that help in classification. The list of time domain features extracted based on the RR interval variability parameter is given in Table 4.

Abbreviations for the statistical features representing RR interval variability are defined as listed below:

Table 4 Time domain parameters for RR interval variability

1. Number of beats detected	11. Mean of heart rate
2. Interquartile range	12. Median of heart rate
3. Standard deviation (SD)	13. SD of heart rate
4. Mean absolute deviation (MAD) from its mean	14. Mean PR interval
5. Mean epoch	15. Median of the PR interval
6. pNN50$_1$ 7. pNN50$_2$	16. SD of PR interval
8. SDSD measure	17. Mean of QT interval
9. RMSSD (root mean square of standard difference	18. Median of QT interval
10. QRS complex area	19. IQR of QT interval

(i)   The NN50 measure (Variant1) and NN50 (Variant2) are computed as the number of pairs of adjacent RR intervals, where the first interval exceeds the second interval by more than 50 ms, or vice versa, respectively.

(ii) pNN50$_1$ and pNN50$_2$: Measures divided by the total number of RR intervals.

(iii) The SDSD measure: Measure of deviation between adjacent RR intervals.

(iv) The RMSSD: Root means the square value of differences between the adjacent RR intervals.

(v)   IQR is the difference between 75th and 25th percentile of RR value distribution.

(vi) MAD is the Mean of Absolute values obtained by the subtraction of the mean RR interval values from all the RR interval values in an epoch.

The AHI is computed by counting the number of apnea episode occurrences in the form as described below:

(a)   The heart rate variability pattern and RR interval variability can be tracked in the time domain as (i) Repeated episodes of Bradycardia followed by an episode of tachycardia (ii) continuous episodes of tachycardia (iii) Bradycardia episode followed by abrupt tachycardia.

(b)   Spectrogram of heart rate showing a pattern of growing amplitude and decreasing frequency is tracked by analyzing the RR interval variability in the frequency domain.

(c)   Patterns with increased low-frequency (LF) band spectral power around 0.02 Hz and decreased HF band spectral power is observed in spectral characteristics.

The time domain methods are computationally simple, but they lack the ability to discriminate between sympathetic and parasympathetic contributions of HRV. Application of FFT HT and WT can provide the spectral features of ECG parameters [24–27]. The set of spectral features extracted from the ECG signal in the proposed work is given in Table 5.

Table 5 Frequency domain parameters for RR interval variability (in addition to 19 features of Table 4)

20. High-frequency (HF) power	26. Power of spectrogram of HRV
21. Low-frequency (LF) Power	27. Spectrogram of heart rate
22. LF power/HF power ratio	28. Power spectral density (PSD) of RR intervals
23. Spectrogram of S wave amplitude	29. PSD of R wave maximum
24. Spectrogram of QRS complex pulse energy	30. Apnea–hypopnea index (AHI)
25. Pulse energy	

1.10 Feature Reduction Using Principal Component Analysis (PCA)

The parameters need to be extracted for the entire length of the ECG record for processing and analysis. Even a single minute of ECG features can be a large set of values based on the sampling rate used during the data acquisition process. The values that are captured can be either in the time domain and frequency domain based on the techniques used for ECG signal processing and analysis. Hence, a statistical estimate of these time and spectral features can be computed to reduce the data size and represent data in a more interpretive manner. Feature reduction and dimensionality reduction techniques help in reducing the requirement of memory used for storing the data. Also, the time required for computation and training is lesser after using feature reduction and dimensionality reduction. Hence, the data interpretation and visualization are easier. There are several feature reduction and dimensionality reduction techniques reported in the literature [30, 31]. In the proposed work, the PCA technique has been employed.

Principal component analysis (PCA)
Features extracted from the ECG signal may have inherent redundancy. The need is to retain the features that are statistically significant, having discriminatory characteristics. Hence, the PCA is employed for dimensionality reduction of feature set, which reduces the overall computation cost-effectively. Each of the principal components is a linear combination of original variables and is arranged in the descending order of variance of the dataset. The first principal component explains the maximum variance in the dataset, whereas the second principal component explains the variance of the remaining dataset, having no correlation with the first principal component. The same is true for all the remaining components.

The input and output of the PCA stage are described as follows:

(i) PCA function in Matlab software takes the input as the feature data matrix of size $n \times p$, where n is the number of observations for each record $= 420$ (7 h * 60 min) and p is the number of features $= 30$.

(ii)   The output of PCA is the $p \times p$ matrix, where p number of principal components arranged in the order of decreasing variance. Hence, the size of the output matrix from the PCA stage is 30×30, from which the dominant principal components are chosen for computation of the performance measures of the proposed method. Then, the optimal number of PCA components required to achieve the maximum accuracy is determined. Thus, the method is employed to yield improved accuracy and reduced execution time or computation cost.

1.11   Classification Using ANN Classifiers

The proposed work employs two artificial neural networks (ANN) classifiers, namely, ANN trained by using Levenberg–Marquardt (LM) algorithm (ANN-LM) and ANN trained by using scaled conjugate gradient (SCG) algorithm guided by k-means clustering (ANN-SCG). The result of classification is that the input ECG signal recording as either belonging to the apnea category or Normal category. The classification process of ANN uses the two stages, namely, the training stage and testing stage. The database records are segregated into three sets, namely, training set, validation test, and testing set so that they are nonoverlapping. Hence, the 35 records are used for training of ANN out of a total of 70 records in the database. The remaining 35 data records are used for testing and validation. The training stage is used for training of neural network with minute-wise data from each of the train datasets and the train labels taken from their respective expert annotations [32–36].

1.11.1   ANN Trained with LM Algorithm

The inputs to the network are the PCA components, and the training of ANN is done using LM optimization algorithm. LM algorithm uses second-order numerical optimization technique combines the advantages of Gauss–Newton and steepest Descent algorithm. LM converges faster-consuming $O(N^2)$ operations and hence require larger storage, N being the total number of weights in the backpropagation method. The algorithm efficiency and stability are high but at the cost of high computation time. It provides a numerical solution to the problem of minimizing a nonlinear function. LM is one more optimal way of training the neural networks resulting in minimized error value by finalizing the weights and bias values while using LM optimization process [34–36].

The performance of the network is estimated in terms of mean of squared errors (MSE). Here the training is supported with validation and test vectors. The training of the network is stopped if the network performance fails to improve or remains the same for max_fail epochs in a row. The performance plots of ANN_LM classifier are analyzed for training and testing. The training stage is used for training of neural network with minute-wise data from each of the train datasets and the train labels taken from their respective expert annotations [9–11].

1.11.2   K-Means Clustering and ANN Trained with SCG Algorithm (ANN_SCG)

In this approach, the feature vectors in principal component space are fed to K-means ++ algorithm for clustering of data points based on their distance measure from the cluster centroids. Further, the clustered data is used as input to ANN, which gets trained with scaled conjugate gradient (SCG) algorithm. K-means clustering helps in organizing the input data samples into a finite number of clusters, which ensures data reduction. Instead of the data samples, the inputs are in the form of clusters identified by the respective cluster index and the cluster centroid location. The outputs of the network are recognition rate of training data, the recognition rate of test data, the produced class labels of training data, RMSE of training data. The performance plots of ANN classifiers are analyzed for training and testing. Scaled conjugate gradient (SCG) algorithm. The scaled conjugate algorithm follows the optimization using numerical investigation and the conjugate gradient method. Bigger the error, bigger is the scale of the gradient to ensure faster convergence. SCG does not require any user-dependent parameters and hence saves time [34, 35].

1.11.3   Artificial Neural Network (ANN)

ANN is a layered structure of neurons used in information processing. The basic artificial neural network architecture is as shown in the model shown in Fig. 10. The layers are input layer, hidden layer, and output layer.

Generally, nonlinear functions such as Tan, Sigmoid, and logistic functions are used, as shown in Fig. 11. The summation of the product of individual signals and

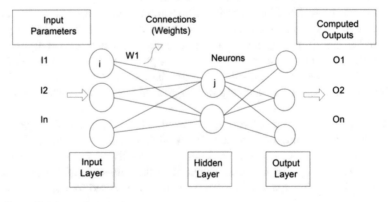

Fig. 10 Artificial neural network

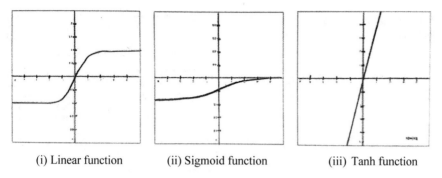

(i) Linear function   (ii) Sigmoid function   (iii) Tanh function

Fig. 11   General forms of activation functions

their weight value is used to decide on the output of the neuron using the activation function as given in Eq. (2) [34–36].

$$\text{net}_i = \sigma \sum_{j=1}^{N} w_{ij} x_j + \mu_i \tag{2}$$

where net_i = output of ith layer, x_j = inputs, N = number of input neurons, w_{ij} = link weights of the network, μ_i = threshold values for hidden neurons, and σ = activation function.

The optimum number of neurons in the hidden layer is decided based on the number of neurons in the input and output layer as $\sqrt{nm}$, if n number of input layer neurons and m is the number of output layer neurons as per a thumb rule [34–36].

Formation of Training, Validation, and Testing Datasets

ANN has three sequential stages of operation, Training, Validation, and Testing. There are designated and independent datasets used for training, validation, and testing of ANN. ANN employed for solving any real-time classification problem needs can be configured for the suitable network type, training algorithm, selection of parameters like initial weight, learning rate, momentum rate, network structure, etc. Training of ANN has to be carried out using a sufficient number of datasets [34–36].

1.11.4   Backpropagation Algorithms

The backpropagation algorithm is a supervised training algorithm, where the error is fed back to the input of the training algorithm to adjust the weights and thus reduce error, as shown in Fig. 12. The popular training algorithms are (i) gradient descent, (ii) resilient backpropagation, (iii) conjugate gradient, (iii) quasi-Newton, (iv) Levenberg–Marquardt algorithm, and (v) scaled conjugate gradient algorithm.

Fig. 12 Backpropagation
algorithm

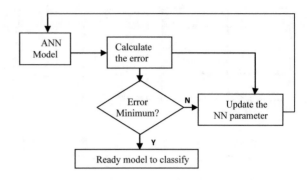

Fig. 13 Graph of MSE
versus bias weights of ANN

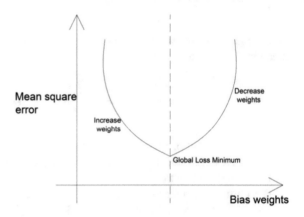

While designing ANN, there is a need to initialize weights of the connection links randomly for the first time. Hence, the training of the ANN can be done by fine-tuning the weights in such a way that the error becomes minimum using iterations involving trial and error. The adjustment of weight is done in such a way that the error is minimized by figuring out the direction in which the error gets reduced. The point is refereed to be Global loss minimum as shown in Fig. 13 as shown. The parameters like rate of convergence, number of epochs taken by the network to converge and the Mean squared error are used to compare the performance of different ANN's trained with different training algorithms [34–36].

1.12 Performance of ANN Model

The performance of the ANN model is estimated using measures like mean absolute error (MAE), mean squared error (MSE), root mean squared error, and normalized root mean squared error (NMSE). The most popular choice is MSE [28, 29] as given in Eq. (3).

$$\text{MSE} = \sum_{i=1}^{n} \frac{(y_p - y_o)^2}{n} \tag{3}$$

where n = number of datasets used for training the network, y_p = mean of predicted value, and y_o = experimental or target value.

Performance Measures of Classification
The performance measures of classification are measured in terms of accuracy (Acc), sensitivity (Se), specificity (Sp), and precision (Pr) in percentages and F-measure lying in the range from 0 to 1. Low-power behavior is estimated in terms of execution time by profiling of the algorithm. Profile time is a tool available in Matlab software that measures the time spent by a program so that the parts of the program, which are computationally intensive can be identified for the required corrections [28, 29].

2   Literature Survey

2.1   Sleep Apnea Diagnosis and Treatment Methods

There are standards developed from various internationally recognized organizations with expert groups like American Academy for Sleep Medicine (AASM), All India Institute of Medical Sciences (AIIMS), New Delhi, and other international bodies to provide guidelines for therapists and researchers working in the area of sleep disorders. Indian Society for Sleep Research (ISSR) is active since 1992, working toward creating awareness and promoting research in the area of sleep and sleep medicine by creating a forum for the exchange of thoughts by several experts in the area [1–6, 32–42].

2.2   Sleep Apnea Detection Methodologies in the Literature

The sleep apnea detection methodologies have evolved from years to analyze the different physiological signals as single or in combination. Among them, the important ones are breathing pattern, nasal airflow, ECG, EEG, SaO$_2$, limb movement, gathered from the human being using a set of sensors [43–59]. Some of the physiological parameters in single or combination used for sleep apnea study along with extracted feature parameters are tabulated as examples in Table 6.

Table 6 Physiological parameters used for sleep apnea study

Author	Physiological parameters	Feature extraction
Liu et al. [39], Lin et al. [40]	EEG	The measure of a shift from delta waves to theta and alpha waves
Chazal et al. [41], Almazaydeh et al. [42], Bsoul et al. [43]	ECG	Cyclic variations in heart rate, RR interval features for short duration epochs
Almazaydeh et al. [44], Canosa et al. [45], Marcos et al. [46], Burgos et al. [47]	SpO2	Oxygen de-saturation level, the linear combination of spectral and nonlinear features used with PCA
Zhao et al. [48], Andrew et al. [49]	Snoring	Formant frequency estimation

2.3   Sleep Apnea Detection Systems for Home Use

Polysomnography (PSG) is a sophisticated and lengthier lab test performed at a hospital with a sleep lab facility to detect different types of sleep disorders. Home testing devices help the patient to get the testing done at ease and leisure at an affordable cost [43–59]. AASM (2009) approves the use of sleep monitoring devices as screening tools. Table 7 provides a list of some of the real-time OSA monitoring devices.

The survey of the work carried out by researchers on ECG signal preprocessing, QRS complex detection, and feature extraction are discussed in further sections.

2.4   Detection Methods of QRS Complex in Literature

The methods available for QRS complex enhancement are categorized under time domain, frequency domain, and transform domain approaches. Among the time domain approaches for QRS enhancement, amplitude threshold, first derivative method, second derivative method, and digital filter are used popularly. Mathematical morphology filtering, empirical mode decomposition (EMD) and filter bank methods are popular among frequency domain methods. Hilbert transform method and wavelet transform constitute the transform domain methods. Among the different QRS detection methods available, threshold-based method, artificial neural networks, hidden Markov models, matched filters, syntactic method, zero crossings, and singularity method are used by researchers. The combination of different QRS enhancement and QRS detection methods can be experimented based on the application need and their performance in terms noise sensitivity, the number of parameters to be handled and the level of complexity [12–27]. The summary of the time domain methods used for QRS detection is given in Table 8 [60–78]. Pan–Tompkins method [18, 19] and

Table 7 Real-time OSA monitoring devices

Parameters	System features	Communication modules	Type of processing
Multiparameter input	ApneaLinkTM Plus [56] for respiratory effort, pulse, oxygen saturation, and nasal airflow	Data storage facility	Off-line (commercial)
	Alice Night One from Philips [58], Respironics, 3 sensors, 7 channels for airflow, snore, thorax effort, SpO2, pulse rate, body position, and PPG	USB, wireless	Online (commercial)
	HealthGear uses blood oximeter to measure blood oxygen level [59]	Bluetooth and a smartphone	Online
ECG only	Apnea MedAssist [43], using an Android operating system-based smartphone using single lead ECG sensor, an SVM classifier	Bluetooth connectivity	Online (commercial)

Table 8 Summary of time domain methods

Author	ECG analysis with MITDB	Performance (%)
Pan and Tompkins [18, 19]	Slope, amplitude, the width of ECG, adaptive threshold	Acc = 99.3
Szczepanski et al. [62]	Voltage values and time distribution values	Acc = 99.4
Melgarejo et al. [28]	POLLING function on multi-lead ECG	Acc: 99.5, Se: 99.8, Sp: 99.5
Lu et al. [63]	I derivative + differentiation + squaring + moving average integration	Adaptive threshold Acc: 99.72, Sp: 99.72

a similar method proposed by Lu et al. [63], both being time domain approaches and computationally efficient with a competent detection accuracy over 99% were adopted for QRS detection of ECG signals in one of the proposed approaches.

Table 9 Summary of methods in the transform domain

Author	ECG-based method/features extracted from MITDB	Classifier and performance measures[a] (%)
Benitez et al. [22]	First differential + HT method	Acc = Se = Sp = 99.9
Ulusar et al. [64]	HT for fetal QRS complex using multiple ECG channels, 156 datasets	Adaptive rule-based decision, Acc: 98.9
Bsoul et al. [65]	Wavelet transformation	Acc: 99.8
Zidelmal et al. [67]	Power spectrum of energy levels, WT	Acc: 99.8 Se: 99.6
Kaur and Rajni [75]	Hybrid linearization + Kalman filter + WT	Se: 99.9, PPR: 99.97
Park et al. [76]	WT + modified Shannon energy envelope	Se = 99.9 = PPR, Acc: 99.8

[a]*Acc* Accuracy, *Se* Sensitivity, *PPR* Positive prediction rate

There is significant work carried out in the QRS detection of ECG signal by using transform domain by several researchers, some of which are summarized in Table 9 [20–27, 61–78].

Based on the survey, the list of different QRS enhancement and QRS detection methods are tabulated, as shown in Tables 10 and 11, where the summary of their performance in response to noise, number of response configuration parameters, and the order of computation complexity are presented [15–27, 63, 64].

Among the transform domain approaches, HT and WT methods are quite competitive, while being used to emphasize QRS complex and accurately detect the position and amplitude of QRS complexes. Similarly, the summary of decision-making used in QRS detection process is given in Table 11 based on the categories made, namely, threshold, neural network, fuzzy logic, neuro-fuzzy logic, hidden Markov model (HMM) and matched filter [79–90]. QRS complex being of the distinct and dominant wave shape, with definite bandwidth, can be easily detected using the threshold technique as compared to all other waves of ECG signal. The proposed methods employ threshold method for detection of a true QRS complex against a false QRS complex.

2.5   Feature Reduction Methods as Per Literature

Some of the works proposed by researchers using feature reduction methods are presented in Table 12 [60, 77, 78].

Table 10 QRS enhancement methods

Technique	Algorithm to emphasize QRS complex	Performance	Order of complexity
Amplitude threshold [16, 17, 61, 62]	First(I) derivative + amplitude threshold	Lower accuracy for lengthy ECG	Lower
I Derivative Method [16–19]	I derivative + threshold, I derivative + digital filter + threshold, morphology filter + I derivative, I derivative + HT + threshold, I derivative + WT + threshold	High-frequency noise problem	Lower
Digital filter [20–25, 60–65]	I derivative + digital filter + threshold, band-pass filter (BPF) + I derivative + threshold, BPF + HT + threshold	SNR improved with filter type, order	Lower
EMD [20, 21]	EMD + threshold, EMD + singularity + threshold, HPF + EMD + threshold	Improves SNR	Higher
HT [22, 23, 68, 69]	I derivative + HT + threshold, BPF + HT + threshold, WT + HT + threshold	Effective de-noising	Higher
WT [20, 21, 65, 67, 68, 70]	WT + threshold, I derivative + WT + zero crossing + threshold, WT + HT + threshold	Effective de-noising	Lower

2.6   Classifiers for Sleep Apnea Detection as Per Literature

Some of the classifiers proposed in the literature on sleep apnea detection methods experimented with Physionet.org's ECG Apnea database (PNDB) are summarized in Table 13 [32–59, 79–87].

Table 11 QRS detection methods

Technique	Steps in algorithm	Performance
Threshold [16, 17, 61, 62]	The threshold applied to last stage	Signals with low SNR and fixed threshold cause inaccurate detection
Neural networks (NN) [39–42, 79, 80]	WT + NN, NN + matched filter	Highly sensitive to noise, improved performance, need multiple parameters. Iterative training, higher memory storage
Fuzzy and neuro-fuzzy technique [79, 80]	Fuzzy rules, mimicking human decision, fuzzy inputs + ANN	Robust programming needed with a large set of fuzzy rules to build intelligence

Table 12 Summary of feature reduction methods

Author	Feature reduction methods with MITDB	Performance measures (%)
Kanaan et al. [77]	Binary SVM + PCA, binary SVM + kernel PCA	PPR: 95 and Sp: 100
Martis et al. [60, 78]	Three approaches gather principal component (PC) features + ANN + LS-SVM	Acc(max): 98.1, Se: 99.9, Sp: 99.1 PPR: 99.6
	DCT coefficients of ECG + PCA + ANN + LS-SVM + probabilistic neural network	Se: 98.69, Sp: 99.91 Acc: of 99.52

Table 13 Comparison of classifiers used for sleep apnea detection

Classifier	Author	ECG features and PNDB	Performance measures (%)
KNN classifier	Penzel et al. [41]	PSD of HRV and R peak area, Bi-variate AR model	Acc: 100
Fuzzy classifier	Vafaie et al. [85]	A dynamic model of ECG signals	Fuzzy classifier Acc: 93.34% genetic algorithm Acc: 98.7%
K-means + SVM	Shen et al. [81]	MITDB	Acc: 98.9%

3 Motivation and Objectives

3.1 Motivation

Awareness toward sleep apnea in India as well across the world is very poor. Polysomnography (PSG) is the gold-standard test used for detecting sleep apnea problem and requires sophistication in the facilities in the form of the sleep lab, instrumentation systems, and monitoring equipment. The facility is not affordable by poorer sections of society. Even it is a cumbersome test. There is a need for a simplified and reliable screening tool that can be used at primary and secondary health care facilities. The monitoring devices need to be portable and energy-efficient. Apart from using the optimized hardware for data acquisition, the design of efficient algorithms is very important [79–90].

The problem statement is:

To design and develop a low-power, high-performance algorithm for sleep apnea detection based on ECG signal using ANN classifier and experiment it with benchmark ECG signal dataset and also an own dataset.

3.2   Proposed Objectives

(i)   ECG signal-based sleep apnea detection using Hilbert transform method and extract time and spectral features for classification using ANN classifiers using Matlab 2017A.
(ii) ECG signal-based sleep apnea detection using Pan–Tompkins algorithm and extract time and spectral features for classification using ANN classifiers using Matlab 2017A.
(iii) ECG signal-based sleep apnea detection using wavelet transform method and extract time and spectral features for classification using ANN classifiers using Matlab 2017A.
(iv) ECG signal-based sleep apnea detection methods using Hilbert transform, Pan–Tompkins algorithm, and wavelet transform and extract time and spectral features for classification using ANN classifiers using Matlab 2017A.
(v)   To experiment with the proposed algorithms with an own dataset using Matlab 2017A.

4   Results and Discussion

The implementation of all the proposed methods is done using Matlab2017a on a personal computer with configuration, Intel(R) Core(TM), i3CPU, 2.4 GHz, 4 GB RAM, and 64-bit operating system. The ECG signal nighttime recordings are gathered from the publicly available benchmark dataset, namely, ECG Apnea database from PNDB. It consists of 70 nighttime ECG recordings of length from 8 to 10 h, sampled at 100 Hz and the expert annotations on occurrences of QRS beats and apnea events. The database has ECG recordings of apnea patients (a01–a20), borderline-case patients (b01–b05) and control patients (c01–c10). This forms a set of 35 min-wise annotated recordings used for training of ANN classifier. The expert annotations contain labels as either "A" or "N", indicating the presence of apnea or normal condition, respectively, for each minute. Every one minute of ECG record is categorized as either an apnea minute or a normal minute. There is a collection of other 35 ECG recordings X01–X35 of both apnea patients and normal beings, which are used as test datasets. A total of 55 ECG records are chosen from SDMCMSH database of Sleep lab facility, SDM College of Medical Sciences and Hospital, Sattur, Dharwad, Karnataka state, India. The ECG signals are annotated by medical experts at the facility. The ethical committee clearance certificate has been obtained for the usage of the ECG records for the present experimental work [7–10, 91–95].

4.1 Sleep Apnea Detection Using the HT Method and ANN Classifiers

In this section, the sleep apnea detection method using HT-based feature extraction and ANN-based classification is discussed. The QRS detection methodology of ECG signal carried out using HT method, followed by feature extraction of a total of 30 features (19 time domain features and 11 frequency features). Feature reduction technique is applied with PCA to decide on a reduced set of discriminatory features needed for classification. Further, the sleep apnea detection and classification are done using two ANN classifiers, namely, one trained with LM algorithm and the other trained using SCG method. Among the two classifiers, the results of ANN_SCG classifier are better than ANN_LM classifier. The performance measures of HT-based ANN_ classifiers vary in the range from 94 to 99(%). The Hilbert transform is chosen to transform the ECG signal to the frequency domain for feature extraction [22, 23, 64, 68, 69].

4.1.1 Proposed Methodology

The proposed methodology of detection of sleep apnea using ECG signal comprises the following steps: (i) preprocessing to remove noise present in ECG signal using band-pass filter, (ii) QRS complex detection to extract peak references by applying HT, (iii) feature extraction in terms of ECG parameters, (iv) feature reduction using PCA, and (v) classification using two ANN classifiers trained with Levenberg–Marquardt (LM) algorithm and scaled conjugate gradient (SCG) method, respectively [91–94]. The block diagram of the proposed methodology is shown in Fig. 14.

4.1.2 Results of Preprocessing and QRS Complex Detection

The preprocessing implemented using FFT and band-pass filter (BPF) has improved the signal-to-noise ratio (SNR) of the ECG signal by 9% (approximately). The SNR

Fig. 14 Block diagram of the proposed methodology of sleep apnea detection using HT-based method

values are computed using the power of the periodic signal and noise. The sample signal records, namely, ×05, ×08, ×10 and ×15 from the ECG Apnea database are shown in Fig. 15 in their original form and the corresponding resultant signal after applying Hilbert transform (HT).

The performance of R peak detection algorithm experimented with the 35 recordings is presented in terms of average values of total beats, detected beats, accuracy, sensitivity, precision, specificity, F-measure, and execution time in Table 14. The results indicate that the QRS complex detection is effectively done and the performance measures, namely, accuracy, specificity, sensitivity are estimated as 98.7%, 92%, 94%, and 94%, respectively, for all the three cases, namely, normal, borderline apnea, and apnea cases. The F-measure is computed as 0.93. Closer, the F-measure value is to 1, better is the algorithm. It is observed that HT does not contribute to any further improvement of SNR of the ECG signal.

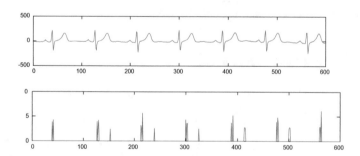

Fig. 15 Sample ECG signal record × 10 and their R peak detected signals [7–10]

Table 14 Average values of performance measures of QRS complex detection using the HT method

Parameters	Average
Total beats	35,968
Detected beats	35,604
Accuracy (%)	98.7
Precision (%)	92
Sensitivity (%)	94
F-measure	0.93
Specificity (%)	94
Execution time (s)	1.42

Fig. 16 ROC curve for FRR
versus FAR for QRS
detection using the HT
method

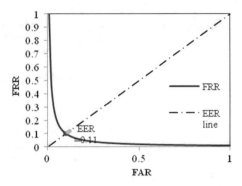

The execution time of the QRS detection algorithm is the time spent in processing the test ECG signal recording and is estimated using the profiler option in Matlab 2017a. The average execution time of the HT algorithm alone is found to be 1.42 s per minute recording as compared with the average total time of 18.04 s spent in the processing of the entire recording. Receiver operating characteristic (ROC) curve obtained by plotting false acceptance rate (FAR) versus false rejection rate (FRR) is shown in Fig. 16. The FAR = 1 − specificity, while the FRR = 1 − sensitivity. The Equal error rate (ERR) corresponds to the point on the ROC curve at which FAR = FRR and the ERR (=FAR = FRR) is of the value 0.11. The system operating at the point of ERR exhibits the optimal performance.

4.1.3   Results of Feature Extraction

The proposed method uses the Hilbert transform to detect QRS complex feature, using which the feature extraction of 30 features gathered minute-wise, belonging to both time domain (19 features) and frequency domain (11 features) is attained (Sect. 1.8). Feature extraction is carried out on all the 35 training ECG records, namely, a01–a20, b01–b05 and c01–c10 records, and the 30 features are computed minute-wise from each of these records. Thus, a total of 14,700 feature vectors of dimension 30 are obtained (35 records * 7 h * 60 min). A feature vector of dimension 30 per minute is used during the training phase of ANN. During the test phase of ANN, the feature extraction process yields 420 feature vectors, each of dimension 30 (1 record * 7 h * 60 min) for every test record. PCA stage is used to improve the efficiency of the feature extraction process by eliminating redundancy in the data [95].

4.1.4   Results of Feature Reduction and Classification Accuracy

There is a need to select an optimum number of PCA components in order to select the reduced and significant feature set to be given to the classifier. Hence, the experimentation has been done to estimate the classification accuracy and average execution time by using the different number of PCA components for sleep apnea detection in case of both ANN-LM and ANN-SCG algorithms and the results are tabulated in Table 15. It is observed that ANN-SCG requires only 20 PCA components to attain accuracy of 99.2% with 6.5 s of execution time, whereas ANN-LM algorithm requires 25 PCA components to attain accuracy of 98.7% with execution time of 15.5 s. The use of a higher number of PCA components does not yield any appreciable improvement in the values of accuracy. Thus, the ANN-SCG algorithm outperforms ANN-LM algorithm in terms of reduced computational cost. Lower the number of inputs to the classifier, lower is the complexity of the classifiers with a reduced number of neurons. The reduction of the complexity of the classifier leads to a reduction of computation time and thus making the algorithm power-efficient [95].

The plots of the accuracy and the execution time attained for the different number of PCA components used are shown in Figs. 17 and 18, respectively, for both the classifier algorithms.

Table 15 Classifier accuracy and execution time computed for different number of PCA components [95]

No of PCA components	Classification accuracy (%)		Execution time (s)	
	ANN-LM	ANN-SCG	ANN-LM	ANN-SCG
5	94.5	95.2	12	3.9
10	97.4	98.3	15	3.7
15	98.1	99	16	5.2
20	98.2	99.2	15.2	6.5
25	98.5	99.2	15.5	6.9
30	98.5	99	20	7.8

Fig. 17 Plot of accuracy versus number of PCA components for both ANN classifiers

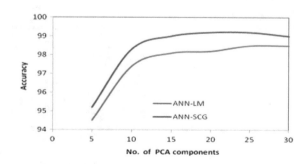

Fig. 18 Plot of execution (profile) time versus number of PCA components for both the ANN classifiers

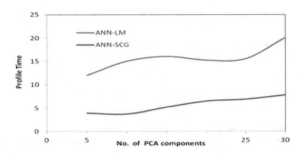

4.1.5 Results of Classification Using ANN-LM and ANN-SCG Algorithms

The classification results of ANN-LM and ANN-SCG algorithms tested for the records of ECG Apnea database are tabulated in Table 16. For ANN-LM algorithm, the average values of Acc, Pr, Se, and Sp are found to be 98.5%, 95.2%, 95%, and 94%, respectively, with 25 PCA components. For ANN-SCG algorithm, the average values of accuracy, precision, sensitivity, and specificity are found to be 99.2%, 96%, 97%, and 98%, respectively, with 20 PCA components, which indicate a marginal increase in comparison with that of ANN-LM. But in terms of execution time and MSE, there is a significant decrease achieved by 58% and 83%, respectively, for ANN-SCG algorithm as compared to that for ANN-LM algorithm. The F-measure, which is the geometric mean of Precision and Sensitivity (Recall), is 0.965 for ANN-SCG algorithm, which is also marginally higher as compared to that of ANN-LM algorithm. The inherent nature of fast convergence behavior of ANN-SCG algorithm has significantly contributed to the decrease of MSE and execution time [95].

Receiver operating characteristic (ROC) curves obtained by plotting FAR vs. FRR for both the ANN classifier algorithms are shown in Fig. 19. It is observed from tabulated results and ROC curves that the ANN-SCG algorithm supported by K-means clustering is proved to be a better classifier as compared to ANN-LM algorithm.

Table 16 Comparison of performance of ANN classifiers

Measures of performance	ANN-LM algorithm	ANN-SCG algorithm
No of PCA	25	20
Accuracy (%)	98.5	99.2
Precision (%)	95.2	96
Sensitivity (%)	95	97
Specificity (%)	94	98
F-measure	0.95	0.965
MSE	0.24	0.04
Execution time (s)	15.5	6.5

Fig. 19 ROC for classification using ANN-LM and ANN-SCG algorithms

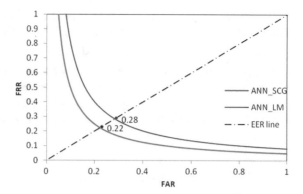

The performance plots on error performance, validation performance, and regression, obtained during ANN training and testing are analyzed. The significant improvements are observed in terms of execution time and mean square error because of the faster convergence behavior of SCG method supported with K-means clustering. The results indicate that the ANN classifier trained with SCG algorithm guided by K-means clustering outperforms ANN classifier trained using the LM algorithm in terms of computational cost. These results are significant and have the potential for application to sub-classification of sleep apnea cases with greater accuracy and reduced time cost [95].

4.1.6 Comparison of the Proposed Method with Other Methods

The performance comparison of the proposed method with different methods in the literature for sleep apnea detection based on ECG signal analysis are tabulated in Table 17 in terms of accuracy (Acc), sensitivity (Se), and specificity (Sp) as reported

Table 17 Comparison of performance of the proposed HT-based ANN classifier with other methods in the literature

Methods	Features	Classifier	Performance measures (%)
Proposed method	Time–frequency features, AHI + HT and PCA	ANN-LM algorithm	Acc: 98.5, Se: 95, Sp: 94
		ANN-SCG + K-means clustering	Acc: 99.2, Se: 97, Sp: 98
Khandoker et al. [50]	Wavelet	Support vector machines	Acc: 93%
Varon et al. [83]	Time domain features principal components	Least squares support machines using an RBF kernel	Acc > 90%

in the literature [27–29, 95]. The dataset used for experimentation for the methods (Table 17) is the ECG Apnea database of PNDB [7–10].

Future Scope

The proposed method yielded the results that are quite encouraging in terms of higher values of performance measures as well the satisfactory execution time of the algorithm. HT-based feature extraction method, followed by ANN classifier trained using ANN_SCG algorithm is one of the good choices for detection of sleep apnea.

4.2 Sleep Apnea Detection Using PT Method and ANN Classifiers

In the present section, again, the PT method is used for QRS detection, followed by the extraction of the same set of 30 features used by HT method, as described in Sect. 1.8. Further, PCA is applied to the feature set to reduce the number of features to avoid redundancy and reduce computation time. The experimentation was carried out to determine the optimal number of principal components representing the feature vectors to attain the desired classification accuracy. Finally, the classification is carried out using the two classifiers ANN-LM and ANN_SCG and the results are compared with that of HT-based ANN_LM and ANN_SCG classifiers as demonstrated in Sect. 4.1.6 [95, 96].

4.2.1 Proposed Methodology

The proposed methodology comprises the following steps: (i) Preprocessing the input ECG signal for noise removal, (ii) QRS detection process using Pan–Tompkins algorithm to locate QRS complexes in ECG signal, (iii) Feature extraction to gather ECG parameters, keeping QRS complex as the reference, (iv) Feature reduction using principal component analysis (PCA), and (v) classification using ANN to classify the input ECG signal record as either belonging to the apnea category or normal category. The block diagram of the proposed methodology is shown in Fig. 20.

Fig. 20 Proposed PT-based method of sleep apnea detection

4.2.2   Results of Preprocessing

The digital filtering using band-pass filter (BPF), followed by differentiation operation used in the implementation of PT algorithm has improved the SNR by 12%, approximately.

4.2.3   Results of QRS Detection

The average performance measures are estimated in terms of accuracy (Acc), sensitivity (Se), specificity (Sp), and precision (Pr), which are found to be 94%, 95%, 93%, and 92%, respectively. The performance measures of the proposed Pan–Tompkins algorithm-based QRS detection experimented with X01–X35 records of PNDB are tabulated along with the performance measures attained using HT method demonstrated in for comparison in Table 18. The performance measures of the proposed method, Acc, Se, Sp, and Pr are considerably lower than that of the HT method. But the execution time of the PT method is quite lower, as an advantage over the HT method. Hence, QRS complex detection using PT method proves to be highly computationally efficient. F-measure, which is the harmonic mean of precision and sensitivity, is found to be 0.92 for the proposed method as compared to 0.93 value computed for the HT method [95, 96].

The receiver operating characteristics (ROC) curve of QRS detection is shown in Fig. 21, as a plot of false rejection rate (FRR) versus false acceptance rate (FAR), where FRR = 1 − Sensitivity and FAR = 1 − Specificity for both the methods. The equal error rate (EER), the desired operating point on the ROC curve with equal values of FRR and FAR, is found to be 0.176 and 0.11 for the PT method and HT method, respectively. For the HT method, EER is reduced by 37% as compared with the proposed method of QRS detection, which has the advantage of reduced computation cost by 65% as compared with its counterpart [95, 96].

Table 18   Comparison of average values of QRS detection performance measures

Performance measures	Proposed PT method	HT method (Sect. 4.1.3)
Accuracy (%)	94	98.7
Sensitivity (%)	95	94
Specificity (%)	93	94
Precision (%)	92	92
F-measure	0.92	0.93
Execution time (s)	0.5	1.42
Equal error rate	0.176	0.11

Fig. 21 ROC curve of FRR
versus FAR for QRS
detection for proposed PT
and HT method

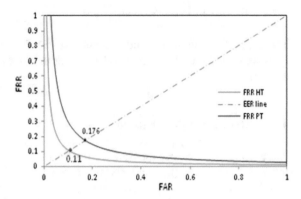

4.2.4 Results of Feature Extraction

Feature extraction is carried out on all the 35 training ECG records, namely, a01–a20, b01–b05 and c01–c10 records, and the 30 features are computed minute-wise from each of these records. Thus, 14,700 feature vectors of dimension size, 30 are obtained (35 records * 7 h * 60 min), i.e., one feature vector per minute, which are fed to PCA stage to select the dominant set of features to be used for training of ANN classifiers. Later, during the test phase, for every test record, the feature extraction process yields 420, minute-wise feature vectors (1 record * 7 h * 60 min) of dimension size, 30 that are fed to PCA stage [95, 96].

4.2.5 Results of Feature Reduction and Classification Accuracy

The experimentation has been done on test ECG records with a different number of PCA components to determine the number of dominant principal components sufficient enough to attain the desired accuracy of classification. The classification accuracy and execution time for a different number of PCA components in case of both the classifiers under the proposed method is summarized in Table 19. The same tabulated data are shown as graphs in Fig. 22 and Fig. 23, respectively. There is no significant improvement in accuracy and execution time observed with further increase in the number of PCA components after 20. Hence, the first 20 dominant features are used with both the classifiers during train and test stage. The accuracy is found to be 95% for ANN_SCG classifier, a significant improvement over 92% found for ANN_LM classifier. But the execution time of ANN_SCG classifier is reduced by a greater margin, taking around 1.2 s against 3.5 s of execution time required for ANN_LM. Hence, by reducing the execution time, the computation cost gets reduced by 66% approximately. The saving attained is due to the reduction of feature set size achieved through PCA by 43% during training and 33% during testing, along with the saving of time achieved by fast converging SCG training algorithm [95, 96].

Table 19 Classification accuracy and execution time versus number of PCA components for ANN_SCG and ANN_LM algorithms

No. of principal components	Proposed ANN classifier with PT method				ANN classifier with HT method in Sect. 4.1.5			
	Accuracy (%)		Execution time (s)		Accuracy (%)		Execution time (s)	
	LM	SCG	LM	SCG	LM	SCG	LM	SCG
5	82	85	0.8	0.5	94.5	95.2	12	3.9
10	88	91	1.9	0.83	97.4	98.3	15	3.7
15	89	94	2.3	0.95	98.1	99	16	5.2
20	92	95	3.5	1.2	98.2	99.2	15.2	6.5
25	92	95	4.5	1.6	98.5	99.2	15.5	6.9
30	92	95	6	2.1	98.5	99	20	7.8

Fig. 22 Accuracy versus no. of PCA components for ANN_LM and ANN_SCG of proposed PT method

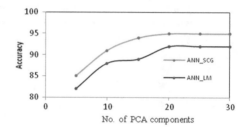

Fig. 23 Execution time versus no. of PCA components for ANN_LM and ANN_SCG of proposed PT method

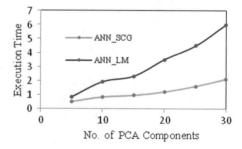

Further, the results are compared with results of the classifier using HT-based feature extraction method proposed in Sect. 4.1.4, which use 25 and 20 principal components for ANN_LM and ANN_SCG methods, respectively. As both the classifiers use the same training algorithms LM and SCG, the results can be compared in a justifying manner. The proposed classifiers both ANN_LM and ANN_SCG require 20 principal components for classification achieving the lower performance measures, but with improved computation cost efficiency as compared with HT-based ANN classifiers proposed in Sect. 4.1.5. It is observed that the ANN_SCG classifier performs better as compared with ANN_LM classier for both PT and HT methods, in terms of each of the performance measures as well as the computation time. This is due to the support of K-means clustering used with SCG method along with its inherent fast convergence nature. While considering the final classification, the HT method performs better in terms of its accuracy measure, whereas the PT method performs well in terms of significant computation efficiency (Table 19) [95, 96].

4.2.6 Results of Classification Using ANN Algorithms

As per the experimentation results got in the PCA stage (Sect. 4.1.5), the proposed method uses 20 principal components for classification of test records. The proposed algorithm uses the train data and train labels for training the ANN classifier utilizing twenty dominant PCA components that represent the features ECG signal and is tested with X01–X35 records of the database. The tabulation of results of detection of apnea minutes for each record is done in Table 20, as percentage values using the total recording time of ECG signal, for both the classifiers, ANN_LM and ANN_SCG.

Expert annotations are available as the number of apnea minutes (in percentage) and also the decision of a record belonging to apnea category as "Yes" and the one belonging to the normal category as "No". A comparison of results of ANN_SCG and ANN_LM algorithms is done with the expert annotations to estimate the number of TP, FN, TN, and FP values, needed to compute the performance measures of classification summarized in Table 21. The test input ECG signal record is decided to be of apnea category if the percentage of detected apnea minutes is more than 10%; otherwise, it is decided to be of Normal category. The summary of the results of Table 20 is shown in Table 21 for comparison of the performance of the proposed method with the HT method described in Sect. 4.1.5 and Sebastian et al. [87]. The performance measures of both the ANN_LM and ANN_SCG classifiers are better than the method proposed by Sebastian et al. [87]. But in comparison with ANN classifier using HT-based feature extraction method proposed in Sect. 4.1.4, the performance measures for the proposed classifiers are quite low [95, 96].

But, the execution time taken by proposed classifiers is quite low, making it computationally efficient as compared with HT method. Accuracy values obtained by the proposed PT-based method reduce by 6.5 and 4.2% for LM- and SCG-based ANN classifiers as compared with that obtained by HT-based method. However, the total execution timings of the proposed PT-based method reduce by 77.4% and 81.5% for LM- and SCG-based classifiers, respectively, as compared with that for HT-based method. The computation time incurred by the proposed PT-based method is significantly lesser than that for HT-based method, but with a significant decrease in the accuracy.

The performance plots of error, validation check, and regression for both the ANN classifiers, ANN_LM, and ANN_SCG are analyzed. The performance measures of the proposed method are compared with some of the other methods reported in the literature, as shown in Table 22 [65, 83, 96–99].

Table 20 Results of sleep apnea detection for X01–X35 records using ANN_LM and ANN-SCG algorithms for PT-based method

ECG record	Expert annotation		ANN-LM		ANN-SCG	
	% Apnea minutes	Decision	% Apnea minutes	Decision	% Apnea minutes	Decision
X01	71.70	Yes	43	Yes	65	Yes
X02	44.25	Yes	65	Yes	52	Yes
X03	2.58	No	5	No	4	No
X04	0.00	No	2	No	12	Yes
X05	63.61	Yes	80	Yes	50	Yes
X06	0.00	No	2	No	2.3	No
X07	47.30	Yes	8	No	52	Yes
X08	62.84	Yes	72.65	Yes	60	Yes
X09	32.87	Yes	18	Yes	45	Yes
X10	18.84	Yes	3	No	15	Yes
X11	3.07	No	1	No	8	No
X12	10.80	Yes	15	Yes	13	Yes
X13	57.44	Yes	29	Yes	62	Yes
X14	88.21	Yes	42	Yes	75	Yes
X15	38.64	Yes	25	Yes	22	Yes
X16	13.59	Yes	37	Yes	15	Yes
X17	0.25	No	4	No	1	No
X18	0.65	No	7	No	3	No
X19	82.92	Yes	87	Yes	79	Yes
X20	52.10	Yes	64	Yes	45	Yes
X21	23.74	Yes	48	Yes	30	Yes
X22	0.83	No	5	No	3	No
X23	22.33	Yes	41	Yes	34	Yes
X24	0.23	No	4	No	2	No
X25	57.59	Yes	65	Yes	61	Yes
X26	66.15	Yes	55	Yes	9	No
X27	97.99	Yes	85	Yes	92	Yes
X28	87.24	Yes	76	Yes	90	Yes
X29	0.00	No	2	No	6	No
X30	61.22	Yes	50	Yes	65	Yes
X31	92.61	Yes	80	Yes	86	Yes
X32	78.97	Yes	55	Yes	62	Yes
X33	1.06	No	23	Yes	5	No
X34	0.84	No	6	No	2	No
X35	0.00	No	2	No	4	No

Table 21 Comparison of performance measures of the proposed method with methods in the HT method and Sebastian et al. [87]

Average performance measures	Proposed classifier		HT-based classifier in Sect. 4.1.6		Sebastian et al. [87]
	ANN-LM	ANN-SCG	ANN-LM	ANN-SCG	
Accuracy (Acc) (%)	91	95	98.5	99.2	83
Sensitivity (Se) (%)	91	94	95	97	92
Specificity (Sp) (%)	92	91	94	98	60
Precision (Pr) (%)	95	96	95.2	96	85
F-measure	0.93	0.95	0.95	0.97	0.88
Execution time (s)	3.5	1.2	15.5	6.5	–

Table 22 Comparison of proposed methods with state of art literature

Method	Features	Classifier	Performance measures (%)
Proposed method	PT-based QRS complex detection, 30 features, AHI	ANN_LM	Acc: 91, Se: 91
		K-means clustering + ANN_SCG	Acc: 95, Se: 94
HT-based method in Sect. 4.1.4	HT-based QRS detection, 30 features, AHI	ANN_LM	Acc: 98.5, Se: 95
		K-means clustering + ANN_SCG	Acc: 99.2, Se: 97
Varon et al. [83]	Time domain features	Least squares SVM	Acc > 90
Bsoul et al. [65]	Time and spectral features	SVM	Se: 96
Oussama et al. [99]	11 time domain features	ANN	Acc: 96

Test of Significance

The Chi-Square test is performed at a 5% significance level to validate the agreement between the expert annotations and the test results for both the PT and HT-based methods. The classification result using the HT method proves to be in better agreement with the expert annotations as compared to that of classification using the proposed PT-based method [95].

4.2.7 Summary

PCA reduces the extracted feature set size by 43% in the training phase and by 33% in the test phase of classification, thus contributing to a considerable reduction of execution time. ANN_SCG classifier has improved performance over ANN_LM classifier in terms of Acc, Se, Sp, Pr, F-measure, and MSE by a smaller margin. But the speed of execution of ANN_SCG classifier has increased compared to ANN_LM classifier by around 66%, due to the fast converging behavior of SCG algorithm, preprocessing using K-means clustering, and the saving achieved through PCA. ANN_SCG classifier has outperformed ANN_LM classifier as well the reference classifier proposed by Sebastian et al. [87]. The merit of the proposed PT-based method is in the form of significantly reduced computational time cost as compared with HT-based method demonstrated in Tables 16 and 18.

Future Scope
Even though the PT-based feature extraction method followed by ANN classifiers demonstrates the satisfactory execution time, it can not be opted because of lower values of performance measures, which are just above 90%. Through the algorithm, we could demonstrate the implementation of feature extraction process using the time domain approach and justify that the classification results of the extracted features are not adequate enough for accurate sleep apnea detection.

4.3   Sleep Apnea Detection Using WT Method and ANN Classifiers

In Sect. 4.2, ECG signal-based sleep apnea detection algorithm using ANN classifiers is demonstrated, where the PT method is used for feature extraction of 30 features and PCA is used for removing the redundancy of features. The results are encouraging in terms of saving attained in terms of computation efficiency. But the performance measures attained in case of both, QRS complex detection and sleep apnea detection algorithms are lower for the PT method as compared with the HT method demonstrated in Sect. 4.1. Hence, the motivation received from the results of the HT method, a time–frequency domain approach, the proposed method uses the popular wavelet transform (WT), which is also a time–frequency domain approach. The objective here is to harness the unique properties of wavelet functions and the multi-resolution decomposition ability of wavelet transform that can be used for denoising and detection of QRS complex detection process in order to improve feature extraction process [67–70, 74–77].

4.3.1   Proposed Methodology

The proposed methodology involves the steps as shown in Fig. 24, namely, (i) preprocessing to free the input ECG signal from noise, (ii) QRS detection to locate QRS complexes using wavelet transform using Daubechies-db4 wavelet, (iii) feature extraction to gather ECG parameters, keeping QRS complex as the reference,

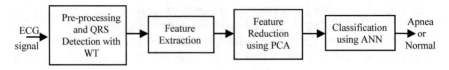

Fig. 24 Proposed methodology of sleep apnea detection

(iv) feature reduction using principal component analysis (PCA), and (v) classification of input ECG signal record using ANN classifiers as either belonging to the apnea category or normal category.

The details on the properties of wavelets and application of WT in preprocessing and QRS complex detection are discussed in Sect. 1.7.2 [97].

4.3.2   Results of Preprocessing

The detail coefficients of the input ECG signal obtained during each decomposition level along with the approximate coefficient, a5 are shown in Fig. 25(i)–(x). The de-noising of ECG signal is achieved using a fixed threshold value and thus free it from noise due to baseline wander, motion artifacts, and power-line interference leading to an improvement of SNR by 15% approximately [97].

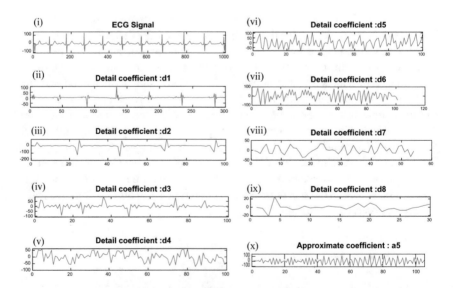

Fig. 25 (i) Input ECG signal, (ii)–(ix) decomposition of ECG signal into detail coefficients $d1$ to $d8$, and (x) approximate coefficient $a5$

4.3.3 Results of QRS Detection

After de-noising operation, the signal is reconstructed using detail coefficients $d3$, $d4$, and $d5$ with the approximate coefficient $a5$ and final detection of peaks are attained by the squaring operation. Thus, an accurate extraction of QRS complex is carried out using WT as shown in Fig. 26, with respect to the de-noised ECG signal. The summary of the results of WT-based QRS complex detection tested with X01–X35 records of the database is given in Table 23. And the results are compared with those of the HT method used for QRS complex detection proposed in Sect. 4.1.3. The proposed WT-based QRS detection method yields considerably higher values of performance features as compared to that of the HT method [96].

There is a marginal increase in computation time observed for the proposed method as compared to the HT method. But, F-measure value is found to be 0.99 for the proposed method, a significant improvement over 0.93 computed for the HT method and 0.92 computed for PT method. The summary of QRS measures tabulated in Table 23 indicates that WT has outperformed the HT method mainly in terms of significantly improved performance measures in the range 1–7% associated with a

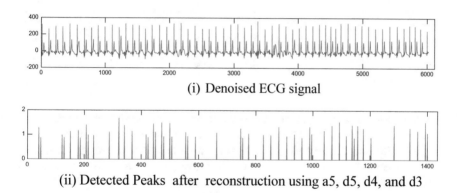

(i) Denoised ECG signal

(ii) Detected Peaks after reconstruction using a5, d5, d4, and d3

Fig. 26 Steps of QRS detection process with the WT method

Table 23 Comparison of QRS detection performance of the proposed WT method and HT method	Average performance measures	Proposed WT method	HT method (Sect. 4.1.3)
	Accuracy (%)	99.7	98.7
	Sensitivity (%)	99.5	94.0
	Specificity (%)	99.6	94.0
	Precision (%)	99.0	92.0
	F-measure	0.99	0.93
	Execution time (s)	1.5	1.42

Fig. 27 ROC curve having FRR versus FAR for QRS detection using WT method

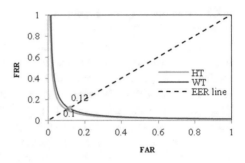

Table 24 Comparison of QRS detection methods in literature with the proposed method (Table 23)

Methods	QRS detection method	Measures[a] (%)
Proposed method	WT + two thresholds	Se: 99.5, Acc: 99.7
Zidelmal et al. [67]	WT + coefficients multiplication + two thresholds	Se: 99.64, Acc: 99.82
Chouakri et al. [101]	WT + histogram + moving average	Se: 98.68, Acc: 97.24
Li et al. [102]	WT + digital filter	Se: 98.89, Acc: 99.94
Martinez et al. [103]	WT, multiple thresholds and zero crossing	Se: 98.8, Acc: 99.86
Gaffari et al. [104]	Hybrid complex WT, threshold	Se: 98.79, Acc: 99.89
Zheng and Wu [105]	Discrete WT + cubic spline, moving average	Se: 98.68, Acc: 99.59

[a]*Se* Sensitivity, *Acc* Accuracy

marginal increase in computation time by around 5%. Receiver operating characteristics (ROC) of QRS detection is found to be very close to each other as shown in Fig. 27, with Equal Error Rate (EER) as 0.12 and 0.11 for the proposed WT method and HT method, respectively. The results of the proposed work are compared with some of the methods in the literature and summarized in Table 24. The proposed method is quite competitive in its performance [95–100].

The comparison of the proposed QRS detection with that of HT method in Sect. 4.1.3 is shown in Table 23. The performance measures of the proposed QRS detection method are compared with that of the other methods in literature as given in Table 24.

4.3.4   Results of Feature Extraction

Feature extraction is carried out on all the 35 training records, namely, a01–a20, b01–b05, and c01–c10 records and the 30 features are computed minute-wise from each of these records for 7 h of time. Thus, the generated set of features contain 4,41,000 features (35 records*7 h*60 min* 30 features), that are fed to the PCA stage to provide the train data to be used for training ANN. Later during the test phase, for every input record, the feature extraction process involves computation of 12,600 features (1 record*7 h*60 min* 30 features) [97].

4.3.5   Results of Feature Reduction and Classification Accuracy

The feature reduction process has been carried out to decide the optimum number of principal components needed for the classification stage to attain a desirable accuracy and are then compared with the results of feature reduction with HT method described in Sect. 4.1.5 as shown in Table 25. In the proposed method, a number of principal components required are 15 to achieve an accuracy of 100% for ANN_SCG as compared with ANN_LM which required 20 principal components to achieve maximum accuracy of 97%. There is a significant saving achieved for ANN_SCG in terms of computation time by 56%, approx., as compared to that of ANN_LM.

The plots of Accuracy and the execution time v/s number of principal components are shown in the Figs. 28 and 29, for both the algorithms ANN_LM and ANN_SCG, which clearly indicate that the ANN_SCG has outperformed ANN_LM. The feature

Table 25 Comparison of attainment of accuracy and execution time against a selected number of principal components, for ANN classifiers

Principal components	Accuracy				Execution time (s)			
	Proposed WT-based method		HT-based method (Sect. 4.1.5)		Proposed WT-based method		HT-based method (Sect. 4.1.5)	
	LM	SCG	LM	SCG	LM	SCG	LM	SCG
05	93	97	94.5	95.2	8.0	5.0	12	3.9
10	95	98	97.4	98.3	9.1	6.4	15	3.7
15	96	100	98.1	99.0	10.5	7.2	16	5.2
20	97	100	98.2	99.2	16.2	8.1	15.2	6.5
25	97	100	98.5	99.2	12.6	9.3	15.5	6.9
30	97	100	98.5	99.0	13.6	9.8	20	7.8

Fig. 28 Plot of accuracy versus no. of PCA components for the proposed WT method

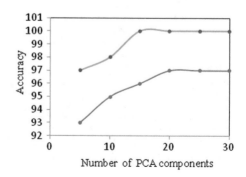

Fig. 29 Plot of execution
time versus no. of PCA
components for the proposed
WT method

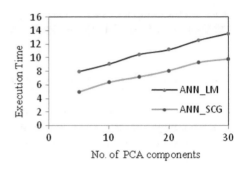

No. of PCA components

reduction process yielded the reduction in number of features (20 features * 420 min
* 1 ECG record) for ANN_LM to reduced number of features (15 features * 420 min *
1 ECG record) for ANN_SCG, thus contributing to significant reduction of execution
time for ANN_SCG in both training and testing of ANN. During ANN training, a
reduced number of features (35 records* 20 features* 420) contribute to execution
time reduction by 40%. While comparing the performance of the proposed method
with that of ANNs trained with LM and SCG in HT method of Sect. 4.1.6, the
classification measures for the proposed method have significantly improved, but
at the cost of a marginal increase in computation time. The comparison of plots
of accuracy and execution time v/s number of PCA components are compared in
Figs. 28 and 29.

4.3.6 Results of ANN Classifiers

The result of classification is: the input ECG signal record belongs to apnea category
or normal category. The comparison of performance measures of the proposed WT-
based method using ANN classifier with that of HT-based method using ANN and
the method in [87] are given in Table 26. For each of the classifier, the TP, TN, FP,

Table 26 Comparison of performance measures of the proposed WT method using ANN with that
of HT method using ANN (Sect. 4.1.6) and the method in [87]

Performance measures	Proposed WT-based method		HT-based method		The method in Sebastian et al. [87]
	ANN-LM	ANN-SCG	ANN-LM	ANN-SCG	
Accuracy (%)	97	100	98.5	99.2	83
Sensitivity (%)	100	100	95	97	92
Specificity (%)	92	100	94	98	60
Precision (%)	95	100	95.2	96	85
F-measure	0.96	1.0	0.95	0.965	0.88
Execution time (s)	11.2	7.2	15.5	6.5	–

and FN are determined, using which the performance measures, namely, Acc, Se, Sp, and Pr are computed. For the proposed method, the performance measures of ANN_SCG classifier prove to be significantly better with values of Acc = 100%, Se = 100%, Sp = 100%, Pr = 100%, F-measure = 1.0 and execution time of 7.2 s as compared with its counterpart ANN_LM classifier.

As compared with the HT method and the method in [87], there is a significant improvement in the performance measures of proposed ANN_SCG, but associated with a marginal increase of computation cost. It is observed that the performance measures of the proposed WT-based method with ANN_SCG perform as the best classifier as compared with the other methods (Table 26). The details on the record-wise classification results of the proposed method using ANN_LM and ANN_SCG experimented with the 35 database records X01–X35 are presented in Table 27 and their comparison results with expert annotations. Each of the classifier performance is tabulated in terms of percentage of apnea minutes present in the entire ECG signal record along with the final decision as either "Yes" or "No" based on the presence and absence of Sleep apnea disorder, respectively. The ECG signal record is decided as belonging to the apnea category, if the percentage of apnea minutes count more than 10%, otherwise it is decided as belonging to Normal category [7–10, 95–97].

Test of Significance

The Chi-Square test is performed at 5% significance level to validate the agreement between the expert annotations and the test results for both the proposed WT-based method and HT-based methods (Sects. 4.1.3 and 4.1.4). The classification result using the proposed WT method, that proves to be in better agreement with the expert annotations as compared to that of HT-based method. The performance of the proposed method for sleep apnea detection is compared with other methods in the literature [67, 101–109] as shown in Table 26 and the proposed method, specifically ANN_SCG classifier proves to be a quite effective technique.

Future Scope

The feature extraction process using WT method followed by ANN classifier trained with Scaled conjugate Gradient algorithm proves to be very efficient in terms of higher values of accuracy, sensitivity, and specificity and reduced values of computation time as compared with HT method (Sect. 4.1) and PT method (Sect. 4.2). Hence, the proposed method can be a good choice for sleep apnea detection.

4.4 Sleep Apnea Detection Using HT-, PT-, and WT-Based Methods on Own ECG Database

In the sections from 4.1 to 4.3, ECG signal feature extraction methods based on PT, HT, and WT methods are proposed for sleep apnea detection. The benchmark database, namely, ECG Apnea from PNDB [7–10] is used for experimentation and the results obtained are quite encouraging. In WT method, wavelet transform used for feature extraction has excelled in its performance which is observed from the

Table 27 Classification of sleep apnea detection with proposed classifiers

ECG record	Expert annotation		ANN-LM Classifier		ANN-SCG classifier	
	% Apnea minutes	Decision[a]	% Apnea minutes	Decision[a]	% Apnea minutes	Decision[a]
X01	71.70	Yes	74	Yes	60	Yes
X02	44.25	Yes	32	Yes	39	Yes
X03	2.58	No	4	No	3	No
X04	0.00	No	3	No	1	No
X05	63.61	Yes	55	Yes	55	Yes
X06	0.00	No	5	No	2	No
X07	47.30	Yes	40	Yes	39	Yes
X08	62.84	Yes	58	Yes	55	Yes
X13	57.44	Yes	50	Yes	52	Yes
X14	88.21	Yes	82	Yes	77	Yes
X15	38.64	Yes	35	Yes	45	Yes
X16	13.59	Yes	16	Yes	15	Yes
X17	0.25	No	2	No	0.5	No
X18	0.65	No	4	No	2	No
X19	82.92	Yes	75	Yes	76	Yes
X20	52.10	Yes	60	Yes	54	Yes
X21	23.74	Yes	20	Yes	19	Yes
X22	0.83	No	4	No	2	No
X23	22.33	Yes	19	Yes	28	Yes
X24	0.23	No	3	No	1	No
X25	57.59	Yes	50	Yes	51	Yes
X28	87.24	Yes	74	Yes	80	Yes
X29	0.00	No	2	No	3	No
X30	61.22	Yes	58	Yes	65	Yes
X31	92.61	Yes	88	Yes	90	Yes
X32	78.97	Yes	74	Yes	70	Yes
X33	1.06	No	5	No	7	No
X34	0.84	No	2	No	2	No
X35	0.00	No	2	No	1	No

[a] Yes—Sleep apnea category, No—Normal category

results of QRS detection performance as compared to PT and HT methods, in terms of increased values of accuracy, sensitivity, and specificity (sections in 4.1.3, 4.2.3 and 4.3.3). The classification carried out using ANN trained with SCG algorithm has outperformed ANN trained with LM algorithm for all the three methods in terms of attained classification performance measures Acc, Se, Sp, F-measure (Sects. 4.1.6, 4.2.6 and 4.3.6) [95–97].

Hence, the proposed work involves the use of only the ANN classifier trained with SCG algorithm on the own dataset for performance comparison among PT, HT, and WT methods. The need is to verify the effectiveness of these algorithms for ECG signal datasets collected from different local hospitals, especially pertaining to Indian subjects. Hence, an own dataset of ECG signal records is collected from a local hospital, namely, Sleep Lab Facility of SDM College of Medical Sciences and Hospital (SDMCMSH), Sattur, Dharwad, Karnataka. The experimentation of proposed methods using such an own dataset makes the study more relevant for demonstrating the robustness of these methods.

4.4.1   SDMCMSH ECG Dataset

The ECG signals gathered through Polysomnography tests conducted at Sleep lab, SDM College of Medical Sciences and Hospital (SDMCMSH), Dharwad, Karnataka, India. The database is named as SDMCMSH ECG database. The ethical committee certification received from SDMCMHS enables the use of real-time records from Hospital as input for the proposed work. There are 55 records of subjects in the age group between 25 and 75 years, out of which 34 are sleep apnea patients 21 are normal subjects. The PSG test is conducted for screening for OSA in patients with suspicious heart-related ailments. The physiological parameters recorded during PSG test are blood oxygen, EEG, breathing patterns, body position, eye movements, ECG and heart rate and rhythms, leg movements, sleep stages, snoring noise, or any unusual movements. Only the ECG signal channel is extracted for processing and analysis in the proposed work, suppressing all the other channels of PSG.

4.4.2   Results of Preprocessing

The SNR improvement is by 10% in the PT method, 7% in HT method, 12% in the WT method by comparing the signal strength before and after ECG signal processing. Hence, the wavelet transform has outperformed in the detection of QRS detection process as compared with the other two methods.

4.4.3   Results of QRS Detection Using PT, HT, and WT Methods

The performance measures of QRS detection are tabulated for each method, namely, PT-, HT-, and WT-based methods tested with ECG signal from SDMCMSH database and later compared the results for each of them with the results obtained using benchmark database ECG Apnea [7–10]. The performance measures like Acc, Se, Sp, F-measure, and Precision are comparatively lower for SDMCMSH database. The reasons are, the high-frequency noise, resulted due to multiple electrode systems of

PSG lead to ECG recording affected with the significant noise level. Even the lack of expertise and experience in recording of parameters during the sleep time of patients caused interruptions in recording and the signal quality was lowered. But the results of WT-based method are still the best among all three methods for experimentation with SDMCMSH database, matching similar performance as with benchmark ECG Apnea database (Table 28).

The ROC plots of FRR against FAR are presented for the PT method for both the ECG Apnea database (PNDB) [9] (Sect. 4.2.3) and SDMCMSH database in Fig. 30. Equal error rate (EER) for SDMCMSH proves a higher value as compared with that of the benchmark ECG Apnea database.

ROC plots for QRS detection using the HT method demonstrated on benchmark dataset (Sect. 4.1.3) and the SDMCMSH database are compared in Fig. 31.

Table 28 Results of preprocessing and QRS detection

Performance measures	PT-based method		HT-based method		WT-based method	
	Bench mark[a]	SDMCMHS[a]	Bench mark[a]	SDMCMHS[a]	Bench mark[a]	SDMCMHS[a]
Accuracy (%)	94	93	98.7	97	99.7	98
Sensitivity (%)	95	92	94	94	99.5	96
Specificity (%)	93	95	94	98	99.6	99
Precision (%)	92	91	92	92	99	96
F-measure	0.92	0.9	0.93	0.93	0.99	0.96
Improved SNR (%)	12	10	9	9	15	13
Execution time (s)	0.5	0.63	1.42	1.84	1.5	1.95

[a]Benchmark database from [9], SDMCMHS database from SDM Hospital, Dharwad, Karnataka

Fig. 30 ROC plot for QRS detection using PT method on the benchmark dataset and local database (Table 28)

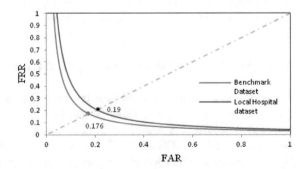

Fig. 31 ROC plot for QRS detection using the HT method on a benchmark dataset and local database

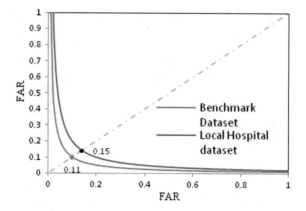

Fig. 32 ROC plot for QRS detection using WT method on a benchmark dataset and local database

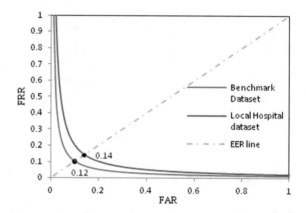

ROC plots for QRS detection using WT method demonstrated on a benchmark dataset and the SDMCMSH database are compared in Fig. 32. The summary of performance comparison of ANN_SCG classifiers with PT, HT, and WT methods is presented in Table 29.

The performance measures of the classifiers exhibit the superior performance with benchmark database as compared with its performance with SDMCMSH database for each of the methods based on HT, PT, and WT. Among the three feature extraction methods, WT-based method outperforms as compared with HT- and PT-based methods. Among the classifiers, ANN_SCG classifier yielded superior results in case of both the databases. Hence, the experimental findings of the proposed methods indicate that the WT-based feature extraction method, followed by feature reduction using PCA and K-means clustering, and finally the classification using ANN trained with SCG has emerged as the best possible solution toward the development of low-power high-performance algorithm for sleep apnea detection.

Table 29 Performance comparison of ANN_SCG classifier with the PT, HT, and WT methods

Classification measures	PT-based method using ANN_SCG (Sect. 4.2.6)		HT-based method using ANN_SCG (Sect. 4.1.6)		WT method using ANN_SCG (Sect. 4.3.6)	
	PNDB[a]	SDMCMSH[a]	PNDB[a]	SDMCMSH[a]	PNDB[a]	SDMCMSH[a]
No of principal components	20	25	20	20	15	20
Accuracy (%)	95	92	99.2	97	100	98
Sensitivity (%)	94	94	97	96	100	97
Specificity (%)	91	91	98	98	100	99
Precision (%)	96	95	96	96	100	97
F-measure	0.95	0.94	0.97	0.96	1.0	0.97
Execution time (s)	1.2	2.3	6.5	7.8	7.2	8.5

[a]Benchmark datasets: ECG Apnea database from PNDB [9], local dataset from SDMCMS Hospital, Sleep lab, Sattur, Dharwad, Karnataka

5   Conclusion and Future Scope

5.1   Conclusion

In the proposed work, it is proposed to automate the process of sleep apnea diagnostics to develop an intelligent assistive technology for sleep apnea diagnostics based on ECG signal characteristics. The objective of the research study is to design and develop a low-power, high-performance algorithm for sleep apnea detection based on the ECG signal. PT algorithm was used for QRS complex detection, which in turn is the basis on which other 30 features are extracted, out of which 29 belong to the time domain, and other 11 of them belong to the frequency domain. Later, using feature reduction and classification technique based on two training algorithms, namely, LM algorithm and SCG algorithm, an efficient method of sleep apnea detection was performed on the ECG Apnea database. The results of the implementation are recorded. Similarly, the second set of experimentation was carried out using HT-based feature extraction to gather the same 30 features followed by feature reduction and classification, as mentioned in the PT-based experimentation. Lastly, the feature extraction strategy was based on the wavelet transform-based method, with the other stages being intact. In each case, the QRS detection performance measures and the classification performance are analyzed. Similar experimentations were carried out using the own dataset, namely, SDMCMSH database to demonstrate the robustness of all the methods and is included in Sect. 4.4.

The summary of the best results obtained for the classification using both the benchmark dataset and the own dataset is presented in Table 29. It is observed that the performance measures of classification using WT-based ANN_SCG classifier are significantly improved over that of HT- and PT-based methods in case of both of the databases. A marginal increase in computation time is incurred for WT method as compared with the HT method. Further, the WT-based feature extraction method, followed by feature reduction using PCA and K-means clustering, and finally the classification using ANN trained with SCG, has emerged as the best possible solution toward the development of low-power high-performance algorithm for sleep apnea detection.

5.2 Future Scope

The outcome of the proposed research work leads to further sleep apnea research by considering the following aspects:

i. Harnessing of architectural features of digital signal processors (DSPs) or use of a customized field programmable generic array (FPGA) device to build a stand-alone system. Design of an autonomous system for sleep apnea detection with the hybrid architectural model of DSP and microcontroller to improve low-power and high-performance characteristics. Extensive use of low-power modes, clock systems, and intelligent peripherals to enhance low-power and high-performance characteristics.
ii. Use of intelligent decision-making using Neuro-Fuzzy classifiers can be further considered for improving the performance of the algorithm.

Acknowledgements Authors are thankful to the Vice-Chancellor and the management of KLE Technological University, Hubballi, for providing the lab facilities, financial support, and encouragement to conduct the proposed research. Authors convey their gratitude to Dr. Shrikanth Hiremath, Pulmonologist, SDM College of Medical Sciences & Hospital for providing the domain knowledge and access to sleep slab facility.

References

1. Jun JC, Chopra S, Schwartz AR (2016) Sleep apnea. Eur Respir Rev 25(139):12–18
2. Consensus and evidence-based INOSA guidelines 2014 (First edition), Writing committee of the Indian initiative on obstructive sleep apnoea. Indian J Chest Dis Allied Sci 57 (2015)
3. Kapur VK, Auckley DH, Chowdhuri S, Kuhlmann DC, Mehra R, Ramar K, Harrod CG (2017) Clinical practice guideline for diagnostic testing for adult obstructive sleep apnea: an American Academy of Sleep Medicine clinical practice guideline. J Clin Sleep Med 13(3):479–504

4. Kligfield P, Gettes LS, Bailey JJ, Childers R, Deal BJ, Hancock EW, Wagner GS (2007) Recommendations for the standardization and interpretation of the electrocardiogram: Part I: The electrocardiogram and its technology: a scientific statement from the American Heart Association Electrocardiography and Arrhythmias Committee, Council on Clinical Cardiology; the American College of Cardiology Foundation; and the Heart Rhythm Society. Circulation 115(10):1306–1324

5. Campbell B, Richley D, Ross C, Eggett CJ (2017) Clinical guidelines by consensus: recording a standard 12-lead electrocardiogram. An approved method by the Society for Cardiological Science and Technology (SCST). http://www.scst.org.uk/resources/SCST_ECG_Recording_Guidelines_2017

6. Thomas RJ, Shin C, Bianchi MT, Kushida C, Yun C-H (2017) Distinct polysomnographic and ECG spectrographic phenotypes embedded within obstructive sleep apnea. Sleep Sci Pract 1:11

7. https://physionet.org/physiobank/database/slpdb/

8. https://sleepdata.org/datasets/shhs

9. https://www.physionet.org/physiobank/database/apnea-ecg/

10. Penzel T, Moody GB, Mark RG, Goldberger AL, Peter JH (2000) The apnea-ECG database. Comput Cardiol 2000:255–258

11. http://www.tcts.fpms.ac.be/~devuyst/Databases/DatabaseApnea/

12. Figoń P, Irzmański P, Jóśko A (2013) ECG signal quality improvement techniques. Przegląd Elektrotechniczny. ISSN 0033-2097, R. 89 NR 4/2013

13. Oweis RJ, Al-Tabbaa BO (2014) QRS detection and heart rate variability analysis. A survey. Biomed Sci Eng 2(1):13–34

14. Elgendi M, Eskofier B, Doko S, Abbott D (2014) Revisiting QRS detection methodologies for portable, wearable battery operated and wireless ECG systems. PLoS ONE 9(1):e84018

15. Gacek A, Pedrycz W (2014) ECG signal processing, classification and interpretation: a comprehensive framework of computational intelligence. Springer Publishing Company

16. Arzeno N, Poon C, Deng Z (2006) Quantitative analysis of QRS detection algorithms based on the first derivative of the ECG. In: Proceedings of the 28th annual international conference of the IEEE engineering in medicine and biology society, pp 1788–1791

17. Arzeno NM, Deng Z-D, Poon C-S (2008) Analysis of first-derivative based qrs detection algorithms. IEEE Trans Biomed Eng 55(2):478–484

18. Pan J, Tompkins WJ (1985) A real-time QRS detection algorithm. IEEE Trans Biomed Eng BME 32(3)

19. Ahlstrom ML, Tompkins WJ (1985) Digital filters for real-time ECG signal processing using microprocessors. IEEE Trans Biomed Eng 32(9):708–713. ISSN 0018-9294

20. Zhang P, Zhang Q, Konaka S, Akutagawa M, Kinouchi Y (2014) QRS detection by combination of wavelet transform and multi-resolution morphological decomposition. Inf Technol J 13:2385–2394

21. Huang N, Shen Z, Long S, Wu M, Shih H et al (1998) The empirical mode decomposition and Hilbert spectrum for nonlinear and nonstationary time series analysis. Proc R Soc Lond A 903–995

22. Benitez P, Gaydecki A, Zaidi A, Fitzpatrick AP (2001) The use of the Hilbert transform in ECG signal analysis. Comput Biol Med 31:399–406

23. Rabbani H et al (2011) R peak detection in electrocardiogram signal based on an optimal combination of wavelet transform, Hilbert transform, and adaptive thresholding. J Med Signals Sens 1(2):91–98

24. Lyon A, Mincholé A, Martínez JP, Laguna P, Rodriguez B (2018) Computational techniques for ECG analysis and interpretation in light of their contribution to medical advances. J R Soc Interface 15(138):20170821

25. Liu F, Liu C, Jiang X, Zhang Z, Zhang Y, Li J, Wei S (2018) Performance analysis of ten common QRS detectors on different ECG application cases. Feife J Healthc Eng 2018, Article ID 9050812, 8 pp

26. Melgarejo-Meseguer F-M, Everss-Villalba E, Gimeno-Blanes F-J, Blanco-Velasco M, Molins-Bordallo Z, Flores-Yepes J-A, Rojo-Álvarez J-L, García-Alberola A (2018) On the beat detection performance in long-term ECG monitoring scenarios. Sensors 18:1387
27. Daniel WW, Cross CL (2014) Biostatistics: basic concepts and methodology for the health sciences, 10th ed. ISV Paperback, Wiley Student Edition
28. Powers DM (2011) Evaluation: from precision, and F-measure to ROC, informedness, markedness and correlation. J Mach Learn Technol 2(1):37–63. ISSN: 2229-3981; ISSN: 2229-399X
29. Gajowniczek K, Ząbkowski T, Szupiluk R (2014) Estimating the ROC curve and its significance for classification models. Assess Quant Methods Econ 15(2):382–391
30. Lee JA, Verleysen M (2007) Nonlinear dimensionality reduction (Information science and statistics), 2007th ed
31. Smit LI. A tutorial on principal components analysis. http://www.cs.otago.ac.nz/cosc453/student_tutorials/principal_components.pdf
32. Faust O, Acharya UR, Ng EYK, Fujita H (2016) A review of ECG-based diagnosis support systems for obstructive sleep apnea. J Mech Med Biol 16(01):1640004
33. Timus O, Bolat ED (2017) k-NN-based classification of sleep apnea types using ECG. Turk J Elec Eng Comp Sci 25:3008–3023
34. Alsmadi MKS, Omar KB, Noah SA (2009) Backpropagation algorithm: the best algorithm among the multi-layer perceptron algorithm. Int J Comput Sci Netw Secur 9(4):378–383
35. Moller MF (1993) A scaled conjugate gradient algorithm for fast supervised learning. Neural Netw 6:525–533
36. Graessle R, MathWorks. Using MATLAB in medical device research and development. https://in.mathworks.com/videos/using-matlab-in-medical-device-research-and-development
37. Prasad CN (2013) Obstructive sleep apnea-hypopnea syndrome—Indian scenario. Perspect Med Res 1(1)
38. Shyamala KK, Khatri B (2016) Study on clinical profile of obstructive sleep apnea (OSA). Sch J App Med Sci (SJAMS) 4(6C):2074–2083. https://doi.org/10.21276/sjams.2016.4.6.43 (Online)
39. Liu D, Pang Z, Lloyd S (2008) A neural network method for detection of obstructive sleep apnea and narcolepsy based on pupil size and EEG. IEEE Trans Neural Networks 19(2):308–318
40. Lin R, Lee R, Tseng C, Zhou H, Chao C, Jiang J (2006) A new approach for identifying sleep apnea syndrome using wavelet transform and neural networks. Biomed Eng Appl Basis Commun 18(3):138–143
41. Chazal P, Penzel T, Heneghan C (2004) Automated detection of obstructive sleep apnoea at different time scales using the electrocardiogram, vol 25, no 4. Institute of Physics Publishing, pp 967–983
42. Almazaydeh L, Elleithy K, Faezipour M (2012) Detection of obstructive sleep apnea through ECG signal features. In: Proceedings of the IEEE international conference on electro information technology (IEEE eit2012), May 2012, pp 1–6
43. Bsoul M, Minn H, Tamil L (2011) Apnea MedAssist: real-time sleep apnea monitor using single-lead ECG. IEEE Trans Inf Technol Biomed 15(3):416–427
44. Almazaydeh L, Faezipour M, Elleithy K (2012) A neural network system for detection of obstructive sleep apnea through SpO$_2$ signal features. Int J Adv Comput Sci Appl (IJACSA) 3(5):7–11
45. Canosa M, Hernandez E, Moret V (2004) Intelligent diagnosis of sleep apnea syndrome. IEEE Eng Med Biol Mag 23(2):72–81
46. Marcos J, Hornero R, Álvarez D, del Campo F, Aboy M (2010) Automated detection of obstructive sleep apnoea syndrome from oxygen saturation recordings using linear discriminant analysis. Med Biol Eng Compu 48:895–902
47. Burgos A, Goni A, Illarramendi A, Bermudez J (2010) Real-time detection of apneas on a PDA. IEEE Trans Inf Technol Biomed 14(4):995–1002

48. Zhao Y, Zhang H, Liu W, Ding S (2011) A snoring detector for OSAHS based on patient's individual personality. In: 3rd international conference in awareness science and technology (iCAST), pp 24–27
49. Andrew K, Tong S et al (2008) Could formant frequencies of snore signals be an alternative means for the diagnosis of obstructive sleep apnea? Sleep Med 9:894–898
50. Khandoker H, Karmaker K, Palaniswami M (2008) Analysis of coherence between Sleep EEG and ECG signals during and after obstructive sleep apnea events. In: Proceedings of 30th IEEE international conference on engineering in medicine and biology society (EMBS 2008), pp 3876–3879
51. Xie B, Minn H (2012) Real time sleep apnea detection by classifier combination. IEEE Trans Inf Technol Biomed 16(3):469–477
52. Heneghan C, Chua CP, Garvey JF, De Chazal P, Shouldice R, Boyle P, McNicholas WT (2008) A portable automated assessment tool for sleep apnea using a combined Holter-oximeter. SLEEP 31(10)
53. Alvarez D, Hornero R, Marcos J, Campo F, Lopez M (2009) Spectral analysis of electroencephalogram and oximetric signals in obstructive sleep apnea diagnosis. In: Proceedings of the 31st IEEE international conference on engineering in medicine and biology society (EMBS 2009), Sep 2009, pp 400–403
54. Angius G, Raffo L (2008) A sleep apnoea keeper in a wearable device for continuous detection and screening during daily life. Comput Cardiol 433–436
55. Shochat T, Hadas N, Kerkhofs M et al (2002) The SleepStripTM: an apnoea screener for the early detection of sleep apnoea syndrome. Eur Respir J 19:121–126
56. ApneaLink^{TM}. https://www.resmed.com
57. Stuart M (2010) Sleep apnea devices: the changing of the guard. Startup J 15(10):1–8
58. Philips Alice Night One. https://www.usa.philips.com/healthcare/product/HC1109289
59. Oliver N, Mangas F (2007) HealthGear: automatic sleep apnea detection and monitoring with a mobile phone. J Commun 2(2):1–9
60. Martis RJ, Acharya UR, Adeli H (2014) Current method in electrocardiogram characterization. Comput Biol Med 48:133–149, 0010-4825
61. Elgendi M, Jonkma M, De Boer F (2009) Improved QRS detection algorithm using dynamic thresholds. Int J Hybrid Inf Technol 2(1)
62. Szczepanski A, Saeed K, Ferscha A (2010) A new method for ECG signal feature extraction. In: Bolc L et al (eds) ICCVG 2010, Part II. LNCS, vol 6375, pp 334–341
63. Lu AX, Pan M, Yu Y (2018) QRS detection based on improved adaptive threshold. J Healthc Eng Article ID 5694595, 8 pp
64. Ulusar UD, Govindan RB, Wilson JD, Lowery CL, Preissl H (2009) Adaptive rule-based fetal QRS complex detection using Hilbert transform. In: 31st annual international conference of the IEEE EM Minneapolis, Minnesota, USA, 2–6 September 2009
65. Bsoul AAR, Ji SY, Ward K, Najarian K (2009) Detection of P, QRS, and T components of ECG using wavelet transformation. In: 2009 ICME international conference on complex medical engineering, Tempe, AZ, pp 1–6
66. Chatlapalli S, Nazeran H, Melarkod V, Krishnam R, Estrada E, Pamula Y, Cabrera S (2004) Accurate derivation of heart rate variability signal for detection of sleep disordered breathing in children. In: Proceedings of 26th annual international conference of the IEEE engineering in medicine and biology society, vol 1, no 5, pp 38–41
67. Zidelmal Z, Amirou A, Adnane M, Belouchrani A (2012) QRS detection based on wavelet coefficients. Comput Methods Programs Biomed 107(3):490–496
68. Farahabadi, A, Farahabadi E, Rabbani H, Mohammad PM (2012) Detection of QRS complex in electrocardiogram signal based on a combination of Hilbert transform, wavelet transform, and adaptive thresholding. https://doi.org/10.1109/bhi.2012.6211537
69. Kohli SS, Makwana N, Mishra N, Sagar B (2012) Hilbert transform based adaptive ECG R-peak detection technique. Int J Electr Comput Eng (IJECE) 2(5):639–643
70. Barmase S, Das S, Mukhopadhyay S (2013) Wavelet transform-based analysis of QRS complex in ECG signals. CoRR abs/1311.6460

71. Rodrígueza R, Mexicanob A, Bilac J, Cervantesd S, Ponceb R (2015) Feature extraction of electrocardiogram signals by applying adaptive threshold and principal component analysis. J Appl Res Technol 13:261–269
72. Xia Y, Han J, Wang K (2015) Quick detection of QRS complexes and R-waves using a wavelet transform and K-means clustering. Bio-Med Mater Eng (IOS Press) 26:S1059–S1065
73. Mou JR, Sheikh MRI, Huang X, Ou KL (2016) Noise removal and QRS detection of ECG signal. J Biomed Eng Med Imaging 3:4
74. Yochum M, Renaud C, Jacquir S (2016) Automatic detection of P, QRS and T patterns in 12 lead ECG signal based on CWT. Biomed Signal Process Control (Elsevier)
75. Kaur H, Rajni R (2017) Electrocardiogram signal analysis for R-peak detection and denoising with hybrid linearization and principal component analysis. Turk J Electr Eng Comput Sci 25:2163–2175
76. Park JS, Lee SW, Park U (2017) R peak detection method wavelet transform and modified Shannon energy envelope. J Healthcare Eng Article ID 4901017, 14 pp
77. Kanaan L, Merheb D, Kallas M, Francis C, Amoud H, Honeine P (2011) PCA and KPCA of ECG Signals with binary SVM classification. 978-1-4577-1921-9/11/$26.00 ©2011 IEEE 344 SiPS
78. Martis RJ, Rajendra Acharya U, Lim CM, Suri JS (2007) Characterization of ECG beats from cardiac arrhythmia using discrete cosine transform in PCA framework. Knowl-Based Syst 45(2013):76–82
79. Avci C, Bilgin G (2013) Sleep apnea detection using adaptive neuro-fuzzy inference system. Engineering 5:259–263
80. Elif Derya Übeyli (2009) Adaptive neuro-fuzzy inference system for classification of ECG signals using Lyapunov exponents. Comput Methods Programs Biomed 9(3):313–321
81. Shen C-P, Kao W-C, Yang Y-Y, Hsu M-C, Yuan-Ting W, Lai F (2012) Detection of cardiac arrhythmia in electrocardiograms using adaptive feature extraction and modified support vector machines. Expert Syst Appl 39(2012):7845–7852
82. Avci C, Akba A (2012) Comparison of the ANN based classification accuracy for real time sleep apnea detection methods. Biomed Eng
83. Varon C, Testelmans D, Buyse B, Suykens JAK, Van Huffel S (2013) Sleep apnea classification using least-squares support vector machines on single-lead ECG. In: 35th annual international conference of the IEEE engineering in medicine and biology society (EMBC), Osaka, pp 5029–5032
84. Rachim VP, Li G, Chung WY (2014) Sleep apnea classification using ECG-signal wavelet-PCA features. Bio-Med Mater Eng 24:2875–2882
85. Vafaie MH, Ataei M, Koofigar HR (2014) Heart diseases prediction based on ECG signals classification using a genetic-fuzzy system and dynamical model of ECG signals. Biomed Signal Process Control 14:291–296
86. Atri R, Mohebbi M (2015) Obstructive sleep apnea detection using spectrum and bispectrum analysis of single-lead ECG signal. Med Physiolog Measur (Institute of Physics and Engineering) 36(9)
87. Canisius S, Ploch T, Gross V, Jerrentrup A, Penzel T, Kesper K (2008) Detection of sleep disordered breathing by automated ECG analysis. In: 30th annual international IEEE EMBS conference, Vancouver, British Columbia, Canada, 20–24 August 2008. 978-1-4244-1815-2/08/$25.00 ©2008 IEEE
88. Vajda S, Santosh KC. (2017) A fast k-nearest neighbor classifier using unsupervised clustering. In: Santosh K, Hangarge M, Bevilacqua V, Negi A (eds) Recent trends in image processing and pattern recognition. RTIP2R 2016. Communications in computer and information science, vol 709. Springer, Singapore. https://doi.org/10.1007/978-981-10-4859-3_17
89. Bouguelia MR, Nowaczyk S, Santosh KC, Verikas A (2018) Agreeing to disagree: active learning with noisy labels without crowdsourcing. Int J Mach Learn Cyber 9:1307–1319. https://doi.org/10.1007/s13042-017-0645-0
90. Santosh KC, Lamiroy B, Wendling L (2013) DTW-radon-based shape descriptor for pattern recognition. Int J Pattern Recognit Artif Intell (World Scientific Publishing) 27(3). https://doi.org/10.1142/s0218001413500080.hal-00823961

91. Bali JS, Nandi A (2013) Design issues of portable, low-power and high performance ECG measuring system. Int J Eng Sci Innov Technol (IJESIT) 2(4):469–475. ISSN: 2319-5967

92. Bali JS, Nandi AV (2016) An experience, using software based tools for teaching and learning mathematically intensive signal processing theory concepts. In: 2016 IEEE 4th international conference on MOOCs, innovation and technology in education, Madurai, 2016, pp 100–104. https://doi.org/10.1109/mite.2016.029

93. Bali JS, Nandi AV (2017) ECG signal based power aware system for obstructive sleep apnea detection. In: 2017 international conference on recent trends in electrical, electronics and computing technologies, Warangal, pp 59–63. https://doi.org/10.1109/icrteect.2017.43

94. Bali JS, Nandi AV (2017) Simplified process of obstructive sleep apnea detection using ECG signal based analysis with data flow programming, vol 2. In: ICTIS 2017. Smart innovation, systems and technologies. Springer, Cham, vol 84, pp 165–173. https://doi.org/10.1007/978-3-319-63645-0_18

95. Bali JS, Nandi AV, Hiremath PS (2018) Performance comparison of ANN classifiers for sleep apnea detection based on ECG signal analysis using hilbert transform. Int J Comput Technol 17(2):7312–7325. https://doi.org/10.24297/ijct.v17i2.7616

96. Bali JS, Nandi AV, Hiremath PS, Patil PG (2018) Detection of sleep apnea in ECG signal using Pan-Tompkins algorithm and ANN classifiers. COMPUSOFT 7(11):2852–2861

97. Bali JS, Nandi AV, Hiremath PS, Patil PG (2018) Detection of sleep apnea from ECG signals using WT and ANN classifiers. IPASJ Int J Electr Eng (IIJEE) 6(11):1–14

98. Sadr N, de Chazal P (2014) Automated detection of obstructive sleep apnoea by single-lead ECG through ELM classification. Comput Cardiol. https://doi.org/10.13140/2.1.3881.3446, 2014

99. Oussama BM, Saadi BM, Zine-Eddine HS (2016) Extracting features from ECG and respiratory signals for automatic supervised classification of heartbeat using neural networks. Asian J Inf Technol 15(1):5–11

100. Greenwood PE, Nikulin MS. A guide to chi-squared testing. Wiley Series in Probability and statistics, 1st ed

101. Chouakri SA, Bereksi-Reguig F, Taleb-Ahmed A (2011) A QRS complex detection based on multi wavelet packet decomposition. Appl Math Comput 217:9508–9525

102. Li C, Zheng C, Tai C (1995) Detection of ECG characteristic points using wavelet transforms. IEEE Trans Biomed Eng 42:21–28

103. Martinez JP, Almeida R, Olmos S, Rocha AP, Laguna P (2004) A wavelet based ECG delineator: evaluation on standard databases. IEEE Trans Biomed Eng 51:570–581

104. Ghaffari A, Golbayani H, Ghasemi M (2008) A new mathematical based QRS detector using continuous wavelet transform. Comput Electr Eng 34:81–91

105. Zheng H, Wu J (2008) Real-time QRS detection method. In: Proceedings IEEE 10th international conference real-time QRS detection method, e-health networking, applications and services, HealthCom 2008, pp 169–170

106. Mittal M, Goyal LM, Hemanth DJ, Sethi JK (2019) Clustering approaches for high-dimensional databases: a review WIREs Data Min Knowl Discov (Wiley) 1–14

107. Mittal M, Sharma RK, Singh VP, Agarwal R (2019) Adaptive threshold based clustering: a deterministic partitioning approach. Int J Inf Syst Model Des (IGI Global) 10(1):42–59

108. Mittal M, Goyal LM, Kaur S, Kaur I, Amit Verma D, Hemanth J (2019) Deep learning based enhanced tumor segmentation approach for MR brain images. Appl Soft Comput 78:346–354

109. Jude Hemanth D, Anitha J, Son LH, Mittal M (2018) Diabetic retinopathy diagnosis from retinal images using modified hopfield neural network. J Med Syst 42(12):247

Chapter 6
Medical Image Processing in Detection of Abdomen Diseases

Kirti Rawal and Gaurav Sethi

1 Introduction

The abdomen diseases that occur in the society are tumor, cyst, calculi, and stone. The major causes of these diseases are poor lifestyle, unhygienic diet, and the water which is available for drinking is very dirty. In this world, 13% of deaths are due to the abdomen diseases like tumor [1, 2], 28% of deaths are due to calculi in kidney [3], and cyst is also the most common disease in the abdomen [4]. So, in order to prevent the human beings from these diseases as well as to reduce the death rate, it is necessary to detect and diagnose these diseases. As we already discussed that most of the persons are suffering from the abdomen diseases but the availability of doctors to detect and diagnose these diseases is very less. Therefore, medical image processing is necessary to overcome the problem of diagnosing various diseases at an early stage.

Medical image processing is a useful noninvasive tool for detecting and diagnosing variety of diseases. The conventional medical imaging modalities are X-rays, Ultrasound, Computed Tomography (CT), and Magnetic Resonance Imaging (MRI) [5, 6]. The image features are extracted from these modalities by using various feature extraction algorithms which further help the researchers to detect these diseases. Afterward, the feature classification algorithms are used for the early diagnosis of diseases, which improves the diagnostic ability and efficiency of radiologists [7]. The computer-aided diagnosis (CAD) is also used for diagnosing variety of diseases. Thus, effort is made in this chapter to process and analyze the abdominal CT images for tissue characterization.

K. Rawal (✉) · G. Sethi
Lovely Professional University, Phagwara, Punjab, India
e-mail: kirti.20248@lpu.co.in

G. Sethi
e-mail: gaurav.11106@lpu.co.in

© Springer Nature Singapore Pte Ltd. 2020
O. P. Verma et al. (eds.), *Advancement of Machine Intelligence in Interactive Medical Image Analysis*, Algorithms for Intelligent Systems,
https://doi.org/10.1007/978-981-15-1100-4_6

Medical imaging processing is considered as the special branch of computer vision that deals with the interpretation of medical images. Various image modalities have been developed during last decades for the imaging of several parts of body. Thus, medical images become increasingly important in tasks like diagnosis, evaluation of diseases, and treatments. Out of all image modalities, CT scan and MRI are considered as most suitable for abdomen disease diagnosis [8, 9]. However, there are limitations associated with CT and MRI.

1.1 Need of Medical Imaging

Medical imaging is the upcoming field for providing better healthcare services. It is considered to be the most powerful tool for increasing the reliability, accuracy, and reproducibility of disease diagnosis. In this era of digitization, intervention of computers in medical diagnosis is a welcome step. In order to accommodate and process the huge amount of data, various techniques are required. Medical professionals have evinced great interest in development of diagnostic systems supported by the computers. The developments of such type of systems are feasible only because of the various image processing algorithms. With the advent of computers, the diagnosis of patients can be done very easily and efficiently. Further, the doctors are able to visualize various diseases by using the modern medical imaging techniques. In addition to original medical image modalities, such as CT scan or MRI, other imaging modalities such as endoscopy or radiography are also prepared with digital sensors. There are numerous advantages of using digital medical image processing methods, for example, digital data will not change when it is reproduced number of times, and it enables the provision of image enhancement that makes it easier for the radiologist to interpret the diseases.

2 Medical Background

Human abdomen is the most complex and important constituent of the human body. The structure of human abdomen is shown in Fig. 1. It consists of liver, pancreas, kidney, gall bladder, and intestines [8]. Following is the brief overview of functioning of abdomen organs.

2.1 Liver

The liver is one of the most important and largest organs in the human body [10]. It is located in the upper part of the abdomen as shown in Fig. 1. The liver is shown in

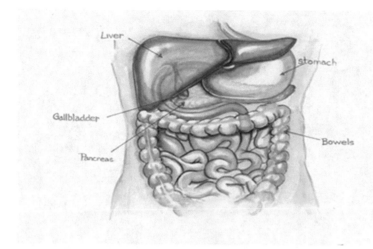

Fig. 1 Human abdomen

Fig. 2. It consists of two lobes, a right lobe and a left lobe, that perform the following functions:

a.   Liver stores the vital nutrients like sugar, vitamins, glucose, and minerals.
b.   Liver plays a crucial role in bile production that helps in digestion of food items.
c.   It helps in blood detoxification and helps in purification of blood by removing unwanted substances.
d.   It produces immune factors and removes bacteria from the bloodstream that helps in fighting diseases and infections.

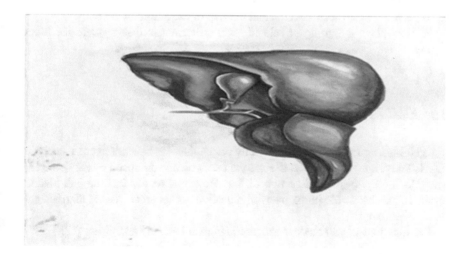

Fig. 2 Human liver

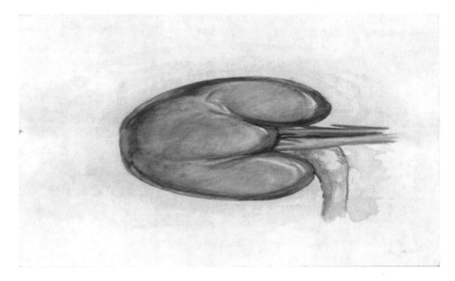

Fig. 3 Human kidney

2.2 Kidney

Kidney is bean-shaped organ of our body as shown in Fig. 3. It is also known as renal or nephron. It is located below the rib cage [11]. The kidneys are very important because they play a pivotal role in keeping the composition of blood stable that is important for proper functioning of the body. Following are the main functions of the kidney in human body [11]:

a. It maintains the composition of body fluids.
b. It prevents waste accumulation and buildup of extra fluid in the body.
c. Kidney plays a significant role in making hormones that regulate the blood pressure.

2.3 Gall Bladder

The gall bladder is considered as the tiny pouch that lies beneath liver as shown in Fig. 4. The primary function is to store the bile which is produced by the liver [12]. The bile is helpful in the digestion of fats. However, removal of the gall bladder results in no observable problems with digestion except small risk of diarrhea and fat malabsorption.

The functions of gall bladder and bile in human body are as follows [12]:

a. It regulates the flow of bile.
b. The bile breaks down the fats in human body.

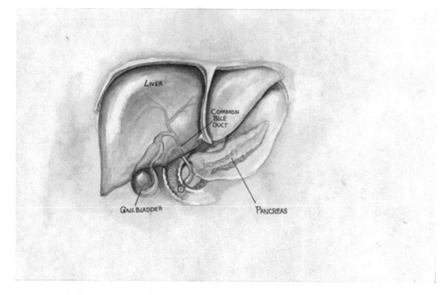

Fig. 4   Gall bladder and pancreas

c. The bile helps in removing the toxins from the human body.
d. The gall bladder helps in digestion of food.

2.4 Pancreas

Pancreas is the organ of human body which is bounded by liver, spleen, and intestines as shown in Fig. 4.

The pancreas is a useful organ of human body for maintaining digestion as well as in controlling the blood sugar levels. Following are the functions of pancreas [13]:

a. It helps in making enzymes that are helpful in digesting proteins, fats, and carbohydrates.
b. The pancreas is composed of exocrine cells that help in the producing enzymes that aid in digestion of food.

2.5 Spleen

Spleen is a fist-sized organ of the lymphatic system as shown in Fig. 5. It is used to purify the blood and benefits in keeping infections at bay and maintains the balance

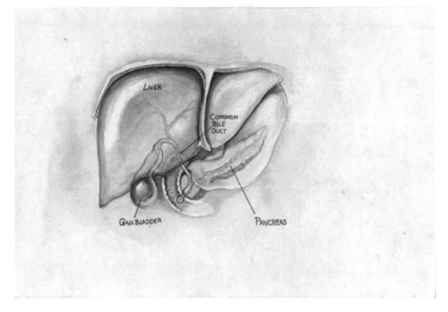

Fig. 5 Human spleen

of body fluid [14]. The spleen also comprises white cells that play a pivotal role in immunity-related diseases. The human spleen performs following functions [14]:

a. It helps in cleaning the impurities from blood.
b. It removes old red blood cells.
c. Spleen stores the blood in case of emergency.

3   Physiological Properties of Abdominal Tissues

The abdominal CT images used in this chapter consist of tumor, cyst, stone, and normal tissues. The physiological properties of abovementioned abdominal tissues are described as follows.

3.1   Tumor

The most credible way to diagnose the tumor is biopsy. It is the process of taking suspected piece of tissues from the body to examine it more closely. But it is painful and often stressful for the patients. In order to avoid unnecessary biopsies, the medical image processing techniques play a vital role in the disease diagnosis. Therefore,

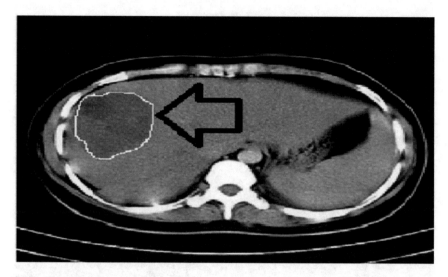

Fig. 6 Tumor (marked by arrow)

there is an increased expectation from the radiologists in diagnosing tumor in the early stages [15].

Many image modalities are used for the diagnosis of tumor. The CT and MRI images have a high sensitivity in the diagnosis of tumor [16]. The potential tumor tissues are shown in Fig. 6 (marked by arrow) which are darker than the normal tissues.

3.2   Cyst

Cyst is the formation of cluster of tissues that can take any place in the human body [17]. These are filled typically with fluid [18] as shown in Fig. 9 (marked by Arrow). The cysts often do not show symptoms of harming the body part. But in some cases, however, pain occurs when it comprises other organs in the neighborhood [19].

Cysts do not need any surgery or treatment. The common treatment, if required, is puncturing of the cyst. But, it is an invasive method that is used to dry the fluid within the cyst and further fill the cyst with solution to make the tissues harder [19]. In order to avoid the invasive surgery, CT scans are used for diagnosis of cysts [20]. The diagnosis of cyst is done with the help of the doctors by visualizing the image of CT scan. As shown in Fig. 7, the impression of cyst is darker as compared to the surrounding or normal tissues. There is a great difference between the contrast of normal tissues and cyst tissues.

Cyst tissues, i.e., the gray level of cyst tissue are much darker than that of normal tissue.

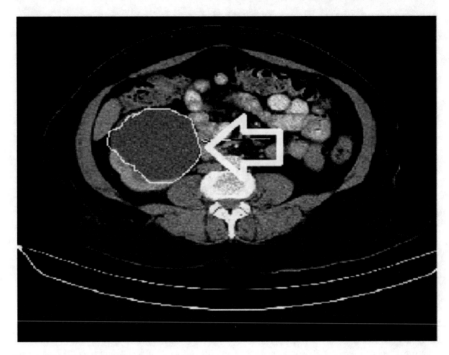

Fig. 7 Cyst (marked by arrow)

3.3   Stone (Calculi)

A stone or calculi is a hard material that is made in the abdomen organ like kidney and gall bladder. A small piece of stone can be easily circulated in the body without causing any discomfort. But, large stone causes certain problems in the urinary tract causing unbearable pain and may block urine. Kidney stone is one of most common disorders prevalent these days. The stone is shown in Fig. 8.

Stone or calculi can be formed when substances like calcium and oxalate in the urine become highly concentrated. One of the main reasons is the consumption of certain food items that promote the stone formation. The texture of stone is completely different from normal, cyst, or tumorous tissues. The main challenge of diagnosis using image processing techniques is the size of the stone. The size of most stones found in kidney and gall bladder is small as compared to other abdomen diseases. The various textures that represent abdomen diseases are classified using various image processing techniques as summarized in Fig. 9.

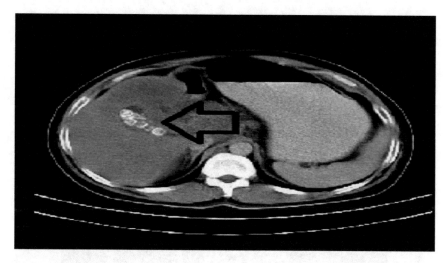

Fig. 8 Stone (marked by arrow)

3.4 Computer-Aided Diagnosis

CAD is the utmost important research area in the medical imaging [21–33]. It is something in which computer assists the specialist in taking decisions. The basic purpose is to generate the computerized output that acts as an automated diagnosis to help the doctors. The intervention of computers in taking decisions helps in improving diagnostic accuracy, increases consistency, and reduces the working load on the radiologists. The simplest example of CAD is shown in Fig. 10. The methodologies involved in designing CAD are (1) detection and segmentation of abnormalities, (2) quantification of image features for abnormalities, (3) optimization of image features for better classification, and (4) classification of diseases using optimized image features.

3.5 Detection and Segmentation of Abnormalities

The first stage of CAD for abdomen tissue characterization is image segmentation. Image segmentation is the process of dividing an image into multiple regions. The region that is required to be segmented is known as the ROI. These ROIs are used as informative inputs for image processing techniques, e.g., feature extraction, selection and, ultimately, classification of diseases. Thus, effective image segmentation is of utmost importance in medical image processing. There are many segmentation techniques available like level set, active contours, thresholding, region growing, and many more.

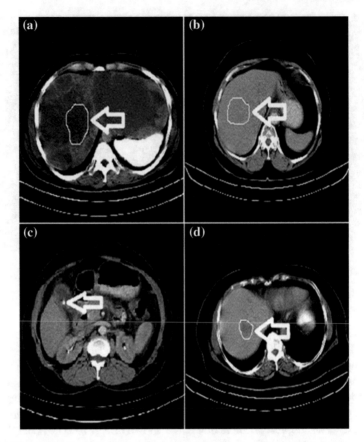

Fig. 9 CT abdominal images showing various kinds of tissues **a** cyst, **b** normal tissues, **c** stone, **d** tumor

3.6 Quantification of Image Features for Abnormalities

From the segmented ROIs and query images, features are required to be computed for extraction of clinically valuable information from the textures. The texture analysis can be considered as supplement to the visual skills of the radiologists that are relevant to the diagnostic problem but may not be visually perceptible. The distinguishing and potent features that characterize a texture are extracted by using different feature extraction methods. The feature extraction methods available are wavelets, curve-lets, GLCM, contour-lets, and DWT [34–40].

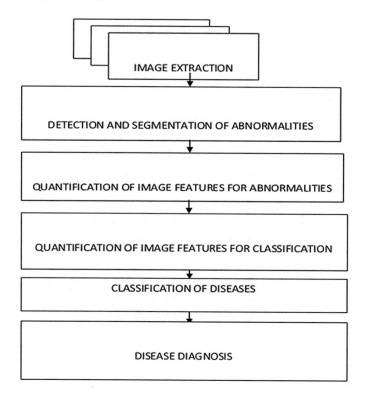

Fig. 10 Computer-aided diagnostic system

3.7   Quantification of Image Features for Classification

These extracted features are now used for classifying abdomen tissues. The process of classification faces challenge of choosing pertinent features from the extracted feature set. This is important as this affects the performance of the classifier. The selection of features from entire feature set is basically an optimization problem, where optimal and near-optimal features are selected in order to enhance the classification accuracy.

3.8   Classification of Diseases

The process of classification is data mining technique that is used to predict the class of data points. The choosing of appropriate classification method is cumbersome and tedious task. The purpose is to use classifier that utilizes the features vectors of query images for the purpose of classification of abdomen tissues. Classification is used to predict the class or membership for data instances. The prime objective of any classifier is to assign a class using feature vectors. The objective of classifier in this research work is to use feature vectors for classifying unknown input query in any one of four classes of tissues (cyst, stone, tumor, and normal).

4 Conclusions

In this chapter, an overview of existing CAD along with its integral components has been discussed. The need and importance of the image processing techniques used in diagnosis of abdomen diseases were also discussed.

The medical background of functioning of various abdomen organs like liver, kidney, pancreas, spleen, and gall bladder was discussed. The chapter summarizes the various abdomen diseases like tumor, cysts and stone (calculi), their texture, and gray-level properties. Various diagnostic systems of different diseases like liver tumor, breast cancer from CT images, and their performance were also surveyed in this chapter.

The image processing techniques are widely used in medical application. Today, there are wide ranges of medical image modalities available for radiologists to go for noninvasive diagnosis of diseases. These image modalities are ultrasound, X-rays, CT scan, and MRI that show the two-dimensional and three-dimensional views of various body organs like heart, abdomen at fairly high resolution. All these image modalities laid the foundation of intervention of computers in the diagnosis of diseases.

References

1. Jemal A, Bray F, Center MM, Ferlay J, Ward E, Forman D (2011) Global tumor statistics. CA Cancer J Clin 61:69–90
2. Burden of non-communicable disease in India report by Cameron Institute (2010)
3. Foster G, Stocks C, Borofsky MS (2009) Emergency department visits and hospital admissions for kidney stone disease
4. Bobadilla JL, Macek MJ, Fine JP, Farrell PM (2002) Cystic fibrosis: a worldwide analysis of CFTR mutations correlation with incidence data and application to screening. Hum Mutat 19:575–606
5. Erkonen WE, Smith WL (2010) The basics and fundamentals of imaging. Philadelphia Wolters
6. Chen E, Chung P (1998) An automatic diagnostic system for CT liver image classification. IEEE Trans Biomed Eng 45:783–794
7. Doi K (2005) Current status and future potential of computer-aided diagnosis in medical imaging. Br J Radiol 78:S3–S19
8. Choi C (2004) The current status of imaging diagnosis of hepatocellular tumor. American Association for the Study of Liver Diseases
9. Freeman R, Mithoefer A, Ruthazer R, Nguyen K, Schore A, Harper A, Edwards E (2006) Optimizing staging for hepatocellular tumor before liver transplantation: a retrospective analysis of the UNOS/OPTN database. American Association for the Study of Liver Diseases, Liver Transplantation
10. http://www.liver.ca/
11. https://www.kidney.org/
12. http://www.healthline.com/

13. www.news-medical.net/
14. www.livescience.com/
15. Llovet JM, Schwartz M, Mazzaferro V (2005) Resection and liver transplantation for hepatocellular tumor. Semin Liver Dis 25:181–200
16. Colli A, Fraquelli M, Conte D (2006) Alpha-fetoprotein and hepatocellular tumor. Am J Gastroenterol 101:1940–1941
17. National Institutes of Health (NIH) Publication, No. 08-4008 November (2007)
18. National Institutes of Health (NIH) Publication, No. 07-4618, February (2007)
19. Battiato S, Farinella GM, Gallo G, Garretto O, Privitera C (2009) Objective analysis of simple kidney cysts from CT images. In: IEEE international workshop on medical measurements and applications, pp 146–149
20. Kak AC, Slaney M (1988) Principles of computerized tomographic imaging. IEEE Press
21. Giger ML, Huo Z, Kupinski MA, Vyborny CJ (2000) Computer aided diagnosis in mammography. Med Imaging Process Anal (SPIE, Bellingham, WA) 2:915–1004
22. Erickson BJ, Bartholmai B (2002) Computer-aided detection and diagnosis at the start of the third millennium. J Digit Imaging 15:59–68
23. Summers RM (2003) Road maps for advancement of radiologic computer-aided detection in the 21st century. Radiology 229:11–13
24. Doi K (2003) Computer-aided diagnosis in digital chest radiography. In: Advances in digital radiography. RSNA categorical course in diagnostic radiology: physics syllabus. RSNA, Oak Brook, IL, pp 227–236
25. Abe H, MacMahon H, Engelmann R, Li Q, Shiraishi J, Katsuragawa S, Aoyama M, Ishida T, Ashizawa K, Metz CE, Doi K (2003) Computer-aided diagnosis in chest radiology: results of large-scale observer tests performed at the 1996–2001 RSNA scientific assemblies. Radio Graph 23:255–265
26. Dodd LE, Wagner RF, Armato SG, McNitt-Gray MF, Beiden S, Chan HP, Gur D, McLennan G, Metz CE, Petrick N, sahiner B, Sayre J (2004) Assessment of methodologies and statistical issues for computer-aided diagnosis of lung nodules in computed tomography: contemporary research topics relevant to the lung image database consortium. Acad Radiol 11:462–475
27. Gur D, Zhang B, Fuhrman CR, Hardesty L (2004) On the testing and reporting of computer-aided detection results for lung tumor detection. Radiology 232:5–6
28. Sethi G, Saini BS, Singh D (2015) Segmentation of cancerous regions in liver using an edge based and phase congruent region enhancement method. Comput Electr Eng 46:78–96
29. Sethi G, Saini BS (2015) Computer aided diagnosis using flexi-scale curvelet transform using improved genetic algorithm. Australas Phys Eng Sci Med 38:671–688
30. Mittal M, Verma A, Kaur I, Kaur B, Sharma S, Meenakshi S, Goyal LM, Roy S, Kim TH (2019) An efficient edge detection approach to provide better edge connectivity for image analysis. IEEE 7(1):33240–33255
31. Mittal M, Goyal LM, Kaur S, Kaur I, Verma A, Jude HD (2019) Deep learning based enhanced tumor segmentation approach for MR brain images. Appl Soft Comput 78:346–354
32. Mittal M, Goyal LM, Jude HD, Sethi JK (2019) Clustering approaches for high-dimensional databases: a review. WIREs Data Min Knowl Discov (Wiley) 9(3):e1300
33. Kaur S, Bansal RK, Mittal M, Goyal LM, Kaur I, Verma A, Son LH (2019) Mixed pixel decomposition based on extended fuzzy clustering for single spectral value remote sensing images. J Indian Soc Remote Sens 47(3):427–437
34. Kaur B, Sharma M, Mittal M, Verma A, Goyal LM, Hemanth DJ (2018) An improved salient object detection algorithm combining background and foreground connectivity for brain image analysis. Comput Electr Eng 71:692–703
35. Jude HD, Anitha J, Son LH, Mittal M (2018) Diabetic retinopathy diagnosis from retinal images using modified hopfield neural network. J Med Syst 42(12):247
36. Jude HD, Popescu DE, Mittal MS, Uma Maheshwari SU (2017) Analysis of wavelet, ridgelet, curvelet and bandelet transforms for QR code based image steganography. In: 14th IEEE international conference on engineering of modern electric systems (EMES) Romanian

37. Mittal M, Goyal LM, Sethi JK, Jude HD (2018) Monitoring the impact of economic crisis on crime in India using machine learning. Comput Econ 53(4):1467–1485
38. Mittal M, Sharma RK, Singh VP, Agarwal R (2019) Adaptive threshold based clustering: a deterministic partitioning approach. Int J Inf Syst Model Des 10(1):42–59
39. Yadav M, Purwar RK, Mittal M (2018) Handwritten Hindi character recognition—a review. IET Image Proc 12(11):1919–1933
40. Garg R, Mittal M, Son LH (2019) Reliability and energy efficient workflow scheduling in cloud environment. Clust Comput 1–15

Chapter 7
Multi-reduct Rough Set Classifier for Computer-Aided Diagnosis in Medical Data

Kavita Jain and Sushil Kulkarni

1 Introduction

The technological growth has led to the humongous data which should be appropriately interpreted and translated. Many a times the data which is collected has lot of issues and before processing it further these issues should be addressed. So basically, given this imprecise chunk of data at hand we try to process it and discover hidden information from it and find meaningful relationships between the attributes which aids us for the classification purpose. Rough Set Methodology (RSM) can handle the imprecision and vagueness in the data quite efficiently and can help with the easy and interpretable rules for the medical diagnosis.

Data is represented in an information table comprising columns which indicates the features of the objects and row indicates the object. Simplifying this table is of extreme importance in many applications. Primary simplification involves doing away with the unnecessary attributes and deletion of identical rows. Similar rows are eliminated as they mean the same decision value and doing away with the redundant attributes saves the computational time. Getting rid of the redundant attributes is also called as dimensionality reduction. Classification performed using the selected feature subset gives approximately the same accuracy as that of the whole feature set. There could be more than one feature subset for the given data set also called as reduct. A reduct is nothing but the necessary part of the given attribute set which helps in discerning the objects. RSM can help us in finding all possible reducts of the given information system. There are various algorithms for constructing the reducts [1].

K. Jain (✉)
University Department of Computer Science, Kalina, Mumbai 400098, India
e-mail: cavita283@gmail.com

S. Kulkarni
Department of Mathematics, Jai Hind College, Churchgate, Mumbai 400020, India
e-mail: sushiltry@gmail.com

© Springer Nature Singapore Pte Ltd. 2020
O. P. Verma et al. (eds.), *Advancement of Machine Intelligence in Interactive Medical Image Analysis*, Algorithms for Intelligent Systems,
https://doi.org/10.1007/978-981-15-1100-4_7

Dimensionality reduction is performed by constructing the discernibility matrix. Discernibility function is computed from this matrix. Simplifying this discernibility function gives us the reducts and these reducts can be used for generating if-then type rules which can be used for the classification purpose [2, 3].

There can be many reducts of the information table. Hence, a classifier can be built based on these reducts called as Multi-Reduct Rough Set Classifier. The outcome of these classifiers is integrated for the final diagnosis, called as ensemble of classifiers. The ensemble model can be homogeneous ensemble model or heterogenous ensemble model. In homogeneous ensemble model all the base learning algorithms are same whereas in heterogenous ensemble model the base learning algorithms are different. There are different meta algorithms that integrate these base classifiers. The most often used algorithms are bagging, boosting, and stacking. The rationale behind ensemble model is to increase the overall performance of the classification. There can be performance improvement if there is diversity in the underlying feature subsets that are reducts. If the base classifiers exhibit the same behavior, then integration of the base classifiers hardly serve the purpose. Diversity of the base classifiers is generally considered in terms of their output. Because of the diversity the error of any classifier is averaged out by the accuracy of the other classifiers.

The basic hypotheses are that if we integrate the base classifiers in right way we can get an improvised or the robust classification model. Although the most popular ensemble algorithms are bagging, boosting, and stacking, many variations of these algorithms are possible, benefiting the given classification problem.

Chapter is arranged into following segments: Sect. 2 elaborates preliminaries of RSM along with real-life data examples. Section 3 talks about the related work of the applications of RSM in the medical field. Section 4 elaborates the concept of dimensionality reduction. Section 5 discourses the outline of the planned model established on the idea of Multi-Reduct Rough Set Classifier and ensemble of classifiers. Finally, Sect. 6 exhibits the concluding remarks.

2   Theoretical Aspects of Rough Set Methodology

Consider the universe of the desired objects. A subset X of U is termed as a concept in the universe.

2.1   Representing Concepts Using an Information System

To understand a concept, it is necessary to consider, namely, the intension indicating the concepts attributes and extension indicating the instances belonging to the concept [4].

An object that belongs to the concept must satisfy all the properties mentioned by the concept's intension. Intension gives the inference to the corresponding extension. In RSM intension and extension are constructed with respect to the given system.

Given system is signified by the tuple:

$$\text{Tuple} = (U, \text{Attr}, \{D_a | a \in \text{Attr}\}, \{V_a | a \in \text{Attr}\}), \tag{1}$$

where

U	Universe of discourse,
Attr	Attributes set,
D_a	Attributes domain, and
$V_a = U \to D_a$	is a description function assigning values.

By definition, $V_a(x) \in D_a$ represents objects value.

Example 1 Let Universe $= \{C_1, C_2, C_3, C_4, C_5, C_6, C_7, C_8\}$, Attr $= \{$Sputum_with_blood, fever, loss_of_appetite, loss_of_weight$\}$, $D_{\text{sputum\_with\_blood}} = \{Yes, No\}$, $D_{\text{Fever}} = \{$Yes, No$\}$, $D_{\text{Loss\_of\_appetite}} = \{Yes, No\}$, and $D_{\text{Loss\_of\_weight}} = \{$Light, Average, Heavy$\}$ as shown in Table 1.

Object C_1 is described as

$$V_{\text{sputum\_with\_blood}}(C_1) = \text{Yes},$$
$$V_{\text{Fever}}(C_1) = \text{Yes},$$
$$V_{\text{Loss\_of\_appetite}}(C_1) = \text{Yes},$$
$$V_{\text{Loss\_of\_weight}}(C_1) = \text{Light}$$

In a table, rows indicate the objects and column indicates the properties which are possessed by the respective objects.

Table 1 The dataset

Conditional attributes					Output class
Case ID	Sputum _With_Blood	Fever	Loss_Of_Appetite	Loss_Of_Weight	Tuberculosis
C_1	Yes	Yes	Yes	Light	No
C_2	No	No	No	Heavy	Yes
C_3	No	No	Yes	Light	Yes
C_4	No	No	No	Average	No
C_5	Yes	Yes	Yes	Light	No
C_6	No	Yes	Yes	Light	Yes
C_7	No	No	No	Average	Yes
C_8	No	Yes	No	Heavy	No

It can be observed in Table 1 that some objects have the same description. For example, objects C_1 and C_5 have the same description. Hence, based on their description one cannot differentiate these two objects. This analysis is the basis of rough set theory [5].

2.2 Definable Concepts and Undefinable Concepts

Some categories that are features of objects can be defined in some knowledge base but may not be definable in some other knowledge base.

From Table 1 suppose if we consider the attribute Loss_Of_Weight = Light will be called as the intension of the concept and its extension is $\{C_1, C_3, C_5, C_6\}$.

A concept is definable if its extension is definable otherwise it is not definable. Hence, if a concept is undefinable, we can approximately define it [6].

2.3 Indiscernibility

Indiscernibility is another elementary notion of RSM. This relation indicates the collection of objects with equivalent values regarding the considered attribute set. It is also termed as an equivalence relation. So, the universe is partitioned by the subsets of attributes symbolized as U/IND(A) [7].

Sets that are indiscernible are termed as elementary units and its unification leads to the definable set [8].

Example 2 Consider the set of case Ids of patients from Table 1. These patients are analyzed with four conditional attributes as indicated in Table 1. For example, a patient having tuberculosis has no sputum with blood, no fever, no loss of appetite, and average loss of weight.

Hence, the set of case Ids of patients can be classified according to sputum with blood, fever, loss of appetite, and loss of weight.

For example, Patients

$C_2, C_3, C_4, C_6, C_7, C_8$—has no sputum with blood
C_1, C_5—has sputum with blood

Patients

C_2, C_3, C_4, C_7—has no fever
C_1, C_5, C_6, C_8—has fever

Patients

C_2, C_4, C_7, C_8—has no loss of appetite
C_1, C_3, C_5, C_6—has loss of appetite

and Patients

C_1, C_3, C_5, C_6—has light loss of weight
C_4, C_7—has average loss of weight
C_2, C_8—has heavy loss of weight

With this kind of grouping, we can define four equivalence relations namely E_1, $E_2, E_3,$ and E_4 which have equivalence classes as

$$U/E_1 = \{(C_2, C_3, C_4, C_6, C_7, C_8), (C_1, C_5)\},$$
$$U/E_2 = \{(C_2, C_3, C_4, C_7), (C_1, C_5, C_6, C_8)\},$$
$$U/E_3 = \{(C_2, C_4, C_7, C_8), (C_1, C_3, C_5, C_6)\},$$
$$U/E_4 = \{(C_1, C_3, C_5, C_6), (C_4, C_7), (C_2, C_8)\}$$

These equivalence classes are elementary concepts for the system.

$$K = (U, \{E_1, E_2, E_3, E_4\}).$$

The common elements of the elementary concepts are called as the basic concepts [2]. As an instance, sets

$$\{C_2, C_3, C_4, C_6, C_7, C_8\} \cap \{C_2, C_3, C_4, C_7\} = \{C_2, C_3, C_4, C_7\},$$
$$\{C_1, C_5\} \cap \{C_1, C_5, C_6, C_8\} = \{C_1, C_5\}$$

are $\{E_1, E_2\}$—basic categories that have no sputum with blood and has no fever, has sputum with blood, and has fever, respectively. Sets

$$\{C_2, C_3, C_4, C_6, C_7, C_8\} \cap \{C_2, C_3, C_4, C_7\} \cap \{C_2, C_4, C_7, C_8\} = \{C_2, C_4, C_7\},$$
$$\{C_1, C_5\} \cap \{C_1, C_5, C_6, C_8\} \cap \{C_1, C_3, C_5, C_6\} = \{C_1, C_5\}$$

are $\{E_1, E_2, E_3\}$—basic concepts that have no sputum with blood, has no fever and has no loss of appetite, has sputum with blood, has fever, and has loss of appetite, respectively.

2.4 Approximations of Sets

Some objects cannot be expressed properly because of insufficient knowledge in hand. Hence, the thought of approximation of sets is used to represent the inaccurate or the imprecise information. This indefinability leads to the very important building block of RSM that is lower approximation (LA) and upper approximation (UA).

LA and UA are the definable sets approximating an undefinable set. These two approximations are formulated using the equivalence relation which gives us the equivalence classes [9]. Figure 1 indicates the Pawlak's Computational model [10].

Fig. 1 Pawlak's
computational model

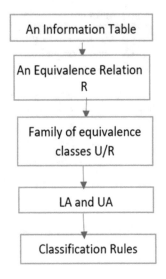

Let IS be the information system $= (U, E)$ and $X \subseteq U$ be an equivalence relation $E \in \text{IND(IS)}$. The two sets called as LA and UA can be described like

$$\underline{E}X = U\{Z \in U/E : Z \subseteq X\}, \tag{2}$$

$$\overline{E}X = \{Z \in U/E : Z \cap X \neq \emptyset\}. \tag{3}$$

The set $\underline{E}X$ comprises all the elements definitely belonging to set X whereas the set $\overline{E}X$ comprises all the elements possibly belonging to X.

The $BN_E(X) = \text{UA} - \text{LA}$ is called as E-boundary of X. This area indicates a set of elements not belonging to X neither the complement based on the given information.

Based on LA and UA, the universe U of elements is divided into the positive, negative, and boundary area as

$$\text{POS}_E(X) = \underline{E}X,$$
$$\text{NEG}_E(X) = \text{Universe} - \underline{E}X,$$
$$\text{BND}_E(X) = \text{E-Boundary area}.$$

The positive area or the LA consists of all the elements which can be surely categorized to X using the available information.

Whereas negative area indicates that the collection of elements which cannot be categorized or classified to this area that is they belong to the complement set of X.

Hence, UA is unification of the positive and borderline regions. Furthermore, X is definable in context to E iff its UA and LA are same else X is termed as E-undefinable.

X is E-definable iff LA $=$ UA,
X is Rough in context to E iff LA $\neq$ UA

Table 2 Equivalence class

Equivalence class	Conditional attribute value set	Decisional attribute value
$E_1 = \{C_2\}$	{No, No, No, Heavy}	Yes
$E_2 = \{C_3\}$	{No, No, Yes, Light}	Yes
$E_3 = \{C_6\}$	{No, Yes, Yes, Light}	Yes
$E_4 = \{C_7\}$	{No, No, No, Average}	Yes
$E_5 = \{C_1, C_5\}$	{Yes, Yes, Yes, Light}	No
$E_6 = \{C_4\}$	{No, No, No, Average}	No
$E_7 = \{C_8\}$	{No, Yes, No, Heavy}	No

Example 3 Consider the universe U of patients in Table 1 described by the four conditional parameters and one output parameter. Table 2 shows the equivalence class for the patients suffering from tuberculosis and not suffering from tuberculosis.

For example in Table 2, considering all the given four conditional attributes (sputum with blood, fever, loss of appetite, loss of weight), then $\{C_1, C_5\}$ is an equivalence class because C_1 and C_5 are not discernible from each other.

In the case of Table 1, for the concept $\{C_2, C_3, C_6, C_7\}$ describing people suffering from tuberculosis, the LA is equal to the set $\{C_2, C_3, C_6\}$ and UA equal to the set $\{C_2, C_6, C_4, C_3, C_7\}$.

2.5 Numerically Characterizing Uncertainty

Uncertainty of the set exists because of the non-empty boundary set. The set's accuracy reduces as the number of members in the boundary set increase. Accuracy measure and Roughness help in numerically characterizing this idea.

Accuracy is given as

$$\alpha(X) = \frac{\text{card LA}}{\text{card UA}}$$

where

$$X \neq \emptyset, \tag{4}$$

Accuracy of set X signifies the completeness of the knowledge about that set. The accuracy value of the set lies in the range of $0 \leq \alpha(X) \leq 1$. Consider $\alpha(X) = 1$ meaning that there are no elements in the boundary region and hence called a definable set and $\alpha(X) \leq 1$ indicates that there are few elements in its boundary area, and hence it is called as an undefinable set [11].

Roughness can be stated as

$$\rho(X) = 1 - \alpha(X), \tag{5}$$

It signifies incompleteness of the given set. Accuracy measure and Roughness are counterpart to each other.

Example 4 Illustrating the above notion with the given information system and equivalence relation, we calculate the accuracy measure for the set of patients suffering from tuberculosis.

$$\alpha(X) = \frac{3}{5}.$$

This numerical value of uncertainty or imprecision is not presumed but calculated with the help of approximations unlike the probability or fuzzy theory. This numerical value indicates the limited knowledge available at hand to classify the object.

3   Related Work

Wang et al. have addressed the issue of microarray-based cancer prediction using α depended degree concept from RSM to choose the highly discriminative features. Wang et al. have taken this approach as attribute reduction in RSM is computationally expensive affair especially for this microarray-based data [12].

In [13], authors have formed a typhoid diagnosis system using a machine learning technique with 18 conditional attributes and 1 decision attribute called as level of severity. LEM2 algorithm was employed for developing the classification model. Total 18 rules were generated using this rough set model for diagnosing typhoid fever. Hybrid approaches for diagnoses of typhoid fever are suggested. Also, authors have pointed the need of a typhoid fever therapy system that can provide a matching effective therapy built on the degree of seriousness.

Rahman et al. have designed a hybrid RSM-based model for prediction of diabetes mellitus and tries to predict the patients with the potential risks. From patient's clinical charts, the rules are mined with the aid of RSM and domain information is employed to assist physicians to identify the patients with the potential risks. The issue of high dimensionality and incomplete values in the dataset is handled by the rough set techniques. The continuous value attributes are converted into discrete value attributes to avoid the intractable rules. For generating the reducts, lattice search method has been applied and rules are generated with the help of LEM2 algorithm.

Two-level of reasoning is done. In first-level diabetes category is predicted and in the second level, it is further concluded whether the case is either normal, borderline, abnormal, risky, etc. [14].

Chen et al. have proposed a diagnostic model for renal disease patients of hemodialysis therapy in Taiwan. Various classification models are dependent on the data at hand. So, authors have proposed this classification model in the context of the healthcare industry. Based on the expert's knowledge the decision attributes are segregated into different categories and the conditional attributes are discretized to improve the overall accuracy. The comprehensible rules were generated for making decision by the RSM LEM2 algorithm which had a low standard deviation compared to the other mentioned methods [15].

In [16], authors have discussed that many expert models have failed to diagnose various medical problems due to inability to handle inconsistencies in data. In order to address this issue, they have proposed Pawlak's rough set concept. Rough set handles inconsistencies in data and generates rules. These generated rules are huge in number. In order to extract the most suitable rules and to avoid redundancies formal concept analysis is being applied and most important rules affecting the decision-making are extracted.

So, authors have done pre-processing of data with the help of rough set and post-processing of data with formal concept analysis. RSM is used for prediction while formal concept analyses are used for description and together it leads to better extraction of knowledge. Authors have taken the case study of heart diseases and have tried to identify the most important rules and attributes which will affect the decision-making.

In [17] authors have discussed presence of imbalanced datasets existing in the healthcare industry and how it affects the performance of the overall classifier at hand. Authors have recommended the application of rough set theory with artificial intelligence as an effective method of classification versus the conventional statistical methods. Data must be discretized first to avoid generation of large decision rules.

In order to tackle imbalanced datasets problem, authors have adopted data level approach that is bootstrap method to change its distribution. Considering, shortcomings of each of undersampling and oversampling methods, authors have integrated both the approaches by making the size equivalent to the average of its classes. Authors have used LEM2 algorithm and generated the interpretable rules which aided in decision-making.

Hassanien et al. have shown the correlation of diabetes to psychological problems among children. In order to address the issue of incomplete knowledge and inconsistent pieces of information authors have used an intelligent technique based on RSM to uncover meaningful relationships. The authors have taken the case study of children having diabetes from Kuwait. Proposed model has two stages that are pre-processing and processing. Pre-processing stage comprises data completion and data discretization. Processing phase comprises relevant feature extraction, feature reduction, and generating rules.

The RSM helps in concluding the strength of the attributes and their power in the classifier. Hassanien et al. have further discussed that the quality of any learning algorithm is based on data discretization.

In order to extract the relevant attributes, it is necessary to identify the core and reducts from the information system. Redundant attributes can be ignored if they do not affect overall accuracy of the information system. Finding all possible reducts is the NP-hard problem. Few reducts which represent the information system are enough. Best reduct can be the reduct with less number of attributes or so.

Authors have done the comparison with other models like neural networks, decision trees, etc. An extension of RSM combining another intelligent system is suggested as the future work [18].

Rafal developed an information system for classifying the patients with genetic susceptibility to DMT1 and indicated the risk factor of patient falling sick. Apart from developing the decision model author also aimed in general to improvise the classifier's performance. The author has used RSM-based methods for classification as underlying data is incomplete, complex, and uncertain.

Rules have been synthesized using reduct/direct approach and LEM2 algorithm. For generating the minimal decision rules using reduct approach author has used the steps like building the discernibility matrix, calculation of reducts, and rules.

Other approach of generation of rules that is LEM2 algorithm is optimal in generating the minimalistic rules. This algorithm is based on local cover computation of information system followed by decision rules from it.

Rafal has first developed the model of diagnosing sickness followed by the model for diagnosis of diabetes. Model is being tested both by the cross-validation and by the expert. The classification accurateness is same with both the methods, but the number of rules inferred by LEM2 algorithm is optimal. Hence, authors have suggested the LEM2 algorithm to be more useful in medical practice [19].

Classical RSM is unable to handle missing values in the given information systems. Zhai et al. have presented RSM-based model on similarity and tolerance relation. For improving the accurateness of the model, the RSM model is defined by similarity relation. Degree of importance of attributes that is weights are calculated to find the more significant attributes. The proposed model reduces the error rate of classification and improvises correctness and granularity [20].

Computation cost with the classical RSM-based model grounded on the equivalence relation for the continuous datasets is too high. To overcome this Kumara et al. have suggested neighborhood RSM for the medical diagnosis. The model uses Euclidean distance as a metric unlike equivalence relation of the classical RSM.

The neighborhood RSM algorithm was implemented on five datasets of UCI repository. Kumara et al. have done the comparative study of neighborhood RSM model with other algorithms like Pawlak's RSM, Neural networks, SVM, etc. Authors have shown that the neighborhood RSM is very effective methodology for the classification [21].

4 Dimensionality Reduction

Reducts are the most important notion of RSM. A reduct is nothing but the set of attributes from the original set which preserves the classification ability of the original dataset. So, to find this reduct we must do away with the redundant attributes called as dimensionality reduction. Obtaining a reduct is like feature selection problem which satisfies certain criterions.

Usually, any information system has multiple reducts. Core is the commonality of whole reducts. In-fact it is used as the base point to compute the reducts. Various heuristics are used for calculation of the minimal reducts [22].

Mainly two methods for calculating the reducts in RSM are discernibility matrix method and second degree of dependency method.

4.1 Discernibility Matrix-Based Dimensionality Reduction

Discernibility matrix characterized for $IT = (U, C \cup d)$ is a symmetric matrix whose items are [23]

$$c_{ij} = \{a \in C | a(x_i) \neq a(x_j)\} \quad \text{for } I, j = 1, \ldots |U|. \tag{6}$$

Consider the example shown in Table 3 for computing the discernibility matrix. The steps for creating the discernibility matrix are

(i)   In the matrix, each entry comprises features varying amongst object i and object j.
(ii) Some entries in the decision-relative matrix, appears to be blank, despite the difference in the feature, is because their corresponding decisions are the same.
(iii) Collecting all singleton items form the core that is attributes which appear in all reducts. Hence, the core is $\{a_2\}$.

Table 3 The dataset

x	a_1	a_2	a_3	a_4	D
x_1	1	1	1	0	1
x_2	0	0	0	2	0
x_3	0	0	1	0	1
x_4	0	0	0	1	0
x_5	1	1	1	0	1
x_6	0	1	1	0	1
x_7	0	0	0	1	0
x_8	0	1	0	2	0

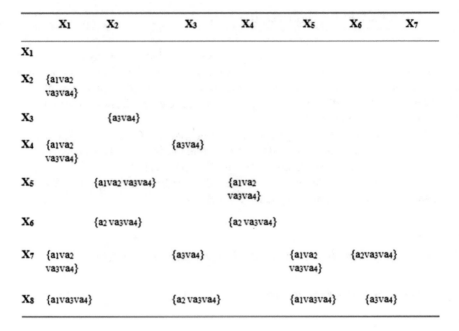

Fig. 2 Discernibility matrix

(iv) Discernibility function is stated which helps in distinguishing the objects from each other.

(v) The minimal reducts are computed from the discernibility function by evaluating its prime implicants [24].

(vi) Discernibility function f_D is calculated with duplicates removed.

The discernibility function of Fig. 2 is

$$f_D(a_1, a_2, a_3, a_4) = \{a_1 \vee a_2 \vee a_3 \vee a_4\} \wedge \{a_1 \vee a_3 \vee a_4\} \wedge \{a_3 \vee a_4\} \wedge \{a_2 \vee a_3 \vee a_4\}$$
$$= \{a_3, a_4\} \text{is an reduct.}$$

4.2 Degree of Dependency

Information Q is derived from information P, if all the elementary classes of Q are outlined by some elementary classes of P. If we can derive Q from P, then it is expressed as

$$P \Rightarrow Q, \tag{7}$$

This type of dependency is called as total dependency. There can be functional dependency that is partial dependency between Q and P. This implies that some information Q can be derived from information P.

Partial dependency is stated as

$$k = \gamma_p(Q) = \frac{\text{card POSp}(Q)}{\text{card}(U)}. \tag{8}$$

The partial dependency formula implies information Q which is derivable with degree $0 \leq k \leq 1$ from information P.

If $k = 1$, it signifies total derivability from P, if $0 < k < 1$ then it signifies partial derivability from P or rough derivability and if $k = 1$, then it signifies total non-derivability from P.

The degree of functional dependency can also be computed from the indiscernibility relation. Furthermore, we can calculate the dispensable and indispensable attributes from it [25].

If the indiscernibility relation for a group of attributes is identical to its power set then, attribute that is part of power set and not part of the set is superfluous. These superfluous attributes are also called as a dispensable attribute.

5   Multi-reduct Rough Set Classifier

5.1   Outline of Proposed Methodology

Outline of proposed model as shown in Fig. 3 is first to generate the multiple reducts based on RSM and at second-step ensemble method is applied on classifiers.

The advantage of ensemble learning is that the error of any classifier is averaged out by the accuracy of the other classifiers.

5.2   Generating Multiple Reducts

There are multiple reducts for any knowledge base. Hence, the classifier can be constructed in varied methods. Hypothetically a classifier prepared dependent on reduct has practically identical execution with the classifier prepared dependent on the whole knowledge base [26].

Since various information subsets can be acquired by applying RSM an ensemble methodology is better for generalizing the performance.

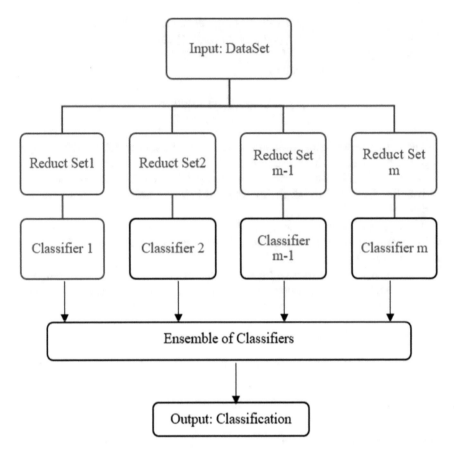

Fig. 3 Proposed methodology

5.3 *Ensemble of Classifiers*

Ensemble methods are nothing but learning algorithms that develop classifiers and then categorize new information by taking the votes of the respective outputs of the learning algorithms. Learning algorithm is often considered as searching a space of assumptions to identify the best assumptions in the space. There are various types of learning algorithms like decision trees, support vector machines, perceptrons, etc.

Reduct classifiers can be combined to improvise the accurateness of the classifier. Numerous techniques to construct ensemble classifiers are bagging, boosting, majority vote, Adaboost, stacking, etc. [27]. Basically, for creating the ensemble of classifiers the training examples are manipulated, input features are manipulated, output targets are manipulated, or randomness is injected [28].

Ensemble learning algorithm works by running the base learning algorithm multiple times.

Ensemble model can be homogeneous ensemble model or heterogenous ensemble model. In homogeneous ensemble model, all the base learning algorithms are same whereas in heterogeneous ensemble model, the base learning algorithms are different. For building a successful ensemble of classifiers it is necessary for the base classifiers to be accurate and diverse.

Example 5 This example demonstrates the importance of accuracy and diversity.

Consider ensemble of four classifiers $\{d_1, d_2, d_3, d_4\}$. Let's say we must classify a new case x.

If all the four classifiers are same then when $d_1(x)$ is wrong, $d_2(x)$ is also wrong, $d_3(x)$ is wrong, and $d_4(x)$ is also wrong.

However, if the errors made by the four classifiers are not correlated then if $d_1(x)$ is wrong, $d_2(x)$ may be correct, $d_3(x)$ may be correct, and $d_4(x)$ may be correct, then bulk vote may correctly tag the given sample.

The basic hypotheses are that if we integrate the base classifiers in right way we can get an improvised or the robust classification model [29].

6 Concluding Remarks

Rough set methodology is a helpful and practical tool for inductive learning and aids in construction of the expert systems. The ease of the model makes it appealing to the research scholars.

The chapter has discussed the theoretical aspects of rough set and methods for dimensionality reduction. The methods mainly are Discernibility matrix-based and degree of dependency.

Further, the chapter has discussed the idea of Multi-Reduct Rough Set Classifier. There can be many reducts of the given information system. The classifiers are constructed based on these reducts and the outcome of these classifiers is integrated for the final classification called as ensemble of classifiers. Ensemble methods give the perfect classifiers by combining the less perfect ones.

The future work is to investigate the methods for generating the minimum discernibility matrix [30–32] and applying heuristic methods to find reducts.

Acknowledgements Indebted to Dr. Sushil Kulkarni, Professor in Mathematics in Jai Hind College and Research Guide at University of Mumbai for his guidance and constant support for conducting my research work. Exclusive thanks to Professor Dr. Pawan Lingras faculty in Mathematics, Saint Mary's University, Canada for giving some insightful comments and directions to my research. Many thanks to Dr. Ambuja Salgaonkar, professor in Computer Science, University of Mumbai for giving support and access to the valuable books and articles.

My heartfelt gratefulness to my husband who has been constant motivation for encouraging me to do the research work. I am grateful to the rock bottom of my heart to my family, pals, and my university colleagues who have backed me directly or indirectly in conducting this research.

References

1. Yao Y, Zhao Y, Wang J (2006) On reduct construction algorithms. On reduct construction algorithms, rough sets and knowledge technology. In: First international conference, proceedings. LNAI, vol 4062, pp 297–304
2. Pawlak Z (1991) Rough sets theoretical aspects of reasoning about data. Theory and decision library D. Springer, Netherlands. ISBN 978-94-011-3534-4
3. Widz S, Slezak D (2012) Rough set based decision support—models easy to interpret. In: Advanced information and knowledge processing. Springer, London. https://doi.org/10.1007/978-1-4471-2760-4_6
4. Yao Y (2015) Rough set approximations: a concept analysis point of view. In: Ishibuchi H (ed) Computational intelligence—Volume I, Encyclopedia of life support systems (EOLSS), pp 282–296
5. Li X (2014) Attribute selection methods in rough set theory. Master's theses and graduate research, San Jose State University
6. Yao Y, Chen Y (2006) Rough set approximations in formal concept analysis. In: Peters JF, Skowron A (eds) Transactions on rough sets V. Lecture notes in computer science, vol 4100. Springer, Berlin, Heidelberg
7. Yao Y, Slezak D (2012) An introduction to rough sets. Rough sets: selected methods and applications in management and engineering. In: Advanced information and knowledge processing. Springer, London. https://doi.org/10.1007/978-1-4471-2760-4_1
8. Pawlak Z, Grzymala-Busse J, Slowinski R, Ziarko W (1995) Rough sets. Commun ACM 38(11)
9. Chan C (2007) Fundamentals of rough set fundamentals of rough set theory and its applications. NUK
10. Pawlak Z (2002) Rough set theory and its applications. J Telecommun Inf Technol
11. Rissino S, Lambert-Torres G (2009) Rough set theory—fundamental concepts, principals, data extraction, and applications. In: Ponce J, Karahoca A (ed) Data mining and knowledge discovery in real life applications. I-Tech, Vienna, p 438. ISBN 978-3-902613-53-0
12. Wang X, Gotoh O (2009) Microarray-based cancer prediction using soft computing approach. Cancer Inform 2009(7):123–139
13. Oguntimilehin A, Adetunmbi AO, Abiola OB (2013) A machine learning approach to clinical diagnosis of typhoid fever. Int J Comput Inf Technol 02(04). ISSN: 2279-0764
14. Ali R, Hussain J, Siddiqi M, Hussain M, Lee S (2015) H2RM: a hybrid rough set reasoning model for prediction and management of diabetes mellitus. Sensors. ISSN 1424-8220
15. Chen Y, Cheng C (2013) Application of rough set classifiers for determining hemodialysis adequacy in ESRD patients. Knowl Inf Syst 34(2):453–482
16. Tripathy B, Acharjya D, Cynthya V (2011) A framework for intelligent medical diagnosis using rough set with formal concept analysis. Int J Artif Intell Appl (IJAIA) 2(2)
17. Chen Y, Cheng C (2012) Identifying the medical practice after total hip arthroplasty using an integrated hybrid approach. Comput Biol Med
18. Hassanien A, Abdelhafez ME, Own H (2008) Rough sets data analysis in knowledge discovery: a case of Kuwaiti diabetic children patients. Adv Fuzzy Syst. https://doi.org/10.1155/2008/528461
19. Rafał D (2011) Accuracy evaluation of the system of type 1 diabetes prediction. In: Rough Sets and knowledge technology. Berlin/Heidelberg, Germany, pp 321–326
20. Zhai Y, Yan R, Huang Z, Guo B (2013) Rough set model based on the dual-limited symmetric similarity relation. Appl Math Inf Sci 7(1L):11–17
21. Kumara S, Inbarani H (2015) A novel neighborhood rough set based classification approach for medical diagnosis. Procedia Comput Sci 351–359
22. Hu K, Lu Y, Shi C (2003) Feature ranking in rough sets. AI Commun (Spec Iss Artif Intell Adv China) 16(1):41–50
23. Jensen R, Jensen Q (2007) Rough set-based feature selection: a review. OAI. https://doi.org/10.4018/978-1-59904-552-8.ch003

24. Yao Y, Zhao Y (2009) Discernibility matrix simplification for constructing attribute reducts. Inf Sci 179(5):867–882
25. Hassanien A, Abdelhafez M, Own H (2008) Rough sets data analysis in knowledge discovery: a case of Kuwaiti diabetic children patients. Adv Fuzzy Syst (Hindawi Publishing Corporation) Article ID 528461, 13 pp. https://doi.org/10.1155/2008/528461 (2008)
26. Hu X (2006) Ensembles of classifiers based on rough sets theory and set-oriented database operations. In: IEEE international conference on granular computing, Atlanta, GA, USA, USA. ISBN No:978-1-5090-9177-5
27. Dietterich T (2000) Ensemble methods in machine learning. In: Multiple classifier systems. Lecture notes in computer science, vol 1857. Springer, Berlin, Heidelberg
28. Rokach L (2005) Ensemble methods for classifiers. In: Maimon O, Rokach L (eds) Data mining and knowledge discovery handbook. Springer, Boston, MA
29. LeDell E (2015) Intro to practical ensemble learning. Group in Biostatistics, University of California, Berkeley
30. Wang J (2001) Reduction algorithms based on discernibility matrix: the ordered attributes method. J Comput Sci Technol
31. Yao Y, Zhao Y, Wang J, Han S (2006) A model of machine learning based on user preference of attributes. In: Proceedings of international conference on rough sets and current trends in computing, pp 587–596
32. Zhao K, Wang J (2002) A reduction algorithm meeting users' requirements. J Comput Sci Technol

Chapter 8
A New Approach of Intuitionistic Fuzzy Membership Matrix in Medical Diagnosis with Application

Nabanita Konwar and Pradip Debnath

1 Introduction

Due to the presence of impreciseness, medical science naturally calls for the use of fuzzy and intuitionistic fuzzy set theory for its modeling. For existent situations, the inexplicit nature of medical documentation and vague informations accumulated for any decision-making requires the utilization of fuzzy and intuitionistic fuzzy concepts. To produce mathematical models that can tackle systems containing elements of uncertainty, Zadeh [1] introduced the notion of fuzzy sets. Another systematic generalization of fuzzy set is intuitionistic fuzzy set. Atanassov [2] introduced intuitionistic fuzzy sets considering both the membership and nonmembership properties of an element. The nonmembership functions have more important roles in case of medical diagnosis. Since then fuzzy set theory and intuitionistic fuzzy set theory have turned into a sturdy area of research in various directions and found numerous applications in the fields of modern sciences and technology including engineering science, graph theory, operations research, artificial intelligence, computer networks, signal processing, robotics, decision-making, pattern recognition, automata theory, medical science, life science, etc. However, decision-making in fuzzy environment was initiated by Bellman and Zadeh [3]. Intuitionistic fuzzy models provide more rigor, pliability, and affinity to the system as compared to the fuzzy models. Sanchez [4, 5] established the diagnostic models in order to characterize the medical knowledge between symptoms and diseases with the help of fuzzy matrices. Esogbue and Elder [6] put forward the idea of fuzzy cluster analysis to develop models for medical

N. Konwar
Department of Mathematics, North Eastern Regional Institute of Science and Technology, Nirjuli 791109, Arunachal Pradesh, India
e-mail: nabnitakonwar@gmail.com

P. Debnath (✉)
Department of Applied Science and Humanities, Assam University, Silchar 788011, Assam, India
e-mail: debnath.pradip@yahoo.com

© Springer Nature Singapore Pte Ltd. 2020
O. P. Verma et al. (eds.), *Advancement of Machine Intelligence in Interactive Medical Image Analysis*, Algorithms for Intelligent Systems,
https://doi.org/10.1007/978-981-15-1100-4_8

185

diagnosis. Thereafter, many mathematicians established a matrix representation of fuzzy soft set for medical diagnosis. Continuing this process, Samuel and Balamurugan [7] initiated the use of intuitionistic fuzzy sets to construct an updated model of Sanchez's approach. Beg and Rashid [8] used intuitionistic fuzzy relations in medical diagnosis. Some work in this direction that deserve attention are Elizabeth and Sujatha [9], Hemanth et al. [10], Khalaf [11], Mittal et al. [12], Shyamal and Pal [13], Sundaresan et al. [14], and Sunitha and Kumar [15].

In the present work, we are going to construct and establish a modified model for medical diagnosis to deal with several features of medical issues using intuitionistic fuzzy membership relations. This work develops a new approach of studying medical diagnosis using intuitionistic fuzzy membership function. First, we construct and establish the procedure for the model and thereafter explain the procedure with the help of some study-based examples. For some relevant fuzzy graph theoretic concepts we refer to Akram and Dudek [16] and Rosenfeld [17].

2 Methods

Here, we put in forward the procedure of modeling a system in medical diagnosis using intuitionistic fuzzy membership matrix. We classify some steps of the procedure which helps us to narrate the whole system accurately.

First, we construct a model in order to verify the diseases accurately through their symptoms. For the system initially we suppose that, symptoms of diseases are denoted by the set S, diseases are denoted by D and patients are denoted by P. Now we classify the steps.

Step 1: First, we transfer all elements of triangular intuitionistic fuzzy (in short, TIF) number matrix into its membership function as shown below:

Membership function of $(a_{ij}, b_{ij}) = ((a_{ijL}, a_{ijM}, a_{ijN}), (b_{ijL}, b_{ijM}, b_{ijN}))$ is defined as

$$(\mu_{ij}, \nu_{ij}) = ((a_{ijL}/10, a_{ijM}/10, a_{ijN}/10), (b_{ijL}/10, b_{ijM}/10, b_{ijN}/10)),$$
$$\text{if } 0 \leq (a_{ijL}, b_{ijL}) \leq (a_{ijM}, b_{ijM}) \leq (a_{ijN}, b_{ijN}) \leq 10.$$

For the whole process, we consider every element as a membership function for the ease of calculation.

Step 2: We construct a TIF Relation V from the set of patients P, considering as membership function to the set of symptoms S, considering as nonmembership function (i.e., on $P \times S$). A TIF Relation V from X_1 to X_2 is a relation of $X_1 \times X_2$ which is specified by a function μ_R called membership function and by the function ν_R called the nonmembership function where the outcome of value of μ_R and ν_R is a TIF number. The pair (V, S) over P is known as patient symptom TIF number matrix.

Step 3: Next, we construct another TIF Relation R from the given list of symptoms S to the collection of diseases D (i.e., on $S \times D$) that establishes connection between symptoms and diseases through membership and nonmembership functions. The pair (R, D) over S is known as symptom–disease triangular intuitionistic fuzzy number matrix.

Step 4: Composition of Intuitionistic Fuzzy Relation:

Suppose A is a TIF set of X_1, construct composition of the TIF Relation $R(X_1 \to X_2)$ with A. Then the relation R is a TIF set B of X_2 and it is denoted by $B = R \cdot A$, and is defined as $R \cdot A(x_2) = (\mu_{R \cdot A}(x_2); \nu_{R \cdot A}(x_2)) = B(x_2)$ where
$\mu_{R \cdot A}(x_2) = \max[\mu_A(x_1)\min \mu_R(x_1, x_2)]$ and $\nu_{R \cdot A}(x_2) = \min[\nu_A(x_1)\max \nu_R(x_1, x_2)]$, $x_2 \in X_2$.

Again suppose $V(X_1 \to X_2)$ and $R(X_2 \to X_3)$ are two TIF relations. The composition $R \cdot V$ is the relation from $X_1 \to X_3$ stated as, $R \cdot V(x_1, x_3) = (\mu_{R \cdot V}(x_1, x_2), \nu_{R \cdot V}(x_2, x_3))$, where $\mu_{R \cdot V}(x_1, x_3) = \max[\mu_V(x_1, x_2) \min \mu_R(y, x_3)]$ and $\nu_{R \cdot V}(x_1, x_3) = \min[\nu_V(x_1, x_2) \max \nu_R(x_2, x_3)]$, $(x_1, x_3) \in X_1 \times X_3$, $x_2 \in X_2$ and $x_3 \in X_3$.

In this procedure, the max product composition $T = R \cdot V$ explains the plight of patients in connection with the ailments as a relation from P to D.

Step 5: Finally, we calculate the score and preciseness of the values with the help of defined functions for each T to perform the medical diagnosis of the concerned patients. The decision is made about the ailment of the patient strongly from the maximum of score values. It helps us to take a strong confirmation about the diseases.

In the next section, we provide examples in support of this new approach. These examples elaborate the above procedure mathematically which is expected to support medical sciences to diagnose disease more easily and specifically.

3   Results and Examples

Here, we describe our main results step-by-step with suitable numerical examples.

Example 1 First, we consider some patients of a hospital such as: P_1 (Gudu), P_2 (Johy), P_3 (Barbee), P_4 (Puli), and P_5 (Rosy) along with their symptoms: S_1 (high fever), S_2 (headache), S_3 (vomiting), S_4 (loose stools), and S_5 (body pain). Also consider the feasible diseases relating to the above symptoms are d_1 (Migraine), d_2 (Dengue), and d_3 (Typhoid).

Step 1: Initially, we consider all elements as the membership function and then we elaborate the procedure from Step 2 onwards.

Step 2: Assume the set of all patients $P = \{P_1, P_2, P_3, P_4, P_5\}$ as the universal set and set of all symptoms $S = \{S_1, S_2, S_3, S_4, S_5\}$ as the set of parameters. Then construct the TIF Relation $V(P \to S)$ as given in Table 1.

Step 3: Assume the set of all symptoms $S = \{S_1, S_2, S_3, S_4, S_5\}$ as the universal set and set of all diseases $D = \{d_1, d_2, d_3\}$ as the set of parameters. The TIF number

Table 1 TIF Relation $V(P \to S)$ between patients and symptoms

V	P_1	
S_1	(0.57, 0.67, 0.77, 0.75, 0.87)	(0.07, 0.17, 0.15, 0.17, 0.19)
S_2	(0.32, 0.47, 0.49, 0.57, 0.67)	(0.09, 0.13, 0.17, 0.27, 0.25)
S_3	(0.05, 0.12, 0.27, 0.28, 0.37)	(0.37, 0.47, 0.45, 0.51, 0.67)
S_4	(0.77, 0.87, 0.82, 0.85, 0.97)	(0.11, 0.14, 0.15, 0.17, 0.19)
S_5	(0.47, 0.57, 0.55, 0.61, 0.65	(0.17, 0.15, 0.25, 0.29, 0.37)
V	P_2	
S_1	(0.26, 0.36, 0.35, 0.46, 0.43)	(0.06, 0.16, 0.12, 0.18, 0.26)
S_2	(0.08, 0.11, 0.15, 0.22, 0.25)	(0.31, 0.43, 0.53, 0.66, 0.73)
S_3	(0.73, 0.75, 0.83, 0.83, 0.86)	(0.12, 0.14, 0.16, 0.18, 0.11)
S_4	(0.57, 0.54, 0.65, 0.67, 0.71)	(0.09, 0.12, 0.15, 0.17, 0.2)
S_5	(0.04, 0.09, 0.15, 0.19, 0.22)	(0.21, 0.25, 0.31, 0.41, 0.62)
V	P_3	
S_1	(0.7, 0.73, 0.8, 0.87, 0.9)	(0.01, 0.03, 0.04, 0.06, 0.08)
S_2	(0.08, 0.09, 0.11, 0.16, 0.18)	(0.41, 0.51, 0.61, 0.71, 0.81)
S_3	(0.31, 0.42, 0.48, 0.51, 0.55)	(0.18, 0.21, 0.28, 0.31, 0.14)
S_4	(0.53, 0.62, 0.68, 0.7, 0.75)	(0.04, 0.09, 0.15, 0.18, 0.21)
S_5	(0.07, 0.09, 0.13, 0.16, 0.22)	(0.33, 0.53, 0.63, 0.73, 0.75)
V	P_4	
S_1	(0.23, 0.43, 0.53, 0.55, 0.63)	(0.11, 0.14, 0.17, 0.21, 0.23)
S_2	(0.06, 0.11, 0.15, 0.27, 0.31)	(0.23, 0.27, 0.33, 0.43, 0.63)
S_3	(0.64, 0.68, 0.71, 0.75, 0.82)	(0.01, 0.06, 0.011, 0.15, 0.16)
S_4	(0.13, 0.23, 0.33, 0.43, 0.51)	(0.017, 0.18, 0.21, 0.25, 0.31)
S_5	(0.53, 0.55, 0.57, 0.63, 0.68)	(0.06, 0.11, 0.15, 0.19, 0.25)
V	P_5	
S_1	(0.17, 0.13, 0.14, 0.27, 0.24)	(0.34, 0.41, 0.48, 0.52, 0.61)
S_2	(0.45, 0.52, 0.59, 0.62, 0.68)	(0.12, 0.18, 0.21, 0.24, 0.28)
S_3	(0.23, 0.28, 0.35, 0.41, 0.46)	(0.03, 0.05, 0.11, 0.16, 0.22)
S_4	(0.31, 0.35, 0.41, 0.45, 0.51)	(0.17, 0.21, 0.25, 0.31, 0.35)
S_5	(0.27, 0.25, 0.29, 0.3, 0.32)	(0.31, 0.41, 0.53, 0.58, 0.61)

matrix (R, D) is a parameterized family of all TIF number matrix over the set S and determined from skilful medical documentation. Thus, construction of TIF Relation $R(S \to D)$ provides medical knowledge of the diseases and their symptoms as given in Table 2.

Step 4: In this step, we construct the composition $T = R \cdot V$ for describing the plight of patients regarding the ailments as a relation from P to D as given in Table 3.

Step 5: Finally, we evaluate the preciseness of the scores as given in Table 4.

Table 2 Construction of TIF Relation $R(S \rightarrow D)$ for the diagnosis of diseases and their symptoms

R	d_1	
S_1	(0.21, 0.26, 0.31, 0.35, 0.51)	(0.07, 0.09, 0.11, 0.13, 0.18)
S_2	(0.55, 0.61, 0.65, 0.71, 0.75)	(0.08, 0.31, 0.14, 0.18, 0.23)
S_3	(0.14, 0.18, 0.2, 0.23, 0.26)	(0.61, 0.66, 0.68, 0.7, 0.72)
S_4	(0.44, 0.49, 0.51, 0.52, 0.58)	(0.3, 0.32, 0.33, 0.38, 0.41)
S_5	(0.31, 0.41, 0.45, 0.51, 0.55)	(0.03, 0.04, 0.08, 0.11, 0.21)
R	d_2	
S_1	(0.22, 0.25, 0.31, 0.35, 0.41)	(0.11, 0.13, 0.14, 0.21, 0.23)
S_2	(0.03, 0.08, 0.11, 0.18, 0.21)	(0.5, 0.55, 0.61, 0.65, 0.75)
S_3	(0.25, 0.33, 0.41, 0.53, 0.61)	(0.13, 0.15, 0.19, 0.21, 0.23)
S_4	(0.61, 0.65, 0.68, 0.73, 0.75)	(0.05, 0.08, 0.11, 0.15, 0.21)
S_5	(0.44, 0.55, 0.66, 0.68, 0.73)	(0.11, 0.13, 0.15, 0.17, 0.21)
R	d_3	
S_1	(0.03, 0.05, 0.13, 0.15, 0.19)	(0.45, 0.51, 0.55, 0.62, 0.67)
S_2	(0.33, 0.41, 0.48, 0.51, 0.56)	(0.18, 0.21, 0.25, 0.29, 0.33)
S_3	(0.42, 0.49, 0.53, 0.56, 0.62)	(0.01, 0.05, 0.13, 0.16, 0.21)
S_4	(0.15, 0.19, 0.21, 0.23, 0.32)	(0.51, 0.52, 0.55, 0.58, 0.63)
S_5	(0.71, 0.82, 0.85, 0.88, 0.92)	(0.01, 0.03, 0.05, 0.08, 0.09)

From the above table, it is evident that patients P_1 (Gudu), P_2 (Johy), and P_3 (Barbee) suffer from diseases d_2 (Dengue), P_4 (Puli) suffers from d_3 (Typhoid), and P_5 (Rosy) suffers from d_1 (Migraine).

This relation can also be explained using a graph called network as given in Fig. 1.

Example 2 In this example, we consider another medical case called insomnia. Poor sleep quality may be categorized as primary and secondary types of insomnia. Through many survey it is found that almost 55% of people get better sleep by doing yoga. Yoga will improve one's sleep in many ways. In this example we consider the diagnosis of insomnia and investigate it with the help of previously discussed procedures.

Assume that P_1 (Giao), P_2 (Julee), P_3 (Barli), P_4 (Pulon), and P_5 (Richi) are some patients; S_1 (difficulty falling asleep at night), S_2 (waking up during the night), S_3 (not feeling well-rested after a night's sleep), S_4 (waking up too early), S_5 (difficulty paying attention, focusing on tasks or remembering), S_6 (daytime tiredness or sleepiness) and S_7 (irritability, depression or anxiety) are symptoms of diseases and d_1 (insomnia primary type), d_2 (insomnia secondary type) are diseases relating to the above symptoms. Then we construct the example according to the above-described procedure.

Step 1: Initially, we consider all elements as the membership functions and then we elaborate the procedure from Step 2 onwards.

Table 3 Construction of the composition $T = R \cdot V$ for describing the plight of patients regarding the diseases as a relation $P \to D$

T	P_1	
d_1	(0.44, 0.49, 0.51, 0.52, 0.58)	(0.07, 0.11, 0.15, 0.17, 0.19)
d_2	(0.61, 0.65, 0.68, 0.71, 0.75)	(0.05, 0.08, 0.11, 0.15, 0.22)
d_3	(0.43, 0.51, 0.55, 0.61, 0.65)	(0.11, 0.15, 0.25, 0.29, 0.33)
T	P_2	
d_1	(0.44, 0.49, 0.51, 0.52, 058)	(0.07, 0.15, 0.12, 0.18, 0.23)
d_2	(0.51, 0.54, 0.63, 0.67, 0.71)	(0.09, 0.12, 0.15, 0.17, 0.22)
d_3	(0.42, 0.49, 0.53, 0.56, 0.62)	(0.18, 0.21, 0.28, 0.31, 0.43)
T	P_3	
d_1	(0.44, 0.49, 0.51, 0.52, 0.58)	(0.07, 0.09, 0.15, 0.13, 0.18)
d_2	(0.53, 0.62, 0.68, 0.7, 0.75)	(0.05, 0.09, 0.15, 0.18, 0.21)
d_3	(0.33, 0.42, 0.48, 0.51, 0.55)	(0.18, 0.21, 0.28, 0.31, 0.42)
T	P_4	
d_1	(0.33, 0.43, 0.45, 0.51, 0.55)	(0.06, 0.11, 0.15, 0.19, 0.25)
d_2	(0.44, 0.55, 0.66, 0.68, 0.73)	(0.07, 0.13, 0.15, 0.17, 0.23)
d_3	(0.51, 0.55, 0.57, 0.63, 0.68)	(0.01, 0.06, 0.11, 0.16, 0.22)
T	P_5	
d_1	(0.45, 0.52, 0.59, 0.62, 0.68)	(0.12, 0.18, 0.21, 0.24, 0.28)
d_2	(0.31, 0.35, 0.41, 0.45, 0.53)	(0.11, 0.15, 0.19, 0.21, 0.23)
d_3	(0.33, 0.41, 0.48, 0.51, 0.56)	(0.03, 0.05, 0.12, 0.16, 0.22)

Table 4 Description of preciseness of the scores

	P_1	P_2	P_3	P_4	P_5
d_1	0.372, 0.644	0.374, 0.642	0.406, 0.634	0.29, 0.59	0.366, 0.778
d_2	0.56, 0.792	0.458, 0.75	0.522, 0.952	0.462, 0.75	0.226, 0.574
d_3	0.324, 0.76	0.244, 0.796	0.176, 0.728	0.478, 0.694	0.344, 0.568

Step 2: Assume the set of all patients $P = \{P_1, P_2, P_3, P_4, P_5\}$ as the universal set and set of all symptoms $S = \{S_1, S_2, S_3, S_4, S_5, S_6, S_7\}$ as the set of parameters. Then construct the TIF Relation $V(P \to S)$ as in Table 5.

Step 3: Assume the list of all symptoms $S = \{S_1, S_2, S_3, S_4, S_5, S_6, S_7\}$ as the universal set and set of all ailments $D = \{d_1, d_2, d_3\}$ as the set of parameters. The TIF number matrix (R, D) is a parameterized family of all TIF number matrix over the set S and established from skilful medical documentation. Therefore, construction of TIF Relation $R(S \to D)$ provides medical knowledge of the ailments and their symptoms as given in Table 6.

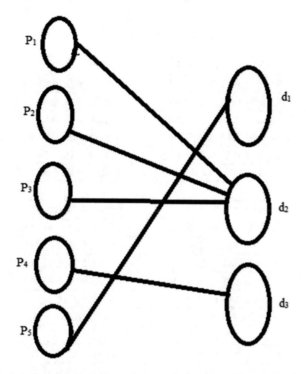

Fig. 1 Intuitionistic fuzzy medical diagnosis network (for three diseases) describing patient and the corresponding disease. Nodes or vertices of the above network denote the patients and diseases, edges denote the confirmation of disease to the patient. Such type of graphical representation improves the application potential of a model. It also provides a quick idea about the model

Table 5 TIF Relation $V(P \rightarrow S)$ between patients and symptoms

V	P_1	
S_1	(0.47, 0.57, 0.67, 0.35, 0.17)	(0.07, 0.16, 0.11, 0.17, 0.09)
S_2	(0.22, 0.37, 0.39, 0.47, 0.57)	(0.08, 0.11, 0.15, 0.24, 0.21)
S_3	(0.05, 0.12, 0.23, 0.27, 0.35)	(0.27, 0.37, 0.45, 0.52, 0.63)
S_4	(0.71, 0.81, 0.72, 0.84, 0.92)	(0.09, 0.13, 0.15, 0.17, 0.2)
S_5	(0.41, 0.53, 0.57, 0.61, 0.64)	(0.17, 0.15, 0.21, 0.26, 0.33)
S_6	(0.11, 0.13, 0.17, 0.21, 0.24)	(0.43, 0.52, 0.55, 0.61, 0.34)
S_7	(0.21, 0.33, 0.57, 0.41, 0.54)	(0.01, 0.53, 0.33, 0.51, 0.44)

(continued)

Table 5 (continued)

V	P_2	
S_1	(0.23, 0.33, 0.35, 0.43, 0.41)	(0.05, 0.15, 0.12, 0.19, 0.23)
S_2	(0.08, 0.11, 0.15, 0.22, 0.25)	(0.31, 0.43, 0.53, 0.66, 0.73)
S_3	(0.63, 0.65, 0.73, 0.73, 0.76)	(0.12, 0.14, 0.16, 0.18, 0.11)
S_4	(0.47, 0.44, 0.55, 0.57, 0.71)	(0.09, 0.12, 0.15, 0.17, 0.2)
S_5	(0.04, 0.09, 0.15, 0.18, 0.23)	(0.21, 0.25, 0.31, 0.41, 0.52)
S_6	(0.23, 0.28, 0.31, 0.35, 0.41)	(0.71, 0.75, 0.72, 0.62, 0.6)
S_7	(0.02, 0.7, 0.11, 0.17, 0.21)	(0.81, 0.84, 0.88, 0.73, 0.75)
V	P_3	
S_1	(0.6, 0.63, 0.7, 0.77, 0.8)	(0.02, 0.04, 0.06, 0.08, 0.03)
S_2	(0.08, 0.09, 0.11, 0.14, 0.17)	(0.31, 0.41, 0.51, 0.61, 0.71)
S_3	(0.41, 0.32, 0.48, 0.52, 0.55)	(0.17, 0.22, 0.27, 0.31, 0.13)
S_4	(0.53, 0.62, 0.68, 0.7, 0.75)	(0.04, 0.09, 0.15, 0.18, 0.21)
S_5	(0.07, 0.09, 0.13, 0.16, 0.22)	(0.33, 0.53, 0.63, 0.73, 0.75)
S_6	(0.11, 0.13, 0.19, 0.22, 0.23)	(0.23, 0.27, 0.31, 0.35, 0.33)
S_7	(0.163, 0.31, 0.5, 0.09)	(0.116, 0.161, 0.21, 0.25, 0.3)
V	P_4	
S_1	(0.33, 0.23, 0.43, 0.55, 0.53)	(0.11, 0.13, 0.17, 0.22, 0.27)
S_2	(0.06, 0.13, 0.15, 0.24, 0.3)	(0.23, 0.26, 0.33, 0.43, 0.53)
S_3	(0.64, 0.68, 0.71, 0.75, 0.82)	(0.01, 0.06, 0.011, 0.15, 0.16)
S_4	(0.13, 0.23, 0.33, 0.43, 0.51)	(0.017, 0.18, 0.21, 0.25, 0.31)
S_5	(0.53, 0.55, 0.57, 0.63, 0.68)	(0.06, 0.11, 0.15, 0.19, 0.25)
S_6	(0.55, 0.58, 0.61, 0.67, 0.6)	(0.31, 0.36, 0.45, 0.49, 0.5)
S_7	(0.77, 0.71, 0.65, 0.61, 0.58)	(0.21, 0.24, 0.31, 0.33, 04)
V	P_5	
S_1	(0.27, 0.24, 0.17, 0.13, 0.14)	(0.34, 0.52, 0.61, 0.41, 0.48)
S_2	(0.59, 0.62, 0.45, 0.52, 0.68)	(0.21, 0.24, 0.12, 0.18, 0.28)
S_3	(0.23, 0.41, 0.46, 0.28, 0.35)	(0.11, 0.16, 0.03, 0.05, 0.22)
S_4	(0.31, 0.41, 0.45, 0.35, 0.51)	(0.17, 0.25, 0.31, 0.21, 0.35)
S_5	(0.27, 0.3, 0.32, 0.25, 0.29)	(0.31, 0.58, 0.61, 0.41, 0.53)
S_6	(0.31, 0.36, 0.44, 0.51, 0.5)	(0.91, 0.85, 0.82, 0.77, 0.66)
S_7	(0.24, 0.17, 0.44, 0.51, 0.11)	(0.14, 0.21, 0.24, 0.11, 0.1)

Table 6 Construction of TIF Relation $R(S \to D)$ for insomnia and its symptoms

R	d_1	
S_1	(0.26, 0.31, 0.21, 0.35, 0.51)	(0.11, 0.13, 0.07, 0.09, 0.18)
S_2	(0.65, 0.71, 0.55, 0.61, 0.75)	(0.14, 0.18, 0.23, 0.08, 0.31)
S_3	(0.23, 0.26, 0.14, 0.18, 0.2)	(0.7, 0.72, 0.61, 0.66, 0.68)
S_4	(0.49, 0.51, 0.52, 0.44, 0.58)	(0.32, 0.33, 0.38, 0.3, 0.41)
S_5	(0.31, 0.51, 0.55, 0.41, 0.45)	(0.03, 0.11, 0.21, 0.04, 0.08)
S_6	(0.5, 0.61, 0.42, 0.31, 0.54)	(0.033, 0.025, 0.11, 0.21, 0.09)
S_7	(0.42, 0.46, 0.24, 0.29, 0.31)	(0.21, 0.02, 0.31, 0.06, 0.33)
R	d_2	
S_1	(0.25, 0.31, 0.35, 0.22, 0.41)	(0.13, 0.14, 0.21, 0.11, 0.23)
S_2	(0.03, 0.18, 0.21, 0.08, 0.11)	(0.5, 0.65, 0.75, 0.55, 0.61)
S_3	(0.25, 0.53, 0.61, 0.33, 0.41)	(0.13, 0.21, 0.23, 0.15, 0.19)
S_4	(0.68, 0.73, 0.61, 0.65, 0.75)	(0.11, 0.15, 0.05, 0.08, 0.21)
S_5	(0.44, 0.73, 0.55, 0.66, 0.68)	(0.11, 0.21, 0.13, 0.15, 0.17)
S_6	(0.35, 0.22, 0.58, 0.53, 0.5)	(0.66, 0.61, 0.35, 0.69, 0.22)
S_7	(0.01, 0.06, 0.09, 0.13, 0.17)	(0.11, 0.21, 0.28, 0.33, 0.31)

Step 4: In this step, we construct the composition $T = R \cdot V$ for describing the plight of patients regarding the ailments as a relation from P to D as given in Table 7.
Step 5: Finally, we evaluate the scores and their preciseness as in Table 8.

From the above table of the score values, it is evident that the patient P_1 (Giao), P_2 (Julee), and P_5 (Richi) are suffering from diseases d_1 (insomnia primary type), P_3 (Barli), and P_4 (Pulon) suffer from d_2 (insomnia secondary type).

This relation can also be presented with the help of a graph called network as given in Fig. 2.

4 Discussion

A new approach of applying intuitionistic fuzzy membership matrix to medical diagnosis has been presented in this chapter. In the first example of Sect. 3, we applied our newly proposed method to the detection of a disease which is the most likely reason for a particular symptom. In our second example, we have seen that our proposed method helps in the diagnosis of different types of insomnia together with the main factor(s) that can be responsible for it.

Table 7 Construction of the composition $T = R \cdot V$ for describing the plight of patients about insomnia as a relation $P \to D$

T	P_1	
d_1	(0.44, 0.58, 0.49, 0.51, 0.52)	(0.07, 0.19, 0.11, 0.15, 0.17)
d_2	(0.71, 0.61, 0.65, 0.68, 0.75)	(0.15, 0.05, 0.08, 0.11, 0.22)
T	P_2	
d_1	(0.44, 0.52, 058, 0.49, 0.51)	(0.07, 0.18, 0.23, 0.15, 0.12)
d_2	(0.51, 0.63, 0.67, 0.54, 0.71)	(0.09, 0.15, 0.12, 0.17, 0.22)
T	P_3	
d_1	(0.49, 0.51, 0.44, 0.52, 0.58)	(0.09, 0.15, 0.07, 0.13, 0.18)
d_2	(0.68, 0.7, 0.53, 0.62, 0.75)	(0.15, 0.18, 0.05, 0.09, 0.21)
T	P_4	
d_1	(0.33, 0.51, 0.55, 0.43, 0.45)	(0.06, 0.19, 0.25, 0.11, 0.15)
d_2	(0.44, 0.68, 0.55, 0.66, 0.73)	(0.07, 0.17, 0.13, 0.15, 0.23)
T	P_5	
d_1	(0.59, 0.62, 0.45, 0.52, 0.68)	(0.21, 0.24, 0.12, 0.18, 0.28)
d_2	(0.31, 0.45, 0.53, 0.35, 0.41)	(0.11, 0.21, 0.23, 0.15, 0.19)

Table 8 Preciseness of scores (for insomnia)

	P_1	P_2	P_3	P_4	P_5
d_1	0.272, 0.541	0.734, 0.426	0.046, 0.364	0.277, 0.5	0.366, 0.578
d_2	0.016, 0.279	0.458, 0.357	0.522, 0.952	0.462, 0.75	0.226, 0.574

5 Conclusion

Procedure that narrates in this work by using triangular intuitionistic fuzzy relation is a new approach for medical issues. The nonmembership functions have more important roles in case of medical diagnosis. The results of the numerical examples provide the best diagnostic conclusions. Therefore, this new way of performing medical diagnosis has been assembled with a much wider concept. To instruct TIF Numbers and its application in many different approaches of medical diagnosis modeling, we successfully applied the score and accuracy functions. Such types of medical diagnosis networks improve the medical diagnosis system technically. By using these intuitionistic fuzzy networks one can also develop more models for different kinds of medical cases like insomnia, etc.

Fig. 2 Intuitionistic fuzzy medical diagnosis network (two diseases). Nodes or vertices of the above network denote the patients and diseases, edges denote the confirmation of disease to the patient

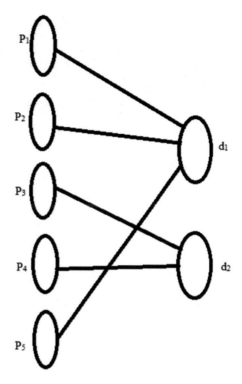

References

1. Zadeh LA (1965) Fuzzy sets. Inf Control 8:338–353
2. Atanassov KT (1986) Intuitionistic fuzzy sets. Fuzzy Sets Syst 20:87–96
3. Bellman R, Zadeh LA (1970) Decision making in a fuzzy environment. Manag Sci 17:144–164
4. Sanchez E (1976) Resolution of composite fuzzy relation equations. Inf Control 30:38–48
5. Sanchez E (1979) Inverse of fuzzy relations, application to possibility distribution and medical diagnosis. Fuzzy Sets Syst 2(1):75–86
6. Esogbue AO, Elder RC (1980) Fuzzy diagnosis decision models. Fuzzy Sets Syst 3:1–9
7. Samuel AE, Balamurugan M (2012) Application of intuitionistic fuzzy sets in medical diagnosis. In: Proceeding of the international conference on mathematical modeling and applied soft computing, vol 1, pp 189–194
8. Beg I, Rashid T (2015) A system for medical diagnosis based on intuitionistic fuzzy relation. Notes IFS 21(3):80–89
9. Elizabeth S, Sujatha L (2013) Application of fuzzy membership matrix in medical diagnosis and decision making. Appl Math Sci 7(127):6297–6307
10. Hemanth JD, Anitha J, Son LH, Mittal M (2017) Diabetic retinopathy diagnosis from retinal images using modified Hopfield neural network. J Med Syst 42(12):247–261
11. Khalaf MM (2013) Medical diagnosis via interval valued intuitionistic fuzzy sets. Ann Fuzzy Math Inform 6(2):245–249
12. Mittal M, Goyal LM, Kaur S, Verma A, Hemanth JD (2019) Performance enhanced growing convolutional neural network based approach for brain tumor segmentation in magnetic resonance brain images. Appl Soft Comput (article in press)
13. Shyamal AK, Pal M (2017) Triangular fuzzy matrices. Iran J Fuzzy Syst 4(1):75–87

14. Sundaresan T, Sheeja G, Govindarajan A (2018) Different treatment stages in medical diagnosis using fuzzy membership matrix. In: National conference on mathematical techniques and its applications, IOP conference series: Journal of Physics, conference series 1000 012094. https://doi.org/10.1088/1742-6596/1000/1/012094
15. Sunitha MS, Kumar V (2002) Complement of a fuzzy graph. Indian J Pure Appl Math 33(9):1451–1464
16. Akram M, Dudek WA (2013) Intuitionistic fuzzy hypergraphs with applications. Inf Sci 218:182–193
17. Rosenfeld A (1975) Fuzzy graphs. Fuzzy sets and their applications. Academic Press, New York

Chapter 9
Image Analysis and Automation of Data Processing in Assessment of Dental X-ray (OPG) Using MATLAB and Excel VBA

Stella, Emmanuel Dhiravia Sargunam and Thirumalai Selvi

1 Introduction

Radiology, the science of imaging of the human body is the mainstay of investigation to arrive at a clinical diagnosis. The reports in this process are a sequence of images that usually follow a repetitive style in normal conditions and vary in pathology. Most of the variations are standard repetitions even in *pathology* (abnormal conditions). In dentistry, stages of tooth development can be used to assess the maturity and age of an individual. In clinical dentistry, this diagnosis is used for treatment planning and in the forensic and legal circle, the stages of tooth development are used to assess the age of the individual. This has also been used to assess the age of the individual to an accuracy of close to a month and thereby giving them a probable date of birth to individuals who are less fortunate to have their date of birth recorded. Most of these individuals are *orphans or victims of conflicts and wars*.

Age estimation plays an important role in *forensic medicine*. The dental age of children can be assessed based on the development of teeth in radiographic images. Age determination is also needed in forensic science where age plays an important role in the identification of the deceased and when *matters of consent, for sports, child labor, or criminal ability* arise [1]. Orthodontists who are specialized dentist use tooth stage to predict the timing of certain treatments. Pediatricians find out whether the dental maturity of a child with the disease has been delayed or advanced

Stella (✉)
Department of Computer Science, Bharathiar University, Chennai, India
e-mail: stellaemmanuel.jc@gmail.com

E. D. Sargunam
Department of Oral and Maxillofacial Surgery, Sri Ramachandra University, Chennai, India
e-mail: emmanuelazariah@hotmail.com

T. Selvi
Department of Computer Science, Government Arts College (Men), Nandanam, Chennai, India
e-mail: sarasselvi@gmail.com

© Springer Nature Singapore Pte Ltd. 2020   197
O. P. Verma et al. (eds.), *Advancement of Machine Intelligence in Interactive Medical Image Analysis*, Algorithms for Intelligent Systems,
https://doi.org/10.1007/978-981-15-1100-4_9

[2]. In addition to this, requests for determining the age of children and juveniles are also increasing in civil legal cases when adoptive parents or the court's demand age estimation for educational or health purposes of children, particularly those adopted internationally [3]. Demirjian and Nolla formulated methods to assess the age of an individual. The most widely used method for the comparison between different populations was first described in 1973 by Demirjian et al. His method is based on the development of seven left permanent mandibular teeth. The age estimation method introduced by Nolla in 1960 comes under age determination during the first two decades of life.

1.1 Overview of Dentition

In human beings, the dentition is of two types *the Primary and Secondary dentition* which is also called as the milk teeth and the permanent teeth, respectively. The jaws are divided into the *upper jaw (Maxilla) and the lower Jaw (Mandible)*. Each jaw is further divided into the right and the left quadrant. Each quadrant has five deciduous teeth (*two incisors, one canine, and two molars*), making it 20 milk teeth. The permanent teeth are eight in each quadrant (*two incisors, one canine, and two premolars* which erupt replacing the deciduous molars and finally *three molars*), making *it 32 permanent teeth.*

The human dentition starts forming at birth and develops at varying rates to complete its eruption close to 18 years of age. At any given point within this timeframe, the teeth are in different stages of tooth formation and assessing them can help us to identify the age of the individual. The growth of the teeth is divided into two stages, *Primary teeth or Milk teeth* and *Permanent teeth* [4]. Age estimation is done using permanent teeth. As early as immediately after birth, the permanent teeth start developing in the bone, the process continues till the last molar teeth erupt. There are some common methods for age estimation [5].

(i) Nolla's method [6]
(ii) Demirjian's method [7]

Both these techniques use an advanced specialized dental radiograph called as *Orthopantomogram* (OPG). This specialized X-ray shows the teeth that have erupted and the tooth that is in various stages of development within the bone as well. An orthopantomogram image of a 6-year-old child has been shown in Fig. 1. The image helps us to see the primary teeth and permanent teeth. When the primary teeth and the permanent tooth are present intraorally in an individual, the dentition is said to be a *Mixed Dentition* [8]. There are various stages of tooth formation that have been classified. The classification is mostly based on the height and width of the tooth being formed and the formation of the different anatomical structures such as enamel, dentin, pulp, cementoenamel junction, roots and completion of the root apices and has been designated stages A-H is shown in Fig. 2. The two techniques of age estimation use the stage of development of the seven permanent teeth in

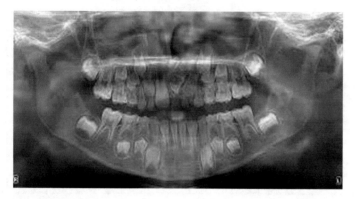

Fig. 1 OPG of 7-year-old boy [15]

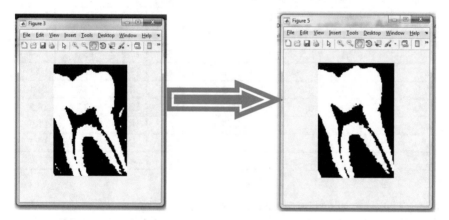

Fig. 2 Extracted image without noise

the mandibular right quadrant of an individual. Once the stage of development of the permanent teeth is identified, the corresponding numerical value from the table (Tables 1 and 2) is calculated. The sum of the values for stages of tooth development for seven teeth in the quadrant gives a value which is then used on a (Table 3) to find the age of the individual.

A study was done on 40 dentists to check the interobserver variability showed *good agreement in central incisors and first molars but poor agreement in premolars and second molars*. These teeth being the most influential tooth stages in the mixed dentition, the result is of concern to the researchers. Done manually, the process is cumbersome. The right stage of development has to be identified and the corresponding values entered and summate to identify the age using the table. Each step can result in an error which can be magnified in the subsequent steps. Automating this process by Image Processing using MATLAB and writing a VB Application for computing the values and arriving at an accurate age will help in eliminating the

Table 1 Mandibular left side 7 teeth scores for girls (Demirjian)

Stage Tooth	O	A	B	C	D	E	F	G	H
M$_2$	0	2.7	3.9	6.9	11.1	13.5	14.2	14.5	15.6
M$_1$				0	4.5	6.2	9	14	16.2
PM$_2$	0	1.8	3.4	6.5	10.6	12.7	13.5	13.8	14.6
PM$_1$			0	3.7	7.5	11.8	13.1	13.4	14.1
C				0	3.8	7.3	10.3	11.6	12.4
I$_2$				0	3.2	5.6	8	12.2	14.2
I$_1$					0	2.4	5.1	9.3	12.9

Table 2 Mandibular left side 7 teeth scores for boys (Demirjian)

Stage Tooth	O	A	B	C	D	E	F	G	H
M$_2$	0	2.1	3.5	5.9	10.1	12.5	13.2	13.6	15.4
M$_1$				0	8	9.6	12.3	17	19.3
PM$_2$	0	1.7	3.1	5.4	9.7	12	12.8	13.2	14.4
PM$_1$			0	3.4	7	11	12.3	12.7	13.5
C				0	3.5	7.9	10	11	11.9
I$_2$				0	3.2	5.2	7.8	11.7	13.7
I$_1$					0	1.9	4.1	8.2	11.8

errors. This automation also can allow us to compare the two techniques, or to add in the corrective factors developed through various epidemiological studies for various human races.

2 Methodology

MATLAB is a scientific programming language and provides strong mathematical and numerical support for the implementation of advanced algorithms. It is for this reason that MATLAB is widely used by the image processing and computer vision community. MATLAB, an abbreviation for "*matrix laboratory*," is a platform for solving mathematical and scientific problems. It can be used to perform *image segmentation, image enhancement, noise reduction, geometric transformations, image registration, and 3D image processing operations* [9]. It is used to retrieve the separate tooth in giving OPG using segmentation. Extract the image without noise to compare with the stored database. Extracted image without noise is shown in Fig. 2.

Table 3 Age value for both Nolla's and Demirjian's method

Nolla's				Demirjian's			
Mandibular teeth of girls		Mandibular teeth of boys		Mandibular teeth of girls		Mandibular teeth of boys	
Age in years	Sum of stages for 7 mandibular teeth	Age in years	Sum of stages for 7 mandibular teeth	Age in months and years	Score	Age in months and years	Score
3	24.6	3	22.3	3	13.7	3	12.4
4	32.7	4	30.3	3.1	14.4	3.1	12.9
5	40.1	5	37.1	3.2	15.1	3.2	13.5
6	46.6	6	43	3.4	15.8	3.4	14
7	52.4	7	48.7	.	.	.	.
8	57.4	8	53.7	.	.	.	.
9	58.4	9	57.9	.	.	.	.
10	64.3	10	61.5	15	99.2	15	97.6
11	66.3	11	64	15.1	99.3	15.1	97.7
12	67.9	12	66.3	.	.	.	.
13	68.9	13	67.8	.	.	.	.
14	69.4	14	69	.	.	.	.
15	69.8	15	69.7	.	.	.	.
16	70	16	70	.	.	.	.
17	70	17	70	16	100	16	98.4

After the process, the system will give the stage of the selected tooth, which can be used to identify the age using Demirjian and Nolla method.

The application is developed using Microsoft Excel VBA (Visual Basic for Application) to find the age of an individual using OPG with the help of two methods Demirjian and Nolla method. Excel VBA helps to automate any process using a programming language. The small programming task can be implemented using VBA [10]. VBA is a Visual Basic Application, a computer programming language that helps to specific processes and calculations [11]. The VBA can be used free of cost by application developers. All the features are used in this application with the help of its build-in function and tools. The stage given by image processing can be given as input to the developed application, which helps to find the age of given OPG.

3 Stages of Analyzing

Each tooth has a different stage, which grows. These can be analyzed using two methods Demirjian and Nolla [12, 13].

3.1 Demirjian Method

In Demirjian's method, the development of teeth is divided into *A to H stages*. Figure 3 shows the stage for Incisors, Canine, Primary Molar, and Molar tooth. Demirjian's gives the score for each tooth according to the stage. The score value also changes according to gender. Tables 1 and 2 give the score value for each tooth. For example, if the second Molar that is M_2 is in stage F, then the score value for M_2 is 14.2 for girls and 13.2 for boys.

Fig. 3 A to H stages of permanent teeth (Demirjian)

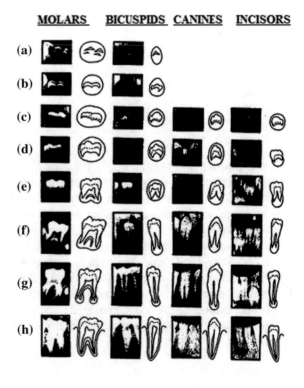

3.2   Nolla Method

In Nolla's method, each tooth is divided into *0–10 stages* as shown in Fig. 4, if *the growth is between any of the two stages, 0.5 is added to the lower stage. If the growth level is almost reached the next stage, 0.7 is added to the lowest stage. If the growth level is yet to reach the middle of stage, 0.2 is added to the lower level.* For example, if the second molar stage is between seventh and eighth stage, then the score value will be 7.5 for second molar tooth. These are the ways to score the stage of the teeth in Nolla's method.

In both the methods, the left seven mandibular teeth are taken to calculate the age. Each seven teeth stage score will be added to get the sum of stages for seven mandibular teeth. Table 3 gives the age for the given sum of stage values for both male and female.

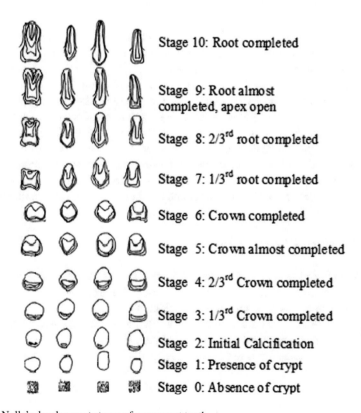

Stage 10: Root completed

Stage 9: Root almost completed, apex open

Stage 8: 2/3rd root completed

Stage 7: 1/3rd root completed

Stage 6: Crown completed

Stage 5: Crown almost completed

Stage 4: 2/3rd Crown completed

Stage 3: 1/3rd Crown completed

Stage 2: Initial Calcification

Stage 1: Presence of crypt

Stage 0: Absence of crypt

Fig. 4 Nolla's development stages of permanent tooth

In a study, the panoramic radiographs of six individuals were given to 40 dentists to evaluate the age using Demirjian and Nolla method. The stage value was fed into an application which is designed using Visual Basic for Application to find the score value of stages and age of the given OPG's. Demirjian's method gave the most accurate value than Nolla's method in assessing the age of the individual. So implementation using image processing was done using Demirjian's technique.

4 Processing Technique

The development of the software program contains two stages:Image processing and calculation of the age using the stage value obtained from image processing. The image processing part was done using MATLAB and the calculation was computerized using Excel VBA.

4.1 *Image Processing*

The image processing process that was used as the stages of tooth development were images and the OPG is usually available in an image format. The idea was to write an algorithm that will identify an image similar to the one selected by the user from a given set of images. The software will run a sequence of image matching to select the most appropriate matching image and this will give the stage of tooth development. SVM Hog classifier is proven to be effective in multivariable classification while selecting a tooth from query image, a separation line need to be drawn first to differentiate the tooth to be assessed. Gabor wavelet transform is used to calculate the shape feature of the selected tooth, along with that, the area, centroid, and diameter of the tooth will be calculated all the extracted features will be trained with its stage labels with SVM classifier and stored in a MATLAB inbuilt data file.

4.1.1 Function of Image Processing

(i)   rgb2gray function is used to convert RGB image or colormap to grayscale.
(ii)   Size function is used to $(d = \text{size}(X))$ return the sizes of each dimension of array X in a vector, d, with ndims(X) elements.
(iii) img2bw function is used to convert image to binary image, based on threshold.
(iv)   imshow function is used to display the image after processing.
(v)   bwareaopen function is used to remove small objects from binary image

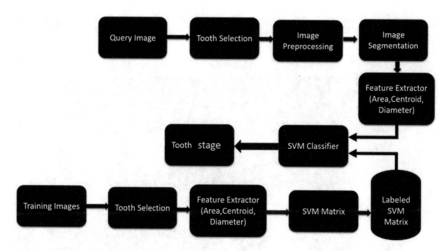

Fig. 5 Block diagram of image processing

A GUI is designed in which the OPG image will be loaded and separate selection buttons are used to extract the left side tooth, when a selection in doing the same process of Gabor wavelet and dimensional features are extracted from the selected image and will be compared to the SVM trained database, an SVM classifier predicts the tooth stage based on highest matching correlation of stage feature value with trained vector. The processing of image using the above and other build-in function overview is shown in Fig. 5. Selection of tooth in the image is shown in Fig. 6. After image processing with all extracted features (Area, Centroid, Diameter) and noise filtering, the image is used to identify the stage. A noise removed image for the selected tooth is shown in Fig. 2.

4.1.2 Training Algorithm

(i) a. Select an image with known stage value and crop individual tooth from the left lower jaw.

  b. Store the cropped tooth image into the respective stage and category folder with its extracted feature vector

  c. Repeat the above training steps for at least 50 images.

(ii) Store the image folder with its stage names in a MATLAB Database ".mat" format

(iii) Develop a GUI to receive user loaded images

(iv) Let the user select the tooth of his choice

(v) Segment the tooth from the original image and store the location details for the image.

(vi) Repeat the user selection for the seven left lower jaw teeth.

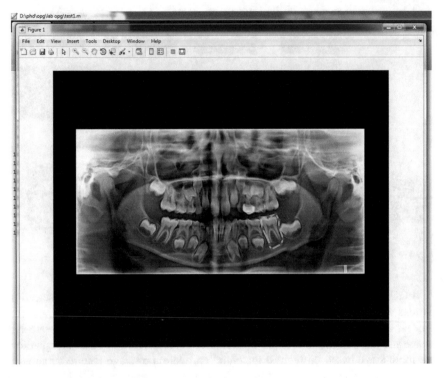

Fig. 6 Selecting the particular tooth to identify the stage

(vii)   After selection, extract the shape, pattern, and size feature vector of the user image and compare the image vector with already trained SVM database.

(viii)   Find the maximum correlation value and predict the best matching stage.

(ix)   Based on the stage prediction, substitute the stages into the "Demirjian" table database and calculate the actual age value.

4.1.3 Coding

In MATLAB, a function is used to retrieve the required information.

```
function [ret_msg,dia,lesion]=feature_detect(handles)
axes(handles.axes3);
ret_msg='None';
dia=0;
lesion=0;
crop=5;
img=handles.img;
 [r,c,dim]=size(img);
    if(dim>=3)
        img=rgb2gray(img);
    end
    img=img(crop:end-crop,crop:end-crop);
    B = im2bw(img, 0.7);
    segmented = imclearborder(B);

    imshow(img);
    segmented=bwareaopen(segmented,50);
    %figure,imshow(B);
    %figure,imshow(segmented);
 [L,N]=bwlabel(segmented);
 [ro,col]=size(segmented);
 for j=1:N
    [r,c]=find(L==j);
    minr=min(r);
    maxr=max(r);
    maxc=max(c);
    minc=min(c);
    msg=sprintf('%d',j);

    %fprintf('\nSize of %d=%d',i,maxc-minc);
    diff=maxc-minc;
    diff_per=diff/col *100;
    if(diff_per>40)
        segmented(minr:maxr,minc:maxc)=0;
    end
end
end
```

```
%figure,imshow(segmented);
%figure,imshow(img);
[L,N]=bwlabel(segmented);
dia_props = regionprops(L, 'EquivDiameter');
area_props = regionprops(L, 'Area');
cent_props = regionprops(L, 'Centroid');
flag=0;
reg=1;
for j=1:N
    d=cent_props(j).Centroid;
    d=d(1);

    [r,c]=find(L==j);
    minr=min(r);
    maxr=max(r);
    maxc=max(c);
    minc=min(c);
    msg=sprintf('%d',j);

    lesion_size=((maxc-minc)*(maxr-minr))/(ro+col)*100;
    area_per=area_props(j).Area/(ro*col)*100;


fprintf('\n%d.Diameter=%f,Area=%f,Centroid=%f',reg,dia_pr
ops(j).EquivDiameter,area_props(j).Area,d(1));
                tmp_msg=sprintf('\n%d . Diameter=%f Area=%f
Centroid=%f',reg,dia_props(j).EquivDiameter,area_props(j)
.Area,d(1));
                ret_msg=[ret_msg tmp_msg];
                flag=1;
                reg=reg+1;
                dia=[dia dia_props(j).EquivDiameter];
                lesion=[lesion lesion_size];
        end
    end
    if(flag==0)
        ret_msg=sprintf('\tCoudlnt Classify\n');
    end
```

4.2   Calculation Using Stage

This process has been done using Visual Basic Application. It has been designed to choose the method of age estimation. Once the method has been chosen, the system asks for the input of the seven teeth stages with the gender of the individual OPG. Both the Demirjian's and Nolla's techniques have a different scoring system for males and females and sex is an important input in the estimation of age. The user can choose the OPG, which has to be evaluated will be displayed on the left side of the application. The stage of tooth identified by the image processing process needs to be selected from the scroll box next to the tooth number. The process can also be synced with the image processing protocol. This will directly enter the appropriate stage to the tooth number. In Demirjan's method, the stage value is A to H, But in Nolla's method, the stage value is 0–10 and additional stage value of 0.5, 0.7, and 0.2 is used to represent the exact stage value. This means if the image similarity between the two stages (50% similarity to two subsequent stages), then it is given an additional value of 0.5. If it is more than 50% of the previous stage, then 0.2 is added and similarly, if it is more than 50% of the higher stage, then 0.7 is added. The MATLAB programming can be trained to incorporate the needs of Nolla's classification also if needed. Once the stages are chosen, activating the "Evaluate the age" button will provide the age of the patient. Program flow diagram is shown in Fig. 7.

Stage value of Central Incisors

+

Stage value of Lateral Incisors

+

Stage value of Canine

+

Stage value of First Premolar

+

Stage value of Second Premolar

+

Stage value of First Molar

+

Stage value of Second Molar

Total stage value

The total stage value is used to find the age with the help of Table 3 provided by Demerijian's and Nolla's method for both male and female. Evaluate age button is used to display the age with the help of programming language code. Figure 8 shows the login method. Figure 9 shows the Demirjian's form and Fig. 10 shows the Nolla's form. Figure 11 shows the data stored in a separate sheet.

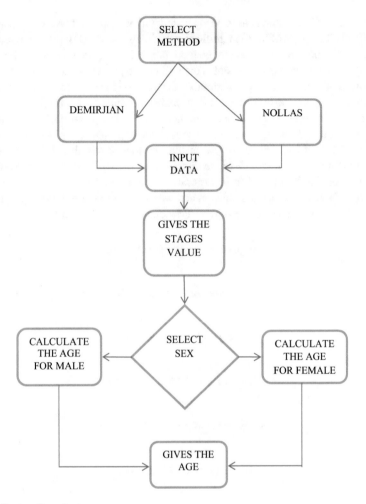

Fig. 7 System flow chart

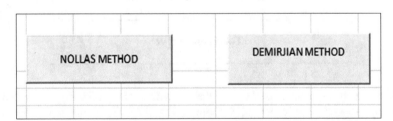

Fig. 8 Select the method (login form)

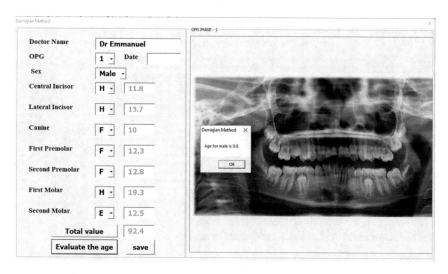

Fig. 9 Demirijan form

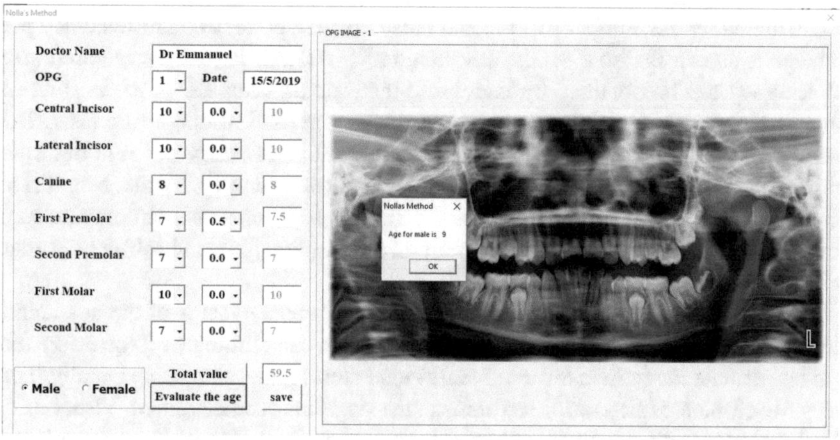

Fig. 10 Nolla form

5 Discussion

The identification of the appropriate image stage was the issue at hand in the assessment of age using radiographs. A trained eye can take asses subtle differences than an untrained eye. So a specialist can be accurate than a novice, even worse is a layman. Advancements in science and technology are aimed at bridging this experience divide and getting the best possible outcome in all situations. The problem being an image-based issue, MATLAB is an appropriate choice given its ease of use and

S.No	Doctor name	opg number	Date	Central Incisor	Lateral Incisor	Canine	First Premolar	Second Premolar	First Molar	Second Molar	Total	Age
1	S.DELPHINE	1	17-05-17	10	10	9	8	7	10	7	62	10
2	S.DELPHINE	2	17-05-17	10	10	9	8	8	10	7	63	9
3	S.DELPHINE	3	17-05-17	10	9	8	7	7	9	6	56	8
4	S.DELPHINE	4	17-05-17	9	9	8	7	7	9	6	56	7
5	S.DELPHINE	5	17-05-17	10	10	9	8	8	10	7	62	9
6	S.DELPHINE	6	17-05-17	9	8	6	6	6	8	2	45	6
7	E.ROJA	1	17-05-17	10	10	9	9	9	10	7	68	13
8	E.ROJA	2	17-05-17	10	10	8	8	7	10	6	59	9
9	E.ROJA	3	17-05-17	10	10	7	7	9	10	6	59	9
10	E.ROJA	4	17-05-17	10	9	8	7	6	10	6	57	7
11	E.ROJA	5	17-05-17	10	9	9	10	8	10	6	62	9
12	E.ROJA	6	17-05-17	7	7	7	6	6	10	6	49	7
13	ROHINI	1	17-05-17	5	7	9	8	8	9	8	60	9
14	ROHINI	2	17-05-17	8	8	8	7	7	10	7	55	7
15	ROHINI	3	17-05-17	10	10	7	7	9	10	6	59	9
16	ROHINI	4	17-05-17	10	9	8	7	6	10	6	57	7
17	ROHINI	5	17-05-17	10	9	9	10	8	10	7	63	9
18	ROHINI	6	17-05-17	7	7	7	6	6	10	6	49	7
19	NIVEDA	1	17-05-17	8	8	8	7	8	9	7	55	8
20	NIVEDA	2	17-05-17	10	10	8	9	7	9	7	62	9
21	NIVEDA	3	17-05-17	9	9	9	9	8	9	4	57	8
22	NIVEDA	4	17-05-17	9	9	9	9	7	9	4	58	8
23	NIVEDA	5	17-05-17	10	10	9	9	9	9	8	64	9

Fig. 11 Stores all data in Excel sheet with stage value

flexibility. The software has all the processes required to isolate and extract data and remove possible irregularities and prepare it for comparison with the existing library of images to identify a closest identical image. In image processing, accuracy is calculated by the number of True and False Positive predicted. The accuracy percentage is shown in Table 4. The advantage of MATLAB is that the required level of accuracy can be obtained by increasing the training samples. In medical terminology, Specificity and Sensitivity are terms that are used in diagnostic tests. This means what kind of accuracy is expected. A sensitive test may not rule out other possibilities (False positive) and can be equated to less accuracy. Mamta Mittal et al. emphasizes the importance of edge enhancement and its subsequent improvement in image identification. Its use is much more in medical imaging and subsequent steps of deep learning [14].

Once the appropriate image is identified, the subsequent steps of the age identification process happen. The process of automation can eliminate all the errors that can be attributed to human error of identifying entering the value mechanically. That virtually eliminates all possible errors after the stage has been identified. Visual Basic Application is again a user-friendly data management tool. It is easily accessible and customizable for our research data management. In the field of medicine, doctors are not adapt to calculations and numerical data management. Automation of these processes using simple software like VBA can potentially enhance the probability of research and its accurate outcome in medical research and implementation. The software is versatile to have *GUI interfaces* to the gender criteria which has a direct

Table 4 Accuracy is calculated using True or False

Training samples	Accuracy (%)	Prediction time (s)
20	32.3	0.07
30	42.7	0.15
40	68.3	0.19

effect on the accuracy of the age. The complexities of the different methods of age estimation like the number of teeth and the inclusion of criteria for intermediate stages can be incorporated in this program. Many researchers have used these two age estimation techniques and have found minor variations in age due to the effect of races (genetic) like Chinese, African Asian, etc., they have devised a correction factor to make these techniques accurate for the particular race. These factors can also be incorporated into this VBA programming.

6 Conclusion

The present study shows that image processing has made estimation of age using Demirjian's technique easier and accurate to about 68%. Eventually, it will also eliminate inter and intra-observer bias as the entire process has been automated. The field of medicine is a unique field where situations can be just a thin line between life and death. *Human errors* have no role to play in this field and every effort is taken to minimize human errors. Human errors are not intentional, they can arise due to distraction due to work pressure and stress to which are healthcare workers are exposed to. The advantages of the machines are that they are free from stress and errors due to repetition. *Machine Learning* is a process where the machines are trained to do a process accurately. Image processing is an integral part of machine learning and this helps in creating *Artificial Intelligence*. In this study, with 40 samples we were able to achieve an accuracy of 68% further training the algorithm with higher number of samples can help in achieving higher accuracy. Similar algorithms in ECG (electrocardiogram), CT scans, MRIs, and radiographs can help in developing a system to alert the physician to abnormal patterns which are critical. Though not a substitute to a physician, these can be a lifesaver to prevent human errors in healthcare system.

References

1. Chandramohan P, Puranik MP, Uma SR (2018) Demirjian method of age estimation using correction factor among indian children: a retrospective survey. J Indian Assoc Public Health Dent 16:72–74
2. Abesi F, Haghanifar S, Sajadi P, Valizadeh A, Khafri S (2013) Assessment of dental maturity of children aged 7–15 years using Demirjian method in a selected Iranian population. J Dent (Shīrāz, Iran) 14(4):165–169
3. Obertová Z, Ratnayake M, Poppa P, Tutkuviene J, Ritz-Timme S, Cattaneo C (2018) Metric approach for age assessment of children: an alternative to radiographs? Aust J Forensic Sci 50(1):57–67
4. Nelson S (2010) Wheeler's dental anatomy, physiology and occlusion, 9th ed. Saunders
5. Willems G (2001) A review of the most commonly used dental age estimation techniques. J Forensic Odontostomatol 19(1):9–17
6. Nolla CA (1960) The development of the permanent teeth. J Dent Child 27:254–266

7. Demirjian A, Goldstein H, Tanner JM (1973) A new system of dental age assessment. Hum Biol 45:211–227
8. https://my.clevelandclinic.org/health/articles/11199-types-of-dental-x-rays
9. https://electronicsforu.com/electronics-projects/image-processing-using-matlab-part-1
10. https://www.excel-easy.com/vba.html
11. Patil V, Gupta R, Rajendran D, Kuntal RS (2019) Formation and designing of "least-cost ration formulation application of cattle" using Excel VBA. Springer, Singapore
12. Thomas D (2014) Age assessment using Nolla's method in a group of Mangalore population: a study on 25 children. J Contemp Med 4(3):121–127
13. Gupta R, Rajvanshi H, Effendi H, Afridi S, Kumar Vuyyuru K, Vijay B, Dhillon M (2014) Dental age estimation by Demirjian's and Nolla's method in adolescents of Western Uttar Pradesh. J Head Neck Physicians Surg 3(1):50–56
14. Mittal M, Verma A, Kaur I, Kaur B, Sharma M, Goyal LM, Roy S, Kim T (2019) An efficient edge detection approach to provide better edge connectivity for image analysis. IEEE Access 7(1):33240–33255
15. http://dmdi.com.au/doc/Maxillofacial_Case_Study_-_7_year_old_male.pdf

Chapter 10
Detecting Bone Fracture Using Transfer Learning

Saurabh Verma, Sudhanshu Kulshrestha, Chirag Rajput and Sanjeev Patel

1 Introduction

Medical imaging domain is witnessing progress in generating enormous medical images as well as developing techniques for interpretation of those images. Specifically, the generic aim of this research is to analyze and identify the disease, as predicted from the medical images having minimal aid from medical experts. Systems such as Computer-Aided Diagnosis (CAD) has proven to be very useful in analyzing a large set of medical data and also to improve the accuracy of interpretation while reducing significant time for diagnosis [1, 2]. There are many imaging techniques available like X-rays, CT scan, ECG, Ultrasound, etc. X-Ray images depend on disease to disease like X-radiography, mammography, CT (Computer Tomography), this varies from individual health conditions. It is believed that X-Rays display prominent results in case of fractures related to bone but it has also been seen that soft tissue injury can't be detected properly using this technology. The X-ray is typically analyzed by doctors physically to deduct the situation to spot the extremity of the fracture [1].

S. Verma · S. Kulshrestha (✉) · C. Rajput · S. Patel
Jaypee Institute of Information Technology, Noida, India
e-mail: sudhanshu.kulshrestha@gmail.com

S. Verma
e-mail: saurabhverma161998@gmail.com

C. Rajput
e-mail: chiragrajput17@gmail.com

S. Patel
e-mail: patelcs01@gmail.com

© Springer Nature Singapore Pte Ltd. 2020
O. P. Verma et al. (eds.), *Advancement of Machine Intelligence in Interactive Medical Image Analysis*, Algorithms for Intelligent Systems,
https://doi.org/10.1007/978-981-15-1100-4_10

1.1 Motivation

Nowadays, all the information about X-ray are digitized and interpreted on machines only. Furthermore, radiographs images and reports help orthopedics doctors and surgeons to determine a proper cure to patient. But some types of fractures are hard to recognize from X-ray manually even for a doctor and this problem of not getting proper treatment, the possibilities for compound fractures in distant future are much probable. A study shows that in UK only, the number of CT examinations increased by 29% from 2012 to 2015, during the time enrollment of new doctors drop back, leaving 9% of concerned orthopedics doctors or surgeons post vacant in 2015 [1]. Likewise, in February 2016, there was an estimate of unattended 200,000 plain radiographs due to lack of doctors or physician availability [1]. These data clearly show the urgent need of a system that will ease the work of doctor as well as the patient. The requirement of an appropriate and convenient decision-making system to assist doctors as well as save time that successively can increase the amount of patients to be treated efficiently.

This article is organized as follows: Sect. 2 presents a study of literature review of existing work in the similar domain. Section 3, describes the proposed methodology and experimental design. It also illustrates the image preprocessing and CNN training in detail. Section 4, discusses the results along with a comparative analysis, future scope, and limitation of the study.

2 Background Study

Kim and MacKinnon [3] retrained the inception v3 network using a radiograph. Their model was trained on 11,112 images after applying the data augmentation technique on initial set of 1389 images. Accuracy attained by them through this technique was 95.4%. Anu and Raman [1] applied image processing for detecting bone fractures. They tested their proposed method on X-ray bone image collection. They have done preprocessing on the data then applied methods of edge detection, segmentation, feature extraction, and classification. The accuracy achieved through this method was 85%. Kaur and Garg [2] used a wavelet approach for Fracture detection in X-Ray images. To find the fracture using X-ray bone images, Multilevel wavelet is used after applying *Gabor* filter. It uses *Hough* transform to find long bones and wavelet decomposition. But it's limitation was that the fracture is found only in horizontal images. OpenCV library is also influential in image processing and information retrieval, as used by authors [4]. Samuel et al. [5] built a system using OpenCV library in combination with *Canny Edge Detection* method to detect the bone fracture. They inferred that the efficiency and accuracy of their proposed model influence the nature of the image. Dimililer [6] developed two-stage system. In its first stage, the fracture images are processed using various image processing techniques to detect their shapes and locations. The second stage is called the classification phase, here a backpropagation neural network is trained and tested on images that have been processed. Cao et al. [7] implemented stacked random forest method to detect fracture from X-rays images. They extracted the features from individual patches in X-ray images using a discriminative learning framework which was named as Stacked Random Forests Feature Fusion. Das [8] implemented first the preprocessing techniques. BAT algorithm has been applied in the preprocessing stage which enhances the image. After that, Self-organizing Map (SOM) is applied to draw segmentation and lastly, K-means clustering is applied to produce objective image. Pan and Yang [9] surveyed Transfer Learning (TL) and explained that TL aims to transfer information among the given sources and targets. In computer vision, TL examples consist of trying to make up for the insufficiency of training samples for some categories by making use of classifiers trained for some other categories. Moeskops et al. [10] trained a CNN model in order to divide six tissues into (i) scanned Magnetic Resonance (MR) brain, (ii) pectoral muscle inside the MR breast images, and (iii) coronary arteries inside the cardiac CTAs. The CNN then try to classify the imaging type, the envision anatomical design, and the classes of tissues. Mini-batch learning was used to train every CNN. The mini-batch used consisted of 210 samples, equally balanced over the tasks of the network. Therefore, the system can be utilized at clinics and hospitals to automatically achieve diverse classification decision without task-specific training. Bharodiya and Gonsai [11] used various image processing techniques to clean and extract useful features of X-ray images to define as fracture or not. First, they used CLHE which, according to them, helped in improving the contrast and also reducing the noise present in the image. Then guided filter to smoothening the image. Then Canny edge detection method was applied which they believe is the best method to detect edges. And then the last step applied was active contour segmentation. This step was performed to identify the portion of X-Ray image where the likelihood of presence of bone fracture is high. Accuracy obtained was 80%. In addition to above explained techniques, Machine Learning is also prevalent in image processing and analysis, as used by Babbar et al. [12]. Hemanth et al. [13] used Neural Network based approach to diagnose diabetic retinopathy disorder from retinal images. In order to analyze images, edge detection and object detection play an important role. Mittal et al. [14] and [15] have worked toward edge detection for better image analysis. Mamta et al. have used deep learning based framework for tumor segmentation in brain images. Table 1 presents an exhaustive comparative study of above explained related work.

2.1 Deep Learning and Transfer Learning

Deep learning [16] is rapidly growing as a key instrument in the applications of artificial intelligence. For topics such as speech recognition, Natural Language Processing (NLP), and Computer Vision (CV), DL has been showing remarkable results. In the specific problem of image classification, deep learning shows excellent results. Image classification problem refers to classify a given image according to possible

Table 1 Comparative study of related work

S. no.	Article	Preprocessing technique	Machine learning technique	Data set size	Accuracy
1	Dimililer [6]	1. Haar Wavelet Transform 2. Invariant Feature Transform (SIFT-Scale Invariant Feature Transform) algorithm	1. Two conventional three-layer back propagation neural networks	100	94.3%
2	Kim and MacKinnon [3]	1. Data augmentation 2. Image processing (red spots removel)	1. CNN training (inception v3 network)	11,112	95.4%
3	Kau et al. [2]	1. Image processing i. Gray scale conversion ii. Histogram equalization iii. Gaussian 2. Smoothing 3. Fuzzy C-means algorithm 4. Gabor filter 5. Hough Transformation 6. Multilevel discreet wavelet transformation	NA	126	89.6%
4	Anu and Raman [1]	1. Gray scale conversion 2. Noise removal median filter 3. Edge detection (sobel edge detector) 4. Segmentation: K-means clustering technique 5. Feature extraction gray-level cooccurence matrix	1. Classification: Decision Tree (DT), Neural Network (NN) and metaclassifier	40	DT and NN-53.25% Meta C gives 85%
5	Kurniawan et al. [5]	1. Noise removal 2. Canny edge detection 3. Shape detection	NA	NA	(Accuracy of 3 iterations respectively) Hairline Fracture 0, 100%, 50% Major detection 50%, 55.6%, 66.7%

(continued)

Table 1 (continued)

S. no.	Article	Preprocessing technique	Machine learning technique	Data set size	Accuracy
6	Cao et al. [7]	NA	1. Stacked Random Forests Feature Fusion (SRF-FF) 2. Fracture localization (subwindow search algorithm) 3. Local patch feature—Schmid texture feature, Gabor texture feature, contextual Intensity feature.	145	81.2%
7	Das [8]	1. Application of Bat Algorithm as an enhancer 2. Application of gamma correction 3. Application of K-means	1. Self-Organizing Map (SOM) Neural Network	NA	NA
8	Moeskops et al. [10]	Nil	1. Convolution neural network contained 32 small (33 voxels) convolution filters for a total of 25 convolution layers	25,000	NA
9	Bharodiya and Gonsai [11]	1. Contrast Limited adaptive Histogram Equalizer (CLHE) 2. Guided filter 3. Canny edge detection 4. Contour segmentation method	Nil	50	80%

categories as per the data set. As far as deep learning is concerned, the problem of classification of an image can be solved, via, TL. Transfer learning [9, 17] is a prominent strategy in computer vision as it grants a developer to create correct and exact models in a more efficient way. With transfer learning, rather than creating the model from beginning, one can use the weights learned while solving different problems. This way by using previous learning, one can avoid starting a model from fresh. In CV, TL is often used through pretrained models. A pretrained model is a structure which is trained on a huge amount of data to determine the similar kind of issue. But, this causes a heavy computational cost during training of such models. Therefore, to avoid such complications, state-of-the-art models from published literature such as VGG19, AlexNet, and ResNet can be used. There are many pretrained model on the web such as VGG, inceptionV3, and many more. However, most of them follow similar steps during TL, as mentioned below:

- Select a pretrained model
- Divide the problem as per the Size Similarity Matrix
- Fine-tune the model.

2.2 Fine-Tuning Pretrained Model

While redeploying the model, we start from deleting the previous classifier from the model and then add a new classifier as per your problem and then fine-tuning takes place through these three methods.

1. *Training of the whole model*—In this situation, we have to train each and every layer of the model from the scratch. It not only requires huge amount of data set but also lot of resources so as to compute large computations.
2. *Train few of the layers and leave remaining frozen*—In cases when a sufficient amount of data set is not available, as well as suitable number of parameters are also not available, there is an option to leave as many as possible number of layers frozen to avoid overfitting in the model. However, if the data set is sufficient enough and the number of parameters is less, model can be enhanced by training more number of layers as per the new task because overfitting isn't a problem.
3. *Freezing of the convolutional base*—This method is applicable in scenarios where there is a well-known trade-off between train and freeze. In this method, convolutional base remains preserved as it was in its initial state. Later, it utilizes the generated result to feedforward to the classifier (Fig. 1).

2.3 Convolutional Neural Network

There are various pretrained TL models based on Convolutional Neural Network (CNN) [19, 20]. CNN surely have gain a lot of popularity in recent years due to its high performance and easiness while training. In general, a CNN model has the following two components:

1. *Convolutional base*—It comprises of convolution and pooling layers whose goal is to generate features map from the image.
2. *Classifier*—It is basically fully connected layer and its goal is to label the image depending on the features coming from preceding layers. Equations (1)–(2) present a generic calculation for various parameters.

$$f(x) = fL(f2(f1(x; w1); w2)), wL) \tag{1}$$

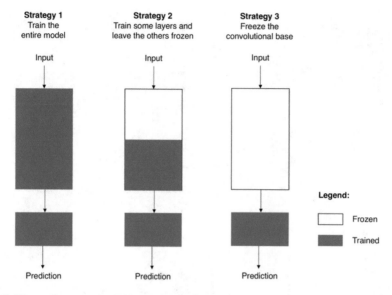

Fig. 1 Fine-tuning strategies [18]

Equation (1) gives feedforward convolution

$$H_{a'b'c'} = \sum_{abc} w_{abcc'} x_{a+a',b+b',c+c'} \tag{2}$$

Equation (2) gives output of convolution

$$H_{abc} = max(0, x_{abc}) \tag{3}$$

Equation (3) illustrates the the nonlinearity function

$$H_{abc} = max(H_{a'b'b'} : a <= a' < a + p, b <= b' < b + p) \tag{4}$$

Equation (2) gives the max pooling function (Figs. 2, 3 and 4).

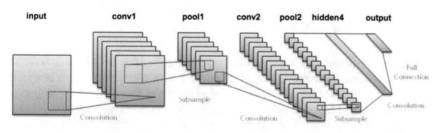

Fig. 2 The figure shows architecture of basic CNN model [21]

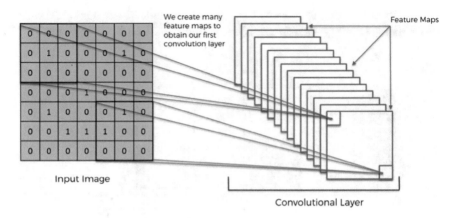

Fig. 3 Intuition of convolution layer [22]

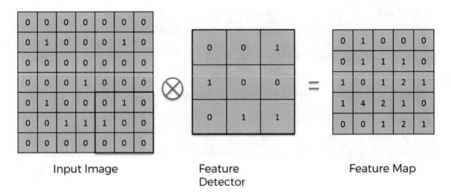

Fig. 4 Matrix calculation [22] of convolution step given in Eq. (2)

3 Proposed Methodology

See Fig. 5

3.1 Image Preprocessing

The data set consists of total 46 plain radiograph images of open fracture in forearm and total 72 plain radiograph images of forearm with no fracture. Each image in the data set is passed through SURF (speeded up robust features) feature extractor which extracts and highlights the important features in the X-ray image that is the fractured part in the bone X-ray image. Figure shows highlighted features in an X-ray images after applying SURF feature extractor. The new image with highlighted

Fig. 5 Flow chart of
working model

features is stored and then data augmentation techniques are applied to each image.
Each image is rotated to different angles and flipped, the resulting data set after image
augmentation consists of 280 training images and 86 testing images. Equations used
in SURF algorithm are

$$S(a, b) = \sum_{k=0}^{a} \sum_{j=0}^{b} I(a, b) . \tag{5}$$

$$H(\rho, \sigma) = \begin{pmatrix} K_{aa}(\rho, \sigma) K_{ab}(\rho, \sigma) \\ K_{ba}(\rho, \sigma) K_{bb}(\rho, \sigma) \end{pmatrix} . \tag{6}$$

Equation (5) is used to filter the image and Eq. (6) is the Hessian matrix $H(\rho, \sigma)$ at
point p and scale where $K_{aa}(\rho, \sigma)$ is calculated as convolution of the second-order
derivative of Gaussian with the image of I(x, y) at the point ρ (Fig. 6).

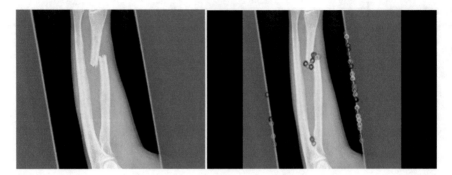

Fig. 6 Left image shows fractured image before SURF Feature extraction and right image shows the extracted feature after the algorithm

3.2 CNN Training

The preprocessed data set is used to train the model using PyTorch. The VGG16 network trained for Imagenet Large Visual Recognition Challenge (ILVRC) using non-radiological images was computationally fitted for determining of the type of fracture. This is done by fine-tuning VGG16 model. A 3×3 convolutional layer heap is used in this step. In addition, depth of other layers is increased incrementally. Here, number 16 represents the count of weighted layers in the architecture of the network. Later, parameters are frozen for all the convolutional layers and are retraining the fully connected part with our data set. This method of training pretrained models for classing new data is called transfer learning. In the training, we are using cross-entropy loss criteria and Adam optimizer to update the parameters. The model is trained for 30 epochs and both training and testing accuracy is calculated at each epoch (Figs. 7 and 8).

The VGG19 architecture consists of 19 layer CNN with 16 convolution layers and three fully connected layers. It uses 3 * 3 small filters in every convolution layer to reduce the number of parameters. VGG19 accepts BGR image with size of 224. The first convolution layer increases the depth and the output becomes 224 * 224 * 64. Another convolution layer is applied right after max pooling the previous output with filter size of 2 * 2 and stride of 2, the output after convolution becomes 112 * 112 * 128 further increasing the depth. After a series of convolutions and max pooling the output becomes 7 * 7 * 512 which is later flattened and two fully connected layers are formed. Finally, after applying softmax, it classifies the 1000 classes of image net challenge. To modify this architecture so as to solve our problem, we are freezing training for all the convolution layers and classification is reduced from 1000 to 2 classes.

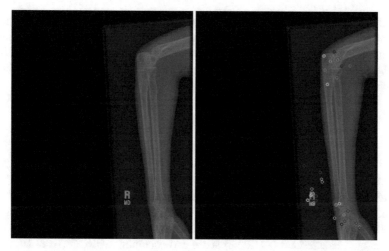

Fig. 7 Left image shows non fractured image before image processing and right image is after image processing

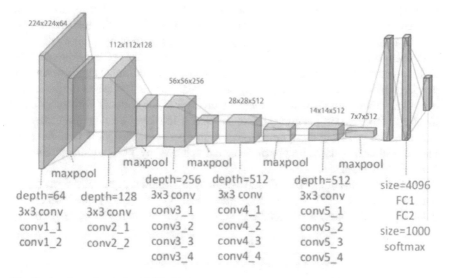

Fig. 8 A schematic of the VGG 19 deep convolutional neural network DCNN architecture trained [23]

4   Result and Discussion

Using preprocessed images with highlighted features using SURF and using transfer leaning technique to adapt the pretrained VGG19 model to classify fracture and no fracture radiological images proved very useful. We are using Area Under the curve of the Receiver Operating Characteristic (AUROC) to evaluate the model. VGG19 scored 0.98 AUROC score on testing images and 0.92 AUC score on training images.

The figure shows performance of other pretrained models on our data set. Although our proposed technique is able classify given number of images with 98.8% accuracy but due to availability of less data for training, it is still vulnerable to overfitting. Furthermore there are certain cases in which feature extraction through SURF is unable to highlight fractured part which later results in poor classification.

4.1 Performance Analysis

To evaluate our proposed method we have used the following pretrained models— AlexNet, DenseNet-121, ResNet152, SqueezeNet, VGG19, and VGG16 with different respective architecture and retrained their fully connected layer for our data set. The main criteria for evaluation are the Area Under the curve of the Receiver Operating Characteristic (AUROC) score and accuracy. AUROC (Area Under the curve of the Receiver Operating Characteristic) basically represents the likeliness of a model classifying 0 as 0 and 1 as 1, higher the AUCROC score, the better the model is at classification. Accuracy is calculated, the total number of two correct predictions that is (True Positive + True Negative) divided by the total number of data in a data set. AlexNet pushed the imagenet classification by a large margin as compared to previous methodologies. Its architecture consists of five convolutional layers and three fully connected layers. AlexNet being a traditional architecture does not perform well enough to classify our small data. As with DenseNet the basic idea behind its architecture is that it connects each layer of the network with every other layer in a feedforward fashion. The ResNet has a 152 layer deep architecture with skip connections. Both these architecture didn't performed as per expectations for the medical data. Finally, best performance is achieved by VGG architecture which was briefly discussed above.

Table 2 presents the AUROC and accuracy of following pretrained models— AlexNet, DenseNet-121, ResNet152, SqueezeNet, VGG19, and VGG16. In testing data set, VGG19 has the AUROC and accuracy as 0.921 and 93.7%, respectively.

Table 2 Comparison of accuracy of proposed technique and others

Model	Training			Testing		
	AUROC[a]	Acc[b]	Loss	AUROC[a]	Acc[b]	Loss
AlexNet	0.791	81.7	0.754	0.81	82.5	0.462
DenseNet-121	0.796	83.5	0.763	0.944	94.1	0.2306
ResNet152	0.805	83.2	0.399	0.912	91.8	0.264
SqueezeNet	0.789	81.4	0.394	0.876	88.3	0.292
VGG19	0.921	93.7	0.1757	0.98	**98.8**	0.064
VGG16	0.903	92.5	0.2179	0.95	96.5	0.165

[a] Area under the curve of the receiver operating characteristic
[b] Accuracy in percentage

Whereas, VGG16 has the second highest AUROC and accuracy as 0.903 and 92.5%, respectively. At the same time, the overall loss for VGG19 and VGG16 is minimum among all other competitors as 0.1757% and 0.2179%, respectively. Similarly, in testing data set, VGG19 has the AUROC and accuracy as 0.98 and 98.8%, respectively. Whereas, VGG16 has the second highest AUROC and accuracy as 0.95 and 96.5%, respectively. At the same time, during testing, the overall loss for VGG19 and VGG16 is minimum among all other competitors as 0.064% and 0.165%, respectively. Hence, it shows that our proposed model outperforms other competitors.

5   Conclusion

The proposed approach detects fracture from the given X-ray images with testing accuracy of 98.8%. Furthermore it is concluded that transfer learning gives significantly comparable results than training model from scratch or even better in some cases. But it is also vulnerable to overfitting due to less training data, and in some cases poor preprocessing leading to poor classification. In our future work, we are trying to introduce better feature extractor and make use of better pretrained models to attain a state-of-the-art solution of the problem, also our solution focuses only on forearm radiograph images so in future work, we are trying to generalize our solution to all fracture types.

References

1. Anu TC, Raman R (2015) Detection of bone fracture using image processing methods. Int J Comput Appl 975:8887
2. Kaur T, Garg A (2016) Bone fraction detection using image segmentation. Int J Eng Trends Technol (IJETT) 36(2):82–87
3. Kim DH, MacKinnon T (2018) Artificial intelligence in fracture detection: transfer learning from deep convolutional neural networks. Clin Radiol 73(5):439–445
4. Sidhwa H, Kulshrestha S, Malhotra S, Virmani S (2018) Text extraction from bills and invoices. In: 2018 international conference on advances in computing, communication control and networking (ICACCCN). IEEE, pp 564–568
5. Kurniawan SF, Putra D, Gede IK, Sudana AKO (2014) Bone fracture detection using opencv. J Theor Appl Inf Technol 64(1)
6. Dimililer K (2017) Ibfds: intelligent bone fracture detection system. Procedia Comput Sci 120:260–267
7. Cao Y, Wang H, Moradi M, Prasanna P, Syeda-Mahmood TF (2015) Fracture detection in X-ray images through stacked random forests feature fusion. In: 2015 IEEE 12th international symposium on biomedical imaging (ISBI). IEEE, pp 801–805
8. Das G (2013) Bat algorithm based softcomputing approach to perceive hairline bone fracture in medical X-ray images. Int J Comput Sci Eng Technol (IJCSET) 4(04)
9. Pan SJ, Yang Q (2009) A survey on transfer learning. IEEE Trans Knowl Data Eng 22(10):1345–1359
10. Moeskops P, Wolterink JM, van der Velden BHM, Gilhuijs KGA, Leiner T, Viergever MA, Išgum I (2016) Deep learning for multi-task medical image segmentation in multiple modalities.

In: International conference on medical image computing and computer-assisted intervention. Springer, pp 478–486

11. Bharodiya AK, Gonsai AM (2019) Bone fracture detection from X-ray image of human fingers using image processing. In: Emerging trends in expert applications and security. Springer, pp 47–53

12. Babbar S, Kesarwani S, Dewan N, Shangle K, Patel S (2018) A new approach for vehicle number plate detection. In: 2018 eleventh international conference on contemporary computing (IC3). IEEE, pp 1–6

13. Jude Hemanth D, Anitha J, Mittal M et al (2018) Diabetic retinopathy diagnosis from retinal images using modified Hopfield neural network. J Med Syst 42(12):247

14. Mittal M, Verma A, Kaur I, Kaur B, Sharma M, Goyal LM, Roy S, Kim T-H (2019) An efficient edge detection approach to provide better edge connectivity for image analysis. IEEE Access 7:33240–33255

15. Kaur B, Sharma M, Mittal M, Verma A, Goyal LM, Jude Hemanth D (2018) An improved salient object detection algorithm combining background and foreground connectivity for brain image analysis. Comput Electr Eng 71:692–703

16. Goodfellow I, Bengio Y, Courville A (2016) Deep learning. MIT Press

17. Raina R, Battle A, Lee H, Packer B, Ng AY (2007) Self-taught learning: transfer learning from unlabeled data. In: Proceedings of the 24th international conference on machine learning. ACM, pp 759–766

18. Towards Data Science. Tl from pre-trained models. https://towardsdatascience.com/transfer-learning-from-pre-trained-models-f2393f124751. Accessed 01 Aug 2019

19. Krizhevsky A, Sutskever I, Hinton GE (2012) Imagenet classification with deep convolutional neural networks. In: Advances in neural information processing systems, pp 1097–1105

20. LeCun Y, Bengio Y et al (1995) Convolutional networks for images, speech, and time series. Handb Brain Theory Neural Netw 3361(10):1995

21. Gopalakrishnan K, Khaitan SK, Choudhary A, Agrawal A (2017) Deep convolutional neural networks with transfer learning for computer vision-based data-driven pavement distress detection. Constr Build Mater 157:322–330

22. SuperDataScience Pty Ltd. CNN: Step 1-convolution operation. https://www.superdatascience.com/blogs/convolutional-neural-networks-cnn-step-1-convolution-operation. Accessed 01 Aug 2019

23. Zheng Y, Huang J, Chen T, Ou Y, Zhou W (2018) Processing global and local features in convolutional neural network (CNN) and primate visual systems. In: Mobile multimedia/image processing, security, and applications, vol 10668. International Society for Optics and Photonics, p 1066809

Chapter 11
GAN-Based Novel Approach for Data Augmentation with Improved Disease Classification

Debangshu Bhattacharya, Subhashis Banerjee, Shubham Bhattacharya, B. Uma Shankar and Sushmita Mitra

1 Introduction

In medical domain, due to scarcity of balanced a need for methods to improve upon the performance of Deep Convolutional Neural Network (DCNN) models [3]. While Transfer Learning and Domain Adaptations are possible solutions [8], there are no pre-trained networks on a large annotated dataset of medical data. This makes us look for data augmentation as a solution [23]. Classical augmentations, using transformations like translation, rotation, scaling, flipping and shearing, have been successful on a large number of datasets. However, in some medical datasets (like the chest X-ray images) some of those transformations, like rotation and flipping, should not be applied to preserve the properties of the annotated data, while the other remaining transformations lead to image duplicity and causes the model to overfit. Ever since Generative Adversarial Networks (GAN) was introduced by Goodfellow et al. [9], it has been extensively studied and used in various fields [12, 18, 30]. The generator of this network can generate the distributions of the original data,

D. Bhattacharya
Chennai Mathematical Institute, Chennai, India
e-mail: dbangshu16@gmail.com

S. Bhattacharya
Heritage Institute of Technology, Kolkata, India
e-mail: sb.skyfall@gmail.com

S. Banerjee · B. Uma Shankar (✉) · S. Mitra
Machine Intelligence Unit, Indian Statistical Institute, Kolkata, India
e-mail: uma@isical.ac.in

S. Banerjee
e-mail: mail.sb88@gmail.com

S. Mitra
e-mail: sushmita@isical.ac.in

© Springer Nature Singapore Pte Ltd. 2020
O. P. Verma et al. (eds.), *Advancement of Machine Intelligence in Interactive Medical Image Analysis*, Algorithms for Intelligent Systems,
https://doi.org/10.1007/978-981-15-1100-4_11

and this method of synthesizing artificial data indistinguishable from original data has been used to good effect in various fields, including data augmentation [1, 4, 7, 14, 15, 24, 26–29]. GANs have already achieved state-of-the-art performance in image generation tasks on various domains. Recently it is also being applied in the medical domain. There have been several studies recently on using GAN to augment medical datasets [6, 22, 27] or using it for segmentation tasks [21]. These augmentations are 'offline' augmentations increasing the dataset size as a method to improve performance. We perform an 'online' augmentation technique keeping the dataset size as constant. For every mini-batch of images, we keep a percentage of original images and the rest are replaced by GAN-synthesized images. This 'keep-prob' parameter was tuned and we used the value 0.7 (i.e. the probability of retaining 70% data and replacing 30% data by online augmentation).

In the present work, the contribution can be summarized as follows: (1) We propose a CNN model, which has optimized hyperparameters to achieve better accuracy for disease classification using X-ray images, where data from the three classes are very small. (2) A schema is proposed for using Deep Convolutional Generative Adversarial Network (DCGAN) model effectively. This approach of online augmenting using DCGAN model-based generated images helped in regularization that minimizes the overfitting and improves the accuracy on the test dataset.

In this chapter, we proposed a novel approach of using the generated images for augmentation of the data based on GAN model to improve the performance of the proposed CNN for disease classification using X-ray images. The rest of the chapter is as follows. Section 2 present the details of the data used, whereas Sect. 3 provides the proposed approach, describing the considered GAN model and CNN model. Section 4 gives the complete details of the experiments and results obtained with analysis. At the end, we provide the conclusions in Sect. 5, with some details about the future work.

2 Chest X-ray Classification

The dataset used is a part of the NIH chest X-ray dataset provided at website[1] [25]. The data used contains images of three classes: Infiltration, Atelectasis and No Findings. The example images are provided in Fig. 1. Our dataset contains 4,215 images of patients having Atelectasis; 9,547 images of patients having Infiltration and 13,762 images of patients having Normal condition. We are splitting the data into training (23,524 images), validation (2,000 images) and testing (2,000 images) sets with each of validation and test data having 500, 500 and 1000 images from Atelectasis, Infiltration and No Findings (NF), respectively. All the images are resized to size 64 × 64.

---

[1] https://nihcc.app.box.com/v/ChestXray-NIHCC/.

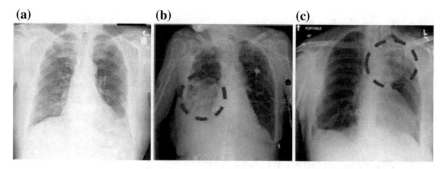

Fig. 1 Visual examples of ordinal images: **a** No Finding, **b** Infiltration and **c** Atelectasis

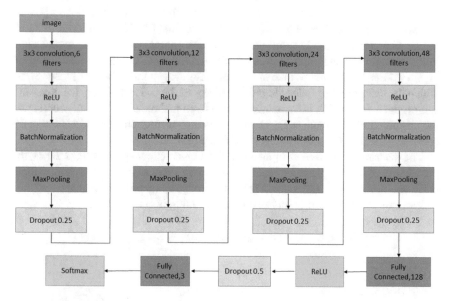

Fig. 2 The proposed model

2.1 CNN Architecture

The network architecture, as shown in Fig. 2, has four convolutional blocks followed by two fully connected blocks. Each convolutional block has a convolutional layer with ReLU activation [5] function followed by Batch Normalization [11], Max Pooling [10] and Dropout layer [5]. Each convolutional layer has a 3×3 kernel with padding ('same'), whereas each maxpooling layer has pool size of 2×2 and has a stride of 2. The kernel of the convolution layer is initialized (by 'he_normal' initialization). The final layer of our model is a softmax layer for classifying images into 3 classes.

Training: The images were normalized to the range 0–1. The loss function used was categorical cross-entropy. Adam optimizer was used with a learning rate of 0.001 [13]. Mini-batches of size 32 were used to train the model for 100 epochs. In an imbalanced dataset like ours, where most examples are from a particular class, the model tends to overfit to a particular class. To combat this, we used the class weights-parameter (weighted loss function) [19], which optimizes the model to handle misclassifications of minority class during backpropagation.

3   Generating Synthetic Chest X-rays

We use a Deep Convolutional Generative Adversarial Network (DCGAN) [20] as a generative model. Generative Adversarial Networks (GANs) comprises two networks, Generator G and Discriminator D, which compete against one another in a minimax approach and improve each other's performance. The discriminator aims to identify if the data fed to it is fake or real, whereas the generator aims to generate data which the discriminator fails to identify as fake. At the end of training, the generator can produce data which is indistinguishable from real data, successfully recreating the original data distribution. DCGANs use deep convolutional neural networks for both the generator and discriminator architecture (Fig. 3).

The discriminator takes as input a chest X-ray sample x and it outputs $D(x)$, the probability of the sample being real. The generator takes as input a 100-dimensional noise vector $z \in [0, 1]$ and maps $G(z)$ to the image I_g (generated image).

The loss function used in this adversarial training is given below:

$$\min_{G} \max_{D} V(D, G) = \mathbb{E}_{x \sim p_{data}(x)} \left[\log D(x) \right] + \mathbb{E}_{z \sim p_z(z)} \left[1 - \log D(x) \right] \quad (1)$$

Fig. 3 DCGAN architecture

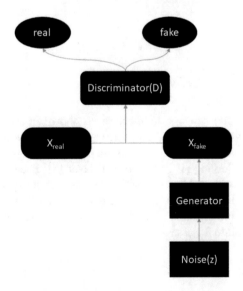

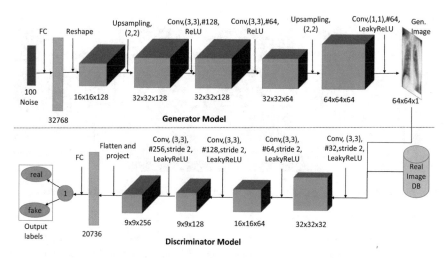

Fig. 4 Generative and discriminative model

where p_{data} is the probability distribution of the original data and p_z is the prior defined on the noise variable [9].

The Discriminator tries to maximize the loss value $V(D, G)$ whereas the generator tries to minimize it. To train the discriminator, the parameters of the generator model are made untrainable, i.e. they cannot be modified by backpropagation and the generator acts as a feedforward network. Similarly for training the Generator, the Discriminator is made untrainable.

Generator Architecture: The Generator takes as input of a 100-dimensional noise vector drawn from normal distribution and outputs a chest X-ray image of shape $64 \times 64 \times 1$ as shown in Fig. 4. There is one fully connected layer, two upsampling layers and three convolutional layers. Each convolutional layer has kernel size of 3×3 with ('same') padding. The inner two convolutional layers are followed by Batch Normalization and ReLU Activation layers. The final convolutional layer is followed by a 'tanh' activation layer.

Discriminator Architecture: The Discriminator takes a typical $64 \times 64 \times 1$ chest X-ray image as input and outputs if it is real or fake. The network comprises four stride convolutional layers followed by a fully connected layer of 1 neuron having a sigmoid activation. Each convolutional layer has a kernel size of 3×3 with stride of 2 and ('same') padding. Each convolution layer is followed by Batch Normalization, LeakyReLU activation and dropout layer.

Training: We trained the DCGAN model for each of the three classes individually. The images were normalized to the range -1 to 1 and then fed into the DCGAN model. We used Adam Optimizer with a learning rate of 0.0002 and momentum of 0.5. We used mini-batches of size 32 and trained for 50, 60 and 70 epochs for

(a)   **(b)**   **(c)**

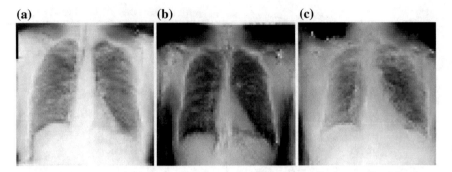

Fig. 5 Visual examples of generated images: **a** No Finding, **b** Infiltration and **c** Atelectasis

classes Atelectasis, Infiltration and NF, respectively, to avoid the overfitting as the data size of three clashes are not same. Figure 5 is an example of image generated by the generator for each of the classes.

4 Experiments and Results

We employed the CNN architecture described in Sect. 2.1, used for training the classifier. Once the generator model of the DCGAN network was trained, we used those models to perform real-time augmentation of the dataset by replacing a part of the data with generated images of the respective classes. We performed this using mini-batches of size 32. Then we used the exact same training procedure as that of CNN training was followed. We compare the performances of the CNN on original dataset and on our online-augmented dataset. The details of online data augmentation algorithm are given in Algorithm 1.

Algorithm 1 Online Data Augmentation algorithm

1: Initialize keep_prob, batch_size;
2: $n \leftarrow$ Calculate number of mini-batches.
3: **for** each mini-batch in n **do**
4:     $prob\_vec \leftarrow$ probability of each image in the mini-batch
5:     **for** each image i in mini-batch **do**
6:         **if** prob_vec[i] $\leq$ keep_prob **then**
7:             pass
8:         **else**
9:             $y \leftarrow$ class of ith entry.
10:             Generator generates image of class y.
11:             Generated image is normalized to the range 0 to 1.
12:             Replace original image by generated image.
13:         **end if**
14:     **end for**
15: **end for**

Fig. 6 Comparison between
CNN and
GAN-augmented-CNN

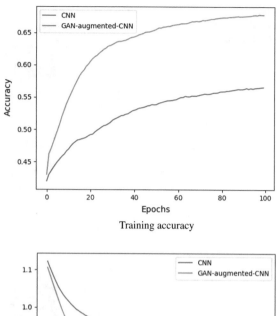

Training accuracy

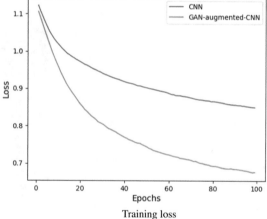

Training loss

As we can see from the training graph shown in Fig. 6, the GAN-augmented CNN model is doing significantly better at classifying chest X-ray images. We are using the total classification accuracy as a metric in evaluating model performance. Accuracy is defined as

$$accuracy = \frac{\sum TP}{\text{Total number of chest X-rays}} \qquad (2)$$

We compare the test set accuracies of the final trained network of the two models (CNN and GAN-augmented-CNN). The confusion matrices of both the model are shown in Fig. 7. The accuracy of the CNN model is reported as 60.3% while the accuracy of the GAN-augmented-CNN model is **65.3%**.

Fig. 7 Confusion matrices of **a** CNN and **b** GAN-augmented-CNN

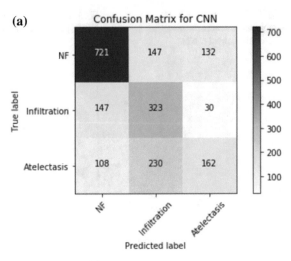

(a)

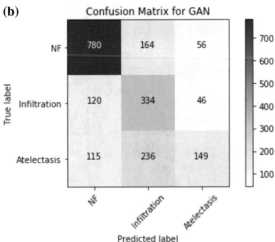

(b)

We repeated the same experiment with three different models. The model with the best results has been shown (model-3). We used Stochastic Gradient Descent (SGD) [13] optimization for one model and had the same architecture as in Section 2.1, but with more filters in each convolution block (32, 64, 128 and 256, respectively). For the other model, we did not use class-weights update during model fitting, i.e. we did not penalize the misclassifications of the lower class data, but the remaining architecture is same as that of Sect. 2.1. We saw that for the SGD optimization, the result obtained was worse than the Adam optimization. As for the model, without using class-weights, the CNN model accuracy was on par with our model CNN performance, but it was mainly overfitting to the classes with more data (majority classes), i.e. No Finding and Infiltration. It failed to classify Atelectasis class which

Table 1 Comparison of models

Model	CNN (%)	GAN-augmented-CNN (%)
Model-1 (SGD)	56.50	58.45
Model-2 (Adam and without class-weights)	60.15	64.35
Model-3 (Adam and class-weights)	60.30	65.30

had low data (minority class). However, the GAN-augmented-CNN model did well to generalize the classifications even without the class-weights parameter. This shows that our method of augmenting data online with generated images reduced overfitting to a large degree and helps in generalizing better results across classes.

Table 1 shows the test set accuracies of the three different models. The consider model (model-3) performed best among all the three models, whereas all the three models with GAN-augmented performed better than plain CNN.

5 Conclusions and Future Work

In this chapter, we have implemented a DCGAN architecture that can be used to create synthetic chest X-ray images for three classes (Infiltration, Atelectasis and No Findings). We proposed a novel schema for using online GAN model effectively. This approach of online augmentation using GAN model-based generated images helped in achieving improved accuracy on the test dataset. This has been demonstrated using three different models; particularly the model-3 has produced reasonably good results for the disease classification. This method shows substantial improvement in results for all the three models in this low and imbalanced data regime. The future work will be concentrated on improving the performance by changing GAN architecture using Conditional GAN (C-GAN) [17] and Wasserstein GAN (W-GAN) [2]. Recently C-GAN and W-GAN studies show that they converge better to Nash equilibrium [16]. Hence, the generated images produced by those architectures when employed in our schema should produce even better results. One of the basic problem of X-ray image analysis for disease classification consist of multi-label classification and associated with different diseases. Our objective will be to extend our schema to such multi-label classification problems. Another reason as to why the results of the Atelectasis class did not come out very well is because it is a condition related to many other diseases. The future work will concentrate on identifying the Atelectasis condition properly with modified GAN architectures.

Acknowledgements The authors (Debangshu Bhattacharya and Shubham Bhattacharya) would like to thank Prof. Dinabandhu Bhandari of Heritage Institute of Technology for his valuable suggestions and guidance. One of the authors, S. Banerjee, acknowledges financial support from the Visvesvaraya PhD Scheme for Electronics and IT by Ministry of Electronics & Information Technology (MeitY), Government of India.

References

1. Ali-Gombe A, Eyad E (2019) MFC-GAN: class-imbalanced dataset classification using multiple fake class generative adversarial network. Neurocomputing. https://doi.org/10.1016/j.neucom.2019.06.043
2. Arjovsky M, Chintala S, Bottou L (2017) Wasserstein GAN. arXiv:1701.07875
3. Banerjee S, Mitra S, Sharma A, Shankar BU (2018) A CADe system for gliomas in brain MRI using convolutional neural networks. arXiv:1806.07589
4. Ben-Cohen A, Klang E, Raskin SP, Soffer S, Ben-Haim S, Konen E, Amitai MM, Greenspan H (2019) Cross-modality synthesis from CT to PET using FCN and GAN networks for improved automated lesion detection. Eng Appl Artif Intell 78:186–194
5. Dahl GE, Sainath TN, Hinton GE (2013) Improving deep neural networks for LVCSR using rectified linear units and dropout. In: 2013 IEEE international conference on acoustics, speech and signal processing (ICASSP). IEEE, pp 8609–8613
6. Frid-Adar M, Diamant I, Klang E, Amitai M, Goldberger J, Greenspan H (2018) GAN-based synthetic medical image augmentation for increased CNN performance in liver lesion classification. CoRR arXiv:abs/1803.01229
7. Gao X, Deng F, Yue X (2019) Data augmentation in fault diagnosis based on the Wasserstein generative adversarial network with gradient penalty. Neurocomputing. https://doi.org/10.1016/j.neucom.2018.10.109
8. Goodfellow I, Bengio Y, Courville A, Bengio Y (2016) Deep learning. MIT Press, Cambridge
9. Goodfellow I, Pouget-Abadie J, Mirza M, Xu B, Warde-Farley D, Ozair S, Courville A, Bengio Y (2014) Generative adversarial nets. In: Ghahramani Z, Welling M, Cortes C, Lawrence ND, Weinberger KQ (eds) Advances in neural information processing systems, vol 27, pp 2672–2680
10. He K, Zhang X, Ren S, Sun J (2015) Delving deep into rectifiers: surpassing human-level performance on imagenet classification. In: Proceedings of the IEEE international conference on computer vision, pp 1026–1034
11. Ioffe S, Szegedy C (2015) Batch normalization: accelerating deep network training by reducing internal covariate shift. arXiv:1502.03167
12. Kahng M, Thorat N, Chau DHP, Viégas FB, Wattenberg M (2018) GAN-Lab: understanding complex deep generative models using interactive visual experimentation. IEEE Trans Vis Comput Graph 25(1):310–320
13. Kingma DP, Ba J (2014) Adam: a method for stochastic optimization. arXiv:1412.6980
14. Li J, He H, Li L, Chen G (2019) A novel generative model with bounded-GAN for reliability classification of gear safety. IEEE Trans Ind Electron 66(11):8772–8781
15. Mao X, Wang S, Zheng L, Huang Q (2018) Semantic invariant cross-domain image generation with generative adversarial networks. Neurocomputing 293:55–63
16. Mescheder L (2018) On the convergence properties of GAN training. arXiv:1801.04406
17. Mirza M, Osindero S (2014) Conditional generative adversarial nets. arXiv:1411.1784
18. Oh JH, Hong JY, Baek JG (2019) Oversampling method using outlier detectable generative adversarial network. Expert Syst Appl 133:1–8
19. Pratt H, Coenen F, Broadbent DM, Harding SP, Zheng Y (2016) Convolutional neural networks for diabetic retinopathy. Procedia Comput Sci 90

20. Radford A, Metz L, Chintala S (2015) Unsupervised representation learning with deep convolutional generative adversarial networks. CoRR arXiv:abs/1511.06434
21. Ronneberger O, Fischer P, Brox T (2015) U-net: convolutional networks for biomedical image segmentation. In: International conference on medical image computing and computer-assisted intervention. Springer, pp 234–241
22. Salehinejad H, Valaee S, Dowdell T, Colak E, Barfett J (2017) Generalization of deep neural networks for chest pathology classification in X-rays using generative adversarial networks. CoRR arXiv:abs/1712.01636
23. Shen D, Wu G, Suk HI (2017) Deep learning in medical image analysis. Annu Rev Biomed Eng 19:221–248
24. Wang D, Vinson R, Holmes M, Seibel G, Bechar A, Nof S, Tao Y (2019) Early detection of tomato spotted wilt virus by hyperspectral imaging and outlier removal auxiliary classifier generative adversarial nets (OR-AC-GAN). Sci Rep 9(1):14, Article no. 4377. https://doi.org/10.1038/s41598-019-40066-y
25. Wang X, Peng Y, Lu L, Lu Z, Bagheri M, Summers RM (2017) ChestX-ray8: hospital-scale chest X-ray database and benchmarks on weakly-supervised classification and localization of common thorax diseases. In: 2017 IEEE conference on computer vision and pattern recognition (CVPR). IEEE, pp 3462–3471
26. Yang W, Hui C, Chen Z, Xue JH, Liao Q (2019) FV-GAN: finger vein representation using generative adversarial networks. IEEE Trans Inf Forensics Secur 14(9):2512–2524
27. Ying X, Guo H, Ma K, Wu J, Weng Z, Zheng Y (2019) X2CT-GAN: reconstructing ct from biplanar X-rays with generative adversarial networks. In: Proceedings of the IEEE conference on computer vision and pattern recognition, pp 10,619–10,628
28. Yu X, Qu Y, Hong M (2018) Underwater-GAN: underwater image restoration via conditional generative adversarial network. In: International conference on pattern recognition. Springer, pp 66–75
29. Zhang M, Zheng Y (2019) Hair-GAN: recovering 3D hair structure from a single image using generative adversarial networks. Vis Inform. https://doi.org/10.1016/j.visinf.2019.06.001
30. Zhou D, Zheng L, Xu J, He J (2019) Misc-GAN: a multi-scale generative model for graphs. Front Big Data 2:10. https://doi.org/10.3389/fdata.2019.00003

Chapter 12
Automated Glaucoma Type Identification Using Machine Learning or Deep Learning Techniques

Law Kumar Singh, Hitendra Garg and Pooja

1 Introduction

Retina is defined as a layered tissue that forms a lining in the interior part of the eye, responsible for converting incoming light into a neurotic signal which is then used for processing the brain's visual cortex. Thereby it is often referred to as an extension for the brain as in Fig. 1. The facility of imaging the retina and developing a methodology to analyze these images is of great interest. Functionality of the retina requires it to see the outside world. So, for image formation to take place, optical transparency of the ocular structures is required. Accessibility for noninvasively imaging retinal tissues and brain tissues is due to the retina being visible from outside [1].

Architecture and functionality of the retina enable it to manifest diseases affecting the eye, as well as those affecting the brain and the circulation system in it. These diseases mainly comprise macular degeneration and glaucoma which are two of the most important causes of blindness in today's world. Retina is also affected by several systematic diseases whose complications result in diabetic retinopathy from diabetes, hypertensive retinopathy from cardiovascular diseases, and multiple sclerosis. We can say that on the one hand, retina is unprotected from systematic and

L. K. Singh (✉) · Pooja
Computer Science and Engineering, Sharda University, Greater Noida, India
e-mail: lawkumarcs@gmail.com

Pooja
e-mail: poojachoudhary80@gmail.com

L. K. Singh
Computer Science and Engineering, Hindustan College of Science and Technology, Mathura, India

H. Garg
Computer Engineering & Applications, GLA University, Mathura, India
e-mail: hitendra.garg@gmail.com

© Springer Nature Singapore Pte Ltd. 2020
O. P. Verma et al. (eds.), *Advancement of Machine Intelligence in Interactive Medical Image Analysis*, Algorithms for Intelligent Systems,
https://doi.org/10.1007/978-981-15-1100-4_12

Fig. 1 Retinal fundus image

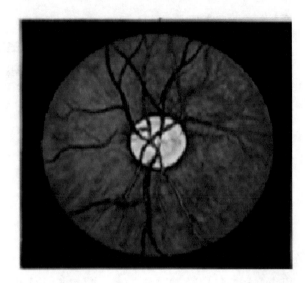

organic specific diseases while, on the other hand, it helps in detecting, diagnosing, and managing complications of hypertension, cardiovascular diseases, and diabetes by enabling imaging of the retina.

1.1 Eye Anatomy

The part of the eye which is visible to us comprises mainly of the cornea which is transparent in nature, the normally white sclera, the normally black pupil, the colored (brown, green, blue, or a mixture of these) iris, and opening in the iris as in Fig. 2. When a ray of light is perceived by the eye, it first passes through the cornea that is responsible for partially focusing the image, after which it passes right through the anterior chamber, the pupil, the lens, thus further focusing the image on the vitreous and finally, it is manifested in the retina. The opaque retinal pigment, called the epithelium, choroid, and sclera, is responsible for supporting the retina. Blood is primarily supplied from choroid to retina, and on the secondary, it is also supplied through retinal vasculature present at the top of the retina [2].

1.2 Retinal Manifestation

Diseases that have origin in the cardiovascular system or the eye tend to manifest themselves in the retina.

Eye imaging and image analysis can be used for studying the following diseases.

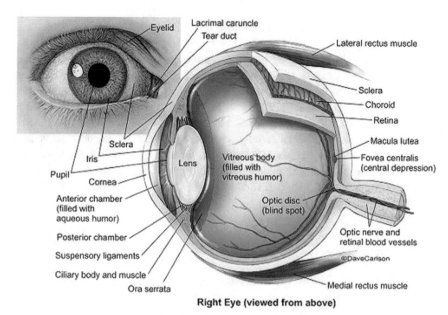

Fig. 2 Eye anatomy

1.2.1 Diabetes

According to the World Health Organization, if a person is diagnosed with fasting plasma glucose more than 7.0 mmol/l then he is said to be suffering from Diabetes mellitus [3]. Actual causes for the disease have not yet been completely understood but it can be said that diabetes can be due to genetic background, sedentary lifestyle and obesity, all leading to an increase in the risk of development of this disease. Change in diet and proper administration of insulin can help in the treatment of the disease. Destruction of small and large blood vessels is caused due to the presence of elevated blood glucose called Hyperglycemia; this further results in damaging the nerve cells, kidney, heart, brain, and eye which is responsible for causing Diabetic Retinopathy which is a further complication of diabetes.

1.2.2 Diabetic Retinopathy

Diabetic Retinopathy, a complex form of diabetes mellitus is the second major factor that leads to progressive loss of eyesight and blindness. There is a rapid increase in patients suffering from diabetic retinopathy. Diabetic retinopathy can be managed by lowering blood sugar through changes in diet, everyday lifestyle, and antidiabetic drugs.

1.2.3   Age-Related Macular Degeneration (AMD)

AMD is responsible for blurring of sharp central vision needed to perform day-to-day activities such as reading, sewing, and driving. It is responsible for affecting the macula that helps us in seeing the fine details. No pain is caused by age-related macular degeneration. AMD is regarded as the cause of 54% of blind people in America. It can be broadly categorized into wet AMD and dry AMD. Dry AMD results in gradual loss of visual activity whereas wet AMD is more destructive as it leads to magnification of maculae's choroid vascular structure which further increases vascular permeability, leading to accumulation of fluid abnormally below or within retinal causing visual impairment. Dietary supplements can be used for slowing the advancement of dry AMD [4].

1.2.4   Glaucoma

Glaucoma is a disorder of the eye that is responsible for damaging the optic nerve that carries all the information from the eye toward the brain. It is the third major factor that leads to blindness in the United States. Optimal treatment and early detection have been proven to reduce the risk for development of glaucoma. Glaucoma is not particularly a part of retinopathy, rather it is apart of neuropathy. Stereo color fundus photography and indirect stereo biomicroscopy can be used for imaging the optic disc two-dimensionally. Cup-to-disc ratio of the retina acts as a structural indicator for detecting the progression and presence of glaucoma in humans [5, 6].

1.2.5   Cardiovascular Disease

These diseases also manifest themselves in the retina in multiple ways. The A/V ratio where A is the diameter of the retinal arteries and V is the diameter of the retinal veins changes as a result of atherosclerosis and hypertension. Risk of stroke and myocardial infarction increases due to the widening of veins and thinning of arteries, which results in the reduction of the A/V ratio. Direct retinal ischemia can also be initiated by hypertension which results in the retinal infarcts to be viewed as cotton wool spots and choroid infarcts to be viewed as deep retinal white spots.

Ironically, the optical characteristics of the eyes that aid information of the image obstructs direct analysis of its retina, i.e., imaging transforms property of the retina to make a focused image make any attempt to acquire a detailed image of the retina from outside with the help of inverse transform.

For centuries, it was believed that when light enters into the eye at a suitable angle, the retina reflects a blurred image which makes the pupil seem red known as a red reflex. There was a need for special techniques to acquire a detailed retinal image [7, 8].

1.3 Types of Glaucoma

Glaucoma can be classified into the following categories.

1.3.1 Open-Angle Glaucoma and Closed-Angle Glaucoma

Open-angle glaucoma is the most common type of Glaucoma and it covers about 90% of all the cases [9]. In this type, there is proper flow of the liquid from outside to inside the drainage canal but a clogging occurs in the flow from inside to outside. It develops over a long period is less painful and also unnoticeable.

Closed-angle glaucoma occurs when the eye's drainage canal gets blocked due to an excessive amount of liquid. The iris becomes less wide and open, in comparison to the normal eye. In this case, the pupil rapidly gets enlarged to a great extent [2]. This type is painful and needs major surgery. It is also called angle-closure glaucoma.

Other types of Glaucoma include the following.

1.3.2 Normal-Tension Glaucoma

Low-tension glaucoma is the other name for this type. In this condition, the IOP is normal but still, the optic nerve gets damaged somehow and therefore loss of visual field occurs.

1.3.3 Congenital Glaucoma

It is common in infants; this is also called children glaucoma. It has two types, i.e., primary congenital glaucoma (occurring due to incomplete or improper development of eye's drainage canal) and secondary congenital glaucoma, resulting from disorders in eye or body, which may or may not be genetic.

1.3.4 Secondary Glaucoma

This type of glaucoma results from some other disease. Pigmentary and neovascular glaucoma are types of secondary glaucoma.

1.3.5 Pigmentary Glaucoma

It occurs due to the presence of pigmented granules present on the backside of the iris, into the aqueous humor. These granules move toward the drainage canals and clog them over a period of time, leading to an increase in intraocular pressure.

1.3.6   Neovascular Glaucoma

It occurs when there is improper formation of blood vessels on the iris and also over the drainage canals. This type mostly has its association with some other diseases, e.g., diabetes.

1.4   Retinal Fundus Images

1.4.1   History

The first method used for image analysis of the retina by Mastsui applied mathematical morphology for vessel segmentation [2]. This approach was implemented on digitized slides of retinal fluorescein angiograms. In the year 1984, Baudoin was the first to propose a retinal image analysis method based on "top-hat" transform that could detect microaneurysms. The introduction of image analysis techniques based on digital filter further gave a boost to research in this sector in the 1990s.

1.5   Current Status of Retinal Imaging

Two major techniques find wide application in medical analysis, namely, Fundus Imaging and Optical Coherence Tomography.

1.5.1   Fundus Imaging

Fundus imaging is the process of representing the 3-D semitransparent tissues of the retina in the two-dimensional imaging plane. The amount of light reflected on the plane represents intensities of the Fundus image.

The following techniques use the method of Fundus imaging:

Fundus Photography

Intensities of an image are formed by light reflected of a specific band.

Color Fundus Photography

Intensities of an image are formed by light reflected of the red, blue and green wavelengths.

Stereo Fundus Photography

Intensities of an image are formed by light reflected from more than one angle of vision. This helps in obtaining z-axis information.

Hyperspectral Imaging

Intensities of an image are formed by light reflected of specific multiple wavelengths.

Scanning Laser Ophthalmoscope

Image intensities are formed by light reflected of laser wavelengths.

Adaptive Optics SLO

Intensities of an image are formed by error-corrected light reflected of laser wavelengths.

Fluorescent Angiography and Indocyanine Angiography

Intensities of an image are formed by light reflected from the fluorescence that is inserted into the bloodstream of the infected patient.

1.5.2   OCT Imaging

Optical coherence tomography works on the principle of estimating the deepness at which a specific kind of backscatter is generated by calculating its time of flight. When transition takes place from one tissue to another, a difference is created in the reflection of the index finger that is responsible for causing backscatters. As it takes more time for light to reach the sensor, backscattering from superficial tissue can be differentiated from those originating from deeper tissues. The interferometer can be used for measuring the differences in time of flight as total thickness of retina lies between 300 and 500 μm. The wavelengths employed for OCT are generally longer than that of visible light. OCT also uses a low coherent light interferometer known as white light interferometer. Illumination of low coherency can be seen as a type of train of overlapped highly correlated "bursts" of light where a unique auto correlogram has been used for labeling each burst. To make our description look intuitive, the term burst is used but we should take note that low coherent light is not pulsed but continuously uniform. The autocorrelation function is used for identifying the burst to calculate delay of light taken after being backscattered from the retina as well as the depth at which backscattering occurred. On optically splitting low coherent light, a mirror is used for reflecting the reference beam kept at a particular distance and the second is made to reflect from the tissues present in the retina; when both flight times are equal to the non zero, cross-interference is said to occur between them, otherwise it is expected to be zero because of the property of low coherency. It is defined as energy or envelope of the nonzero infer gram which is transformed into an intensity representing the amount of backscattering from site of the image.

Wavelengths that can penetrate deep into the retina and choroid tissue are greater than visible light because of which OCT technology relies on infrared light sources such as superluminescent LEDs. A beam splitter is employed for splitting the light beam into two, first a beam that is reflected by the retina representing the sample arm, and the second reflected from the mirror depicting the reference arm. The interferon of energy between these two is converted into image intensities having photosensors called CCD or a CMOS sensor.

Infer gram intensities are used for creating a depth image called a scan by using ultrasound terminology. By employing 3-D imaging, topographic images are created with an A-scan for each in x and y location.

1.6 Deep Convolution Neural Network

Neurons are made up of learnable weights and biases. The function of a neuron-like it takes input then performs a dot product and follows it with a nonlinear activation. Deep learning uses very similar types of neural networks that are convolution neural networks. A single differential point is expressed by the whole network: at the start, a point has raw image pixels and at the endpoint, has a class score. The last layer has also loss function and at the last layer fully connected layer. Convolution Neural Network is a special kind of multilayer neural network. It is the same as an other neural network using backpropagation method as a training method as shown in Fig. 3 [10].

2 Method for Detection of Glaucoma Using Machine Learning Techniques

The method used in machine learning techniques for the detection of glaucoma is shown in Fig. 4.

The following steps are involved in Automatic Detection of Glaucoma using Machine Learning Techniques:

Step 1: Load the dataset of Retinal Fundus Image;
Step 2: Segment the important part of Retinal Fundus Image like Optic Cup and Optic Disc;

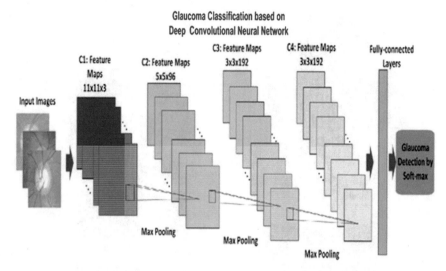

Fig. 3 Deep convolution neural network

Fig. 4 Framework
automatic detection of
glaucoma

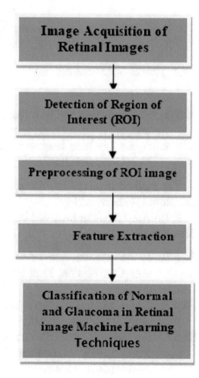

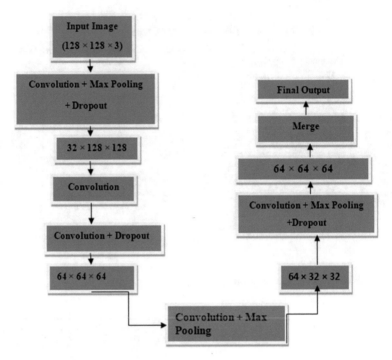

Fig. 5 Flow of the CNN model used

Step 3: Extract import features like statistical or textural features from the Retinal Image;

Step 4: Apply Machine Learning Techniques for classification of Glaucoma and Non-Glaucoma Image.

The method involved in automatic detection of Glaucoma using deep learning techniques is shown in Fig. 5.

Following steps are involved in deep convolution neural network for automatic detection of glaucoma:

Step 1: In the first step, size of an image (128 × 128 × 3) gives input to the system. Convolution, Max Pooling, and Drop Out functions are used by the architecture.

Step 2: Convolution operation is performed on the resultant size of the image (32 × 64 × 64) from Step 1.

Step 3: From Step 2, obtain the size of an image (32 × 64 × 64). At this Step, apply the CONVOLUTION and DROPOUT operation. New matrix is obtained of size (64 × 64 × 64).

Step 4: Again perform CONVOLUTION and DROPOUT operations on the image obtained in STEP 3 and obtain image of size 64 × 32 × 32.

Step 5: At this stage, again perform CONVOLUTION, MAX POOLING, and DROPOUT operation on the image obtained from Step 4 of size (64 × 32 × 32).

Step 6: At this stage, the entire image matrix is to be merged to get the resultant matrix. It is to be passed over the next layer.

2.1 Preprocessing Techniques for Retinal Fundus Images

For the detection of Glaucoma using machine learning techniques, the first step is preprocessing the acquire retinal images. Preprocessing helps in the elimination of unwanted noise from the retinal images as well as in improving the features of an image. It is totally dependent on noise removal, computational time, computational cost, and also on quality of the denoised image.

Blood vessels in the retinal images are in different sizes, shapes, and locations individually. It reflects high level of variation in images that changes reflect the disease in retinal images. So we have to segment the blood vessels and inpaint technique applied to gain the vessel free images as shown in Fig. 6 [11].

In the way of detection of Glaucoma, the main preprocessing task is extracting the region of interest from retinal images. Further, features are extracted from ROI images as shown in Fig. 7.

After blood vessel removal and extraction of the ROI image, retinal image enhancement technique is applied to ROI images. We are able to extract accurate features after enhancing the images [3].

2.1.1 Image Enhancement

Image enhancement is defined as a mechanism of making adjustments in the digital images so that the image becomes more suitable for display and also for making the

(a)   (b)   (c)

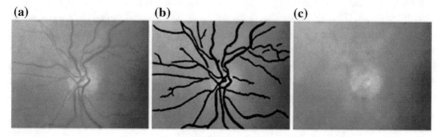

Fig. 6 Result of vessel in painting on **a** original image, **b** image of vessel mask **c** vessel free retinal image

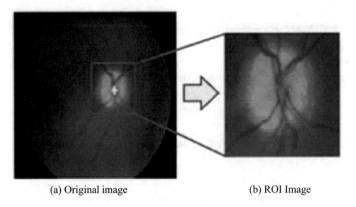

(a) Original image   (b) ROI Image

Fig. 7 **a** Original image **b** region of interest image

image more efficient for performing further analysis [12]. Some of the techniques of image enhancement are mentioned below.

Adaptive Histogram Equalization

Adaptive histogram equalization is employed for improving the contrast levels of retinal images and making it better for further processing. Adaptive histogram equalization method is employed for calculating the total number of obtained histograms where each histogram corresponds to the respective part of an image and it is also used for redistributing the lightness value for the image. It is majorly used for making the definition of edges more prominent corresponding to different regions of an image. In the adaptive histogram equalization method, enhancement of noisy pixels is brought about by drawing a contrast between the background pixel and the required information. The background information represents itself in the form of noisy pixels. Accuracy of histogram equalization is improved by performing the above operation and this also helps in distributing the illumination [13, 14]. Contrasting of the image can be enhanced by the employment of contrast limited adaptive histogram equalization or CLAHE algorithm. The contrasting level of blood vessels is increased by applying this algorithm. Normalization of the gray pixels can be done by stretching the brightness of the image using the CLAHE pixels in the dynamic range which thus helps in eliminating neighboring dark pixel border and image labels. Histogram equalization is for constant enhancement of the images. The main reason for this method is to enhance the contrast between image background and the exudates.

$$h(\upsilon) = \text{round}\left(\frac{\text{cdf}(\upsilon) - \text{cdf}_{\text{min}}}{(M \times N) - \text{cdf}_{\text{min}}} \times (L - 1)\right) \tag{1}$$

where

cdf   minimum nonzero value of cumulative distribution function
$(M \times N)$ Image's number of pixels
L Number of gray levels that are used.

Wiener Filter

Inverse filtering can be defined as a kind of restoration technique. If by using any type of specific filter, blurring of an image takes place, it can be restored to original image by using inverse filtering. This particular method is sensitive to noise. This methodology is used for reducing degradation of an image and thus helps in devising restoration algorithm for different types of degradation methods. Wiener filtering acts as a tradeoff between the noise smoothening method and inverse filtering. This method helps in eliminating blurring effects and noise present in fundus images. Sometimes acquired retinal image has speckle noise. So before extracting the retinal features, removing the speckle noise is necessary. Wiener filter is very helpful for removing speckle noise and regularizing the background [15].

Wiener filter has color image as an input. The channel which has the highest tradeoff is considered as the input image. The contrast value for each channel is calculated in order to select the required color channel. Wiener filter, with the help of low-pass filter, is used for measuring the R, G, and B channels. It can be used for measuring the contrast between low-pass filtering image and the original image.

$$G(u, v) = \frac{H * (u, v) P_{\%}(u, v)}{|H(u, v)|^2 P_{\%}(u, v) + P_{\Omega}(u, v)} \tag{2}$$

where

$H(u, v)$ Fourier transform of point spread function
$P_{\%}(u, v)$ Power spectrum of the signal process
P_n Power spectrum of noise space.

Median Filter

Median filter can be defined as a kind of nonlinear filter. It is mainly employed for eliminating salt and pepper noise. This filter is also used for enhancing the retinal images.

$$y[m, n] = \text{median}\{x[i, j], (i, j) \in w\} \tag{3}$$

where w represents the neighborhood defined by the user, centered around location $[m, n]$ in the retinal image.

Adaptive Median Filters

Adaptive median filter is a type of linear filter in which changing parameters of an image are adjusted by using an optimization algorithm. When compared with standard median filter, it is an advanced method as it performs processing spatially. The adopted methodology is used for classifying the pixel in an image with the help of surrounding neighbor pixels where the size of these pixels is adjustable. Eliminating impulse noise and smoothening other noises is the main purpose of these filters. It helps in reducing excessive thickening and thinning of boundaries as well as all the necessary distortions. While performing filtering operations, these filters tend to reduce the size of filters. The main advantage of these filters is that the impulse response is greater than two; also all the important details are preserved [16].

Gaussian Filter

Fundus images mainly comprise of three bands which are the red band, green band, and blue band. Exudates tend to appear more in the green layer in comparison with the blue and red channels, respectively. In the green layer, optic disc of fundus images also tends to appear brighter. Gaussian smoothening is used for executing the average value of neighboring pixels which is derived on the basis of a Gaussian function. This operation helps in eliminating the effects of illumination and other kinds of noises. It acts as a Gaussian low-pass filter which helps in eliminating the high-frequency component of the image [9].

2.2 Retinal Image Features for Glaucoma Detection

In order to detect glaucoma with the help of images of retina, several important features of the eye such as Cup-to-Disc ratio (CDR), Neuro-Retinal Rim (NRR), Retinal Nerve Fiber Layer (RNFL), Texture feature, statistical features, Haralick texture, wavelet energy, and higher-order spectra are used.

2.2.1 Structural Features

The structure of an object can be referred to as its profile and its physical structure. Structural features are mostly employed for matching and finding the shapes, identifying objects, or calculating measurements of shapes. The structure of an object is mainly determined by the boundary externally present in that object. There are a number of structural changes that occur inside the retina as a result of glaucoma [17].

2.2.2 CDR (Cup-to-Disc Ratio)

CDR is the parameter that can be effectively used for the purpose of detecting glaucoma disease. An increase in intraocular pressure leads to cupping of the optic disc. For normal disc, the CDR should be less than 0.3 but it is higher than 0.3 for glaucoma-affected image. With increase in cup size, the Neuro-retinal Rim is also affected [18].

$$CDR = (Cup\ Area/Disc\ Area) * 2$$

2.2.3 NRR (Neuro-Retinal Rim)

NRR is defined as the region present between the edge of the optic disc and the optic cup. In the condition of glaucoma, ratio of the area occupied by NRR in temporal and nasal regions becomes thicker than that occupied by NRR in superior and inferior regions. Ratio of the area covered in the nasal to the temporal region is thicker than that covered in the inferior to the superior regions. The application of AND operation between the optic cup and optic disc regions provides us the region of Neuro-Retinal Rim.

2.2.4 RNFL (Retinal Nerve Fiber Layer)

Axons of the ganglion cells are responsible for forming the retinal nerve fiber layer. This layer and the retinal layer run parallel to each other. Nerve fibers accumulate at the boundary of the optic disc and then turn at 90 degrees to leave the retina at the optic nerve. It appears similar to a cluster of scratches. In a normal eye, RNFL is commonly seen in the inferior temporal region, followed by regions of the superior temporal, superior nasal, and inferior nasal [19].

2.2.5 Texture Features

Texture may be defined as a repeated occurrence of the same pattern or structure of information at regular intervals. A texture can be understood as superficial properties and characteristics of an object. Such characteristics may include size, density, shape, etc. Texture feature extraction finds application in the fields of remote sensing, medical imaging, and content-based image retrieval. In other words, arrangement of image intensities in space is defined as texture of the image. There may be statistical or structural texture feature. Some of the texture features are listed below:

Structural texture features consist of defining the primitives and consequently, the hierarchy of the arrangement of such primitives in three-dimensional spaces. This allows for easy depiction of the image in the form of symbols. Such features are more suitable for image synthesis as compared to image analysis techniques [1].

2.2.6 Gray Level Cooccurrence Matrix

Gray level Cooccurrence matrix can be used to calculate various statistical texture features. Data about intensity of the image at specific pixel positions can be obtained with the help of GLCM. The following features can be extracted using GLCM [20].

Contrast

Contrast provides a comparison between image intensity of the pixel and its adjacent neighbor. In layman's language, contrast refers to the measure of the difference between color intensity and brightness of objects lying in the same view field.

Homogeneity

It provides a measure of proximity of the pixels to the diagonal of the GLCM. Homogeneity is the measure of similarity between neighboring pixels. Unity Homogeneity is indicative of a diagonal GLCM. Minimal change in the localizing textures suggests larger values of Homogeneity.

$$\text{Homogenity} = \sum_{i=0}^{n-1} \sum_{j=0}^{n-1} \frac{1}{1 + (i - j)^2} P(i, j, d, \theta) \tag{4}$$

Correlation

It provides a measure of the correlation between the pixel and its adjacent neighbor. The values of correlation may vary from -1 (purely negative) to $+1$ (purely positive). Correlation is maximum in the direction of the features, that is, if there exist horizontal textures in an image, the correlation in that direction will be greater than in any other direction [4].

$$\text{Correlation} = \frac{\sum_{i=0}^{n-1} \sum_{j=0}^{n-1} ij P(i, j, d, \theta) - \mu_x \mu_y}{\sigma_x \sigma_y} \tag{5}$$

Entropy

It provides a measure of the degree of dispersion of Gray levels in the GLCM. The word Entropy is taken from the conceptual theory of thermodynamics where it is usually used to characterize the amount of randomness. The value of Entropy is maximum when every element of the GLCM is the same [21].

$$Entropy = \sum_{i=0}^{n-1} \sum_{j=0}^{n-1} P(i, j, \delta, \theta) \log_2 [P(i, j, \delta, \theta)] \tag{6}$$

Energy

It is calculated as the summation of the square of the GLCM. Energy is the measure of the number of times a pixel pair is repeated. A larger value of energy indicates similar pixel values.

$$Energy = \sum_{i=0}^{n-1} \sum_{j=0}^{n-1} [(P(i, j, \delta, \theta))^2] \tag{7}$$

where

$P(I, j, \theta, \delta)$	elements of GLCM matrix
n	Number of pixels
μ	mean
Σ	standard deviation.

3   Discussion of Machine Learning Techniques and the Results

3.1   Dataset

Some datasets of retinal images are publicly available for research purposes. Machine learning approaches are applied to such datasets and performance of machine learning techniques evaluated. The STARE (STRUCTURED ANALYSIS OF THE RETINA) database comprises of 400 raw images which suggest some diseases of the retina such as exudates and hemorrhages that occur in the ONH. Images for DRIVE (DIGITAL RETINA IMAGES FOR VESSEL EXTRACTION) database were collected from a screening program on Diabetic retinopathy in the Netherlands. This database provides 40 images as training data and 20 images as data testing. HRF (HIGH-RESOLUTION FUNDUS IMAGE DATABASE) has 15 healthy patient images, 15 images of patients under the effect of diabetic retinopathy, and 15 of glaucoma patients [5].

3.2 Glaucoma Detection Using K-mean Clustering

K-mean clustering is a type of unsupervised machine learning algorithm. Here, we have data without any well-defined category or group to be more precise, it refers to unlabeled data. The main aim of this algorithm is to detect groups in the collected data where the variable K refers to the total number of groups present in that data. The algorithm tends to perform iteratively where, on the basis of the feature that is provided, the algorithm assigns each data point to one of the K group. So, the detected data points are clustered based on similar characteristics found between the features. CDR is a very important technique for analyzing the risk of glaucoma from fundus images. In this segmentation, OD and OC are performed, further to which CDR is measured. With the help of this methodology, author obtained 92% accuracy for classifying glaucoma [16]. An average square error of 0.02 is achieved for CDR. The database consisted of 73 normal and 27 glaucomatous fundus images with mentioned CDR values. The images were gathered from Armed Forces Institute of Ophthalmology (AFIO), Rawalpindi, Pakistan.

$$J_{km} = \sum_{i=1}^{N} \min_{j=1}^{k} \left\| x_i - m_j \right\|^2 \tag{8}$$

3.3 Detection of Glaucoma Using SVM

SVM or Support Vector Machine is used for mapping multidimensional space into a form of high-dimensional feature space. In this particular feature space, the classifier tries to search for the hyperplane employed for separating the glaucomatous from healthy eyes that maximize the distance of any of the above cases from the hyperplane as in Fig. 8. When the input space is transformed into a feature space, it is called

Fig. 8 SVM input space and feature space

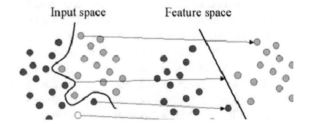

Input space   Feature space

a kernel, mostly linear kernels are used. It is a technique of Supervised Learning; hence, it can be used for both regression and classification.

SVM usually provides imperfect classification between the healthy or normal and glaucomatous eyes. SVM gives correct classification for data where the testing data is not identical but near to the training data. Author worked on a dataset consisting of 47 glaucomatous eyes and 42 healthy eye images. The best results were under SVM where it obtained a sensitivity of 97.9% and 92.5% at a specificity of 80% and 95%, respectively.

Support vector machine is a binary classifier but is also used in multiclass related problems. Classification of glaucoma using support vector machine was performed on 100 fundus images [8]. Consists of 50 normal images and 50 images affected with glaucoma. All the basic steps, preprocessing, feature extraction were performed and SVM was applied for classification. Author found accuracy rate 96%, sensitivity and specificity approximately 100 and 92%.

3.4 Glaucoma Detection Using the Naïve Bayes Technique

The Naive Bayes Classifier is said to be a very efficient and easy to implement classifier which is based on the Bayes theorem. It a simple and easy model designed in order to assign labels from a set having finite values to a vector having feature values. The main benefit of this classification method is that it needs fewer amounts of data to be used for training. Due to this benefit, the required parameters can be easily estimated for the classification procedure. This classification technique categorizes the data into two namely, the training phase and the prediction phase. The training phase makes use of training data in order to estimate the parameter for probability distribution whereas the prediction phase makes use of unseen test data. This method is used for computing posterior probability of the sample belonging to each class. The largest posterior probability is thus used for classifying the data efficiently. After features belonging to the training set are given to the classifier, probabilities for each individual feature as well as probabilities for each class are calculated [6].

$$P(C_i/X) = \frac{P(C_i)P(X/C_i)}{\sum_{j=1}^{k} P(C_j)P(X/C_j)} \tag{9}$$

where

C_i represents the class as glaucomatous or normal
X is used for representing the features obtained from retinal input image
$P(x|C_i)$ represents the class conditional probability
$P(C_i)$ represents the posterior probability
$P(C_i|x)$ represents the maximum posterior hypothesis

A new approach based on this method was used for glaucoma detection where the Histogram oriented features or HOG features were extracted from the region of interest of given fundus image. The Naive Bayes classifier was used for showing the performance of the extracted HOG features against features of the transform domain. The proposed methodology was tested on a publicly available database and was found to give an accuracy of 94% [6].

3.5 Glaucoma Detection Using Artificial Neural Network

An artificial neural network is defined as a processing system based on information that has the same working performance as that of the biological neural network. A neural net is said to be a combination of various simple and easy processing elements known as neurons or nodes. Directed communication links having some weight assigned to them is used for connecting one neuron to another. The weight assigned to each node corresponds to the information needed for solving a problem by the net. Each neuron is associated with an internal state which is known as the activation state or activity level, defined as the functions of inputs it has received. A neuron is used for sending the activation signal to various other neurons. A neural network can be used to perform a specific task by training it. The training of neural network is performed by adjusting the weights associated with each connected node. For a particular input to reach a particular output, the neural network has to undergo various adjustments and training.

The methodology of artificial neural network makes the process of glaucoma detection more accurate and adaptive. The neural network is trained in order to recognize the parameters needed for detection of glaucoma at different stages of its progression. The approach uses two hidden nodes in the hidden layer for ANN-based classification. It achieves accuracy, specificity, and sensitivity as 86.7, 88.9, and 84.4% [22].

3.6 Automatic Detection of Glaucoma Using Deep Learning Techniques

The Deep **Convolution** Neural Network architecture was applied to publicly available dataset from DRISHTI-GS as shown in Fig. 9. It was made up of four convolution layers, six learned layers, and two fully connected layers. For improving the performance, dropout and data augmentation techniques are used in the architecture. For automated detection of glaucoma, this technique focuses on Deep Learning (DL) architecture with a convolution neural network. These DL systems provide a hierarchical representation of images to differentiate between glaucomatous and non-glaucomatous patterns. The proposed architecture consists of four convolution layers, six learned layers, and two fully connected layers. To further improve performance, techniques like dropout and data augmentation strategies are adopted.

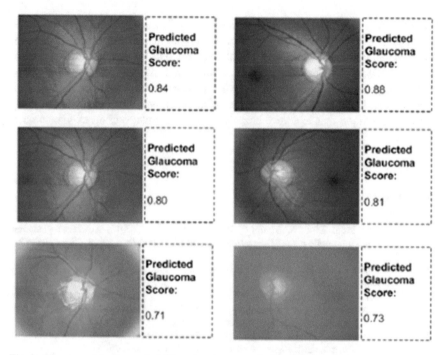

Fig. 9 Glaucoma detection probability using deep convolution neural network

Fig. 10 Histogram showing
the accuracy of the different
architecture of deep learning

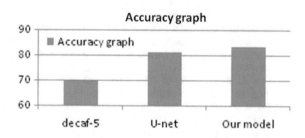

ORIGA and SCES datasets were used for performing the experiments. The AUC values provided by the adopted methodology on ORIGA and SCES are 0.831 and 0.887, respectively [10]. Accuracy can also be analyzed with the help of graph as shown in Fig. 10.

4 Conclusion

Glaucoma is a perilous eye disease if it is not diagnosed at the right time. It poses the possibility of vision loss. Symptom of glaucoma is increasing intraocular pressure in the eye. Different types of image preprocessing methods are introduced that are important for reducing noise from acquired images. The image processing technique is also helpful for segmenting the ROI from the retinal fundus image. Identification of different features that are present in retinal images is also discussed. Automatic detection of glaucoma using machine learning model uses extracted features and classifies healthy or glaucoma affected images. Extraction technique is used for classification of healthy and glaucoma affected images using machine learning techniques. Deep learning technique that is a very effective tool used for detection of glaucoma is elaborated.

References

1. Claro M, Santos L, Silva W, Araújo F, Moura N, Macedo A (2016) Automatic glaucoma detection based on optic disc segmentation and texture feature ex traction. CLEI Electron J 19(2):5
2. Abràmoff MD, Garvin MK, Sonka M (2010) Retinal imaging and image analysis. IEEE Rev Biomed Eng 3:169–208
3. Xu Y, Duan L, Lin S, Chen X, Wong DWK, Wong TY, Liu J (2014) Optic cup segmentation for glaucoma detection using low-rank superpixel representation. In: International conference on medical image computing and computer-assisted intervention. Springer, Cham, pp 788–795
4. Salazar-Gonzalez A, Kaba D, Li Y, Liu X (2014) Segmentation of the blood vessels and optic disk in retinal images. IEEE J Biomed Health Inform 18(6):1874–1886
5. Burgansky-Eliash Z, Wollstein G, Chu T, Ramsey JD, Glymour C, Noecker RJ, Schuman JS (2005) Optical coherence tomography machine learning classifiers for glaucoma detection: a preliminary study. Invest Ophthalmol Vis Sci 46(11):4147–4152
6. Nirmala K, Venkateswaran N, Kumar CV (2017) HoG based Naive Bayes classifier for glaucoma detection. In: TENCON 2017–2017 IEEE region 10 conference. IEEE, pp 2331–2336
7. Zilly J, Buhmann JM, Mahapatra D (2017) Glaucoma detection using entropy sampling and ensemble learning for automatic optic cup and disc segmentation. Comput Med Imaging Graph 55:28–41
8. Dey A, Bandyopadhyay SK (2016) Automated glaucoma detection using support vector machine classification method. J Adv Med Med Res 1–12
9. Balasubramanian M, Bowd C, Weinreb RN, Vizzeri G, Alencar LM, Sample PA, Zangwill LM (2010) Clinical evaluation of the proper orthogonal decomposition framework for detecting glaucomatous changes in human subjects. Invest Ophthalmol Vis Sci 51(1):264–271
10. Chen X, Xu Y, Yan S, Wong DWK, Wong TY, Liu J (2015) Automatic feature learning for glaucoma detection based on deep learning. In: International conference on medical image computing and computer-assisted intervention. Springer, Cham, pp 669–677
11. Youssif AAHAR, Ghalwash AZ, Ghoneim AASAR (2008) Optic disc detection from normalized digital fundus images by means of a vessels' direction matched filter. IEEE Trans Med Imaging 27(1)
12. Yin F, Liu J, Ong SH, Sun Y, Wong DW, Tan NM, Wong TY (2011) Model-based optic nerve head segmentation on retinal fundus images. In: 2011 annual international conference of the IEEE engineering in medicine and biology society. IEEE, pp 2626–2629
13. Yousefi S, Goldbaum MH, Balasubramanian M, Jung TP, Weinreb RN, Medeiros FA, Bowd C (2013) Glaucoma progression detection using structural retinal nerve fiber layer measurements and functional visual field points. IEEE Trans Biomed Eng 61(4):1143–1154
14. Murthi A, Madheswaran M (2012) Enhancement of optic cup to disc ratio detection in glaucoma diagnosis. In: 2012 international conference on computer communication and informatics. IEEE, pp 1–5

15. Tan MH, Sun Y, Ong SH, Liu J, Baskaran M, Aung T, Wong TY (2013) Automatic notch detection in retinal images. In: 2013 IEEE 10th international symposium on biomedical imaging: from nano to macro, San Francisco, CA, USA, 7–11 April 2013
16. Ayub J, Ahmad J, Muhammad J, Aziz L, Ayub S, Akram U, Basit I (2016) Glaucoma detection through optic disc and cup segmentation using K-mean clustering. In: 2016 international conference on computing, electronic and electrical engineering (ICE Cube). IEEE, pp 143–147
17. Bock R, Meier J, Michelson G, Nyúl LG, Hornegger J (2007) Classifying glaucoma with image-based features from fundus photographs. In: Joint pattern recognition symposium. Springer, Berlin, Heidelberg, pp 355–364
18. Kavitha S, Karthikeyan S, Duraiswamy K (2010) Early detection of glaucoma in retinal images using cup to disc ratio. In: 2010 second international conference on computing, communication and networking technologies. IEEE, pp 1–5
19. Tripathi S, Singh KK, Singh BK, Mehrotra A (2013) Automatic detection of exudates in retinal fundus images using differential morphological profile. Int J Eng Technol 5(3):2024–2029
20. Gayathri K, Narmadha D, Thilagavathi K, Pavithra K, Pradeepa M (2014) Detection of dark lesions from coloured retinal image using curvelet transform and morphological operation. IJNTEC 2347–7334
21. Maheswari MS, Punnolil A (2014) A novel approach for retinal lesion detection in diabetic retinopathy images. People 4:6
22. Parfitt CM, Mikelberg FS, Swindale NV (1995) The detection of glaucoma using an artificial neural network. In: Proceedings of 17th international conference of the engineering in medicine and biology society, vol 1. IEEE, pp 847–848

Chapter 13
Glaucoma Detection from Retinal Fundus Images Using RNFL Texture Analysis

Anirban Mitra, Somasis Roy, Ritushree Purkait, Sukanya Konar, Avisa Majumder, Moumita Chatterjee, Sudipta Roy and Sanjit Kr. Setua

1 Introduction

Blindness is a state of being sightless. Some people have some limited ability to see with the help of some special aids or may be able to recognize a slight of light. There are many causes which adverse blindness, one of them is Glaucoma. Glaucoma is a common cause of irreversible blindness globally. The utterance glaucoma originates

A. Mitra (✉) · R. Purkait · S. Konar · A. Majumder
Department of Computer Science and Engineering, Academy of Technology, Adisaptagram
712121, West Bengal, India
e-mail: anirban.mitra.cse@gmail.com

R. Purkait
e-mail: ritushreepurkait@gmail.com

S. Konar
e-mail: sukanya.konar2015@gmail.com

A. Majumder
e-mail: avisa.majumder@gmail.com

A. Mitra · S. Roy · M. Chatterjee · S. Kr. Setua
Department of Computer Science and Engineering, Calcutta University Technology Campus,
JD-2, Sector-III, Salt Lake, Kolkata 700098, India
e-mail: somasis.roy@gmail.com

M. Chatterjee
e-mail: moumitachatterji@gmail.com

S. Kr. Setua
e-mail: sksetua@gmail.com

S. Roy
PRT2L Washington University in St. Louis, 510 South Kingshighway Boulevard, St. Louis, MO
63110-1076, USA
e-mail: sudiptaroy01@yahoo.com

© Springer Nature Singapore Pte Ltd. 2020
O. P. Verma et al. (eds.), *Advancement of Machine Intelligence in Interactive
Medical Image Analysis*, Algorithms for Intelligent Systems,
https://doi.org/10.1007/978-981-15-1100-4_13

from antiquated Greek word, which signifies "clouded or blue-green hue," which means the shade of the eye that isn't dark [1]. Glaucoma is portrayed by loss of retinal ganglion cells, neural rim tissue, and augmentation of the optic cup and peri-papillary Retinal Nerve Fiber Layer (RNFL) loss clinically [2].

A large portion of this mischief is because of colossal strain inside the consideration. The ciliary body of the eye secretes a vitreous liquid called aqueous humor into a space among the iris and focal point called back chamber and depletes it by means of trabecular work network. This discharge and seepage of the liquid is adjusted in the eyes. Glaucoma predominantly emerges when the drainage canal is blocked and there is an expansion in strain which is known as the intraocular pressure harming the optic nerve layer, transmitting visual data from retina to cerebrum. On the off chance that this harm left untreated, it prompts total visual impairment. So the early location of glaucoma is significant and important [3].

There are different kinds of glaucoma, including open-angle glaucoma and angle-closure glaucoma. Open-angle glaucoma is the most well-known kind of glaucoma and it is incessant sort (long span) where the drainage path becomes gradually clogged even though the drainage structure called trabecular meshwork looks normal. As a result the aqueous humor does not drain from the eye properly. Angle-closure glaucoma is less normal and it might be intense (happens all of a sudden) or constant sort (long term). In this kind of glaucoma the space among iris and cornea turns out to be excessively thin with the goal that the liquid isn't depleted appropriately which expands the intraocular weight in the eye [4].

Glaucoma is one of the most commonly known causes behind permanent vision loss, especially if an occurrence of persons over the age of 75 years should occur. In the vast majority of the cases glaucoma have no early manifestations. An individual needs ordinary eye checkup with the goal that it very well may be recognized and analyzed. A past filled with glaucoma or medical issues like diabetes, is inclined to Glaucoma.

The retina is a photosensitive layer of the optic nerve tissue lining in the internal surface of the eyeball. Retinal harms because of different infections can, in the long run, lead to irreversible vision misfortune. As populace maturing has risen as a noteworthy statistic pattern around the world, patients experiencing chorioretinal infections like Age-related Macular Degeneration (AMD) are required to increment later on. Because of glaucoma disease, one of the most affected retinal structures is the retinal nerve fiber layer (RNFL). One of the indications of glaucoma is the vibrant decay of the layer of a retinal nerve fiber (RNFL) due to a loss in cell density. Nerve cell degeneration starts several years before any changes can be registered in the patient's vision. The following figure (Fig. 1) depicts the nature of RNFL in the eye.

Grievously, over the top changes in the RNFL can't be revived by the present remedy. Simply, snappy treatment can stop the development of the contamination. Thus, it is incredibly charming to recognize the disease as fast as time grants. As a general rule, Optical Clarity Tomography (OCT) can study the RNFL thickness. An examination using OCT, however, is truly costly and not yet accessible for the most

Fig. 1 Example of typical
fundus image with
distinctive RNFL loss

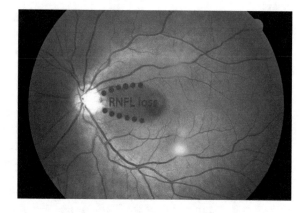

portion. The fundus camera is, of course, deemed a standard and vital crucial device
used at numerous ophthalmic workplaces around the globe.

2   Related Works

Previously various analyzes relied upon fundus pictures as a result of the mod-
erateness of fundus pictures over Optical Coherence Tomography (OCT). Various
noteworthy sorts of research have sought after a novel method to manage surface
examination engaging estimation of the RNFL thickness in for the most part used
shading fundus photographs [5].

Hoyt et al. first showed an undertaking to use fundus cameras to reveal glaucoma
by evaluating the appearance of RNFL. The makers deliberately revealed that the
RNFL configuration's funduscopic signs provide the RNFL loss in the retina with
its most timely target evidence. Accordingly, various inventer attempted this pas-
sionate analysis of fundus photographs. Airaksinen et al. supposedly researched the
RNFL scheme and scored a numerical scale of glaucomatous damage. Peli et al.
used digitized overly differentiating fundus images to perform one of the RNFL
surface's essential self-loader tests. Yogesan et al. conducted a groundwork examination of digitized fundus photographs using surface examination techniques based
on reduced measurement run-length schemes. In addition, energy data on the texture
of the RNFL was further used by Dardjat et al., Lee et al. The RNFL appears to be
darker in fundus images due to glaucomatous hurt. Oliva et al. showed a navigator
focus to look through the lessening RNFL in mechanized shading fundus images. The
paper demonstrates the surface examination self-loader approach based on the RNFL
configuration control assessment. Hayashi et al. used a strategy with the channels of
Gabor to update certain regions with the RNFL framework and to collect these areas
toward the glaucoma region. The article showed significant outcomes, followed by a
social comparison case. The manufacturers expanded the previous plan to examine
and used a larger dataset to perform an assessment. Odstrcilik et al. (2010) made an

altered strategy to texture examination of RNFL subject to Markov self-self-assured fields utilizing the dataset of DCFI with size of 3504 × 2336 pixels (18 normals, 10 glaucomatous) [6]. The standard element of this progression was the highlights capacity to confine among sound and glaucomatous cases is embraced utilizing OCT RNFL thickness measurement.

Muramatsu et al. (2010) arranged a modified system with Gabor stations to upgrade the certain area with RNFL model and grouping of these domains planned to glaucoma revelation. He utilized the dataset of DCFI with a size of 768 × 768 pixels (81 normals, 81 glaucomatous). His examination accomplished the framework appropriate just for the affirmation of central and continuously expansive RNFL incidents imparted by gigantic changes in intensity [7].

Tuulonen et al. (2000) had the option to make a self-loader technique utilizing a microtexture examination of the RNFL design. His structure was capable in demonstrating that adjustments in a miniaturized scale surface of RNFL design are identified with glaucoma harm however there is an absence of little example size [8].

Kolář and Jan (2008) made a programmed strategy to the surface examination of RNFL dependent on fractal measurements. Here, the nearby fractal coefficient was utilized as a component for glaucomatous eye location. Yet, there were issues with the vigorous estimation of this coefficient [9].

Jan et al. (2012) made a programmed strategy to RNFL surface investigation dependent on a mix of power, edge portrayal, and Fourier phantom examination. The capacity of proposed highlights was to arrange RNFL abandons that have been demonstrated by means of correlation with OCT. The examination was done uniquely in a heuristic manner [10].

However, in spite of those profitable examines, the location of glaucoma was not sufficiently acceptable because of the immediate reliance on the fundus picture powers. Glaucoma could be identified appropriately just if the RNFL decay was in a hopeless state having a tremendous visual deficiency as of now. This isn't adequate for early recognition of Glaucoma and anticipation from vision misfortune at a beginning time.

In this way, in our paper, we have attempted to discover the answer for early identification of glaucoma from Retinal Fundus Images utilizing RFNL Texture Analysis. Here, we consider the patients' Retinal Fundus Images and with the assistance of mostly Haralick highlights of surface examination including the run-length feature [11], we would almost certainly distinguish the nearness of glaucoma and even the inclination toward glaucoma in the soonest of stages.

Haralick's highlights dependent on the dim dimension co-event lattice (GLCM) are connected to catch textural designs in the fundus. The target of this work is the determination of the most segregating and discovering the critical surface highlights that can decide the nearness of glaucoma in the eye-picture.

Haralick highlights are measurable highlights that are registered over the whole picture. These estimations are used to depict the general surface of the picture utilizing measures, for example, entropy and whole of fluctuation. Chaddad et al. propose a methodology, in view of Haralick's highlights, to distinguish and group colon disease cells [6]. Here we utilize this essential idea to achieve our point of the undertaking.

3   Proposed Solution

This paper displays another programmed glaucoma discovery framework dependent on Haralick texture features extricated from the retinal fundus image. The proposed framework is cultivated in four phases: picture preprocessing (Extraction of the area of intrigue (ROI) based on OD) [12, 13], include extraction, retinal picture division, and arrangement. The factual examination is utilized to get the best highlights for the arrangement to separate among solid and glaucomatous eyes of the patients. The accompanying areas will portray in detail these stages. All picture examination was accomplished with no learning of patient clinical qualities or status.

The examination of the retina by means of an ophthalmoscope or fundus cameras (vital or advanced) has been used adequately to examine various retinal and eye diseases [1]. The retinal nerve fiber layer (RNFL) can be seen in the same way as the optic plate, macula, and retinal vascular tree, particularly in a non-red light as suggested by Kulwant [2]. This layer creates a stripe-like reference point on the surface, showing the proximity of nerve strands. In fundus images, there was a tendency to separate this layer, which could enhance the finding of glaucoma. Fundamental surface examination for incredible affirmation of RNFL deserts was depicted and pursued by Yogesan et al. [14], the start of 10 low-target digitized fundus pictures. In Tuulonen et al. [8], RNFL's microtexture examination was portrayed in digitized pictures in dull estimation was discussed. The adjacent surface characteristics based on magnificence ability were identified and used as a depiction commitment between glaucoma and normal and visual hypertension. Quantifiable-based processes are an essential tool for surface delineation and are also a promising tool for surface examination by RNFL. These frameworks have three basic classes: techniques subject to estimates of first concern, second sales pieces of data, and encounters with higher demands. The author related a first demand metric here, which relies only on the perspective of the pixel and not on the participation of the pixel. The main clarification behind these immediate observations is that the description of these parameters is directly forward and provides the ground characteristics and their graphic presence a fundamental point of perspective. This assessment consolidates five parameters (features) of Shannon (as described in the information hypothesis): average, standard variation, kurtosis, skewness, and entropy. They are settled from an allocation of energy probability that must be assessed on the basis of the histogram of the image's distinct region. It is possible to discover the definition and representation of these parameters elsewhere.

The reason for Haralick feature is the Gray-level Co-occurrence Matrix, G as the given figure (Fig. 2). This matrix is square with measurement N_g, where N_g is the number of gray level in the picture. Component $[i, j]$ of the grid is produced by checking the occasions a pixel with esteem I is adjoining a pixel with esteem j and after that partitioning the whole framework by the all out number of such examinations made. Every passage is in this manner viewed as the likelihood that a pixel with esteem I will be discovered contiguous a pixel of significant worth j.

$$G = \begin{bmatrix} p(1,1) & p(1,2) \dots \dots & p(1,N_g) \\ p(2,1) & p(2,2) \dots \dots & p(2,N_g) \\ \vdots & \vdots & \vdots \\ \vdots & \vdots & \vdots \\ p(N_g,1) & p(N_g,2) & p(N_g,N_g) \end{bmatrix}$$

Fig. 2 Gray-level co-occurrence matrix G

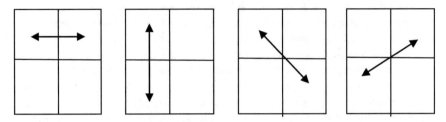

Fig. 3 Four directions of adjacency as defined for calculation of the Haralick texture features. The Haralick statistics are calculated for co-occurrence matrices generated using each of these directions of adjacency

Since adjacency in a 2D, square pixel image (even vertical, left and right diagonals) can be portrayed to occur in each of four directions, four such matrices can be resolved as depicted in the following figure (Fig. 3).

Image texture is a measurement of the spatial variation of gray tone values. Haralick et al. [15] recommended the utilization of gray-level co-occurrence (GLCM). This technique depends on the joint probability distributions of pairs of pixels. GLCM demonstrates how often each gray level occurs as an element of the gray level at a pixel at a specified geometric place [16]. A basic part is meaning of eight nearest-neighbor resolution that characterize various frameworks for various points (0°, 45°, 90°, 135°) (Fig. 4) and separates between the horizontal neighboring pixels.

Haralick then depicted some statistics which can be determined from the matrix of the co-occurrence in order to represent the image's texture. The features will be discussed in the following sections:

Angular Second Moment:

Fig. 4 X3 window definition and spatial relationship for calculating Haralick texture measures

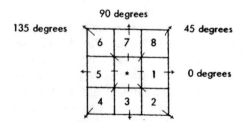

$$\sum_i \sum_j p(i, j)^2 \tag{1}$$

Contrast:

$$\sum_{n=0}^{N_g-1} n^2 \left\{ \sum_{i=1}^{N_g} \sum_{j=1}^{N_g} p(i, j) \right\}, |i - j| = n \tag{2}$$

Correlation:

$$\frac{\sum_i \sum_j (i, j) p(i, j)^2 - \mu_x \mu_y}{\sigma_x \sigma_y} \tag{3}$$

where μ_x, μ_y, σ_x and σ_y are the means and standard deviations of p_x and p_y, the partial probability density functions

Sum of Squares (Variance):

$$\sum_i \sum_j (i - \mu)^2 p(i, j) \tag{4}$$

Inverse Difference Moment (Homogeneity):

$$\sum_i \sum_j \frac{1}{1 + (i - j)^2} p(i, j) \tag{5}$$

Sum Average:

$$\sum_{i=2}^{2N_g} i p_{x+y}(i, j) \tag{6}$$

where x and y are the coordinates (row and column) of an entry in the co-occurrence matrix, and $p_{x+y}(i)$ is the probability of co-occurrence matrix coordinates summing to $x + y$

Sum Variance:

$$\sum_{i=2}^{2N_g} (i - fs)^2 p_{x+y}(i) \tag{7}$$

Sum Entropy:

$$-\sum_{i=2}^{2N_g} p_{x+y}(i) \log\{p_{x+y}(i)\} = fs \tag{8}$$

Entropy:

$$-\sum_i \sum_j p(i, j)\log(p(i, j)) \tag{9}$$

Difference Variance:

$$\sum_{i=0}^{N_g-1} i^2 p_{x-y}(i) \tag{10}$$

Difference Entropy:

$$-\sum_{i=0}^{N_g-1} p_{x-y}(i)\log\{p_{x-y}(i)\} \tag{11}$$

With the observation that, in a coarse texture, relatively long gray level runs would occur more often and that a fine texture should contain primarily short runs, Galloway proposed the use of a run-length matrix for texture feature extraction [17]. For a given image, a run-length matrix $p(i; j)$ is defined as the number of runs with pixels of gray level i and run-length j. Various texture features can then be derived from this run-length matrix. In a more recent study, Dasarathy and Holder [18] described another four feature extraction functions following the idea of joint statistical measure of gray level and run-length, as follows:

Long Run Low Gray-Level Emphasis (LRLGE):

$$LRLGE = \frac{1}{n_r} \sum_{i=1}^{M} \sum_{j=1}^{N} \frac{p(i, j) \cdot j^2}{i^2} \tag{12}$$

Long Run High Gray-Level Emphasis (LRHGE)

$$LRHGE = \frac{1}{n_r} \sum_{i=1}^{M} \sum_{j=1}^{N} p(i, j) \cdot i^2 \cdot j^2 \tag{13}$$

In the proposed methodology LRLGE and LRHGE will be added with some Haralick feature list to fetch the characteristic information of texture for doing the differentiation in between Loss-RNFL and Healthy-RNFL.

3.1   RNFL Analysis

Contrast, Correlation, Homogeneity, Energy, Entropy, Angular Second Moment feature, Long Run Low Gray-Level Emphasis (LRLGE), Long Run High Gray-Level Emphasis (LRHGE), Coarseness, and Busyness are the very important features of a texture to describe its heterogeneous nature. In Fig. 3 we have displayed some sample images from the RNFL-Loss and RNFL-Healthy image set individually. The two very important processes are applied in the Peri-Papillary-Atrophy analysis. To describe the heterogeneous property of a texture, a huge number of features are available. So at first, the Supervised Statistical Linear Discriminate Analysis (LDA) is applied for dimensionality reduction of the feature list. Second, the classification of the Loss-RNFL and the Healthy-RNFL is done using the Support Vector Machine based on the Radial basis function kernel.

Algorithm: Healthy RNFL and Loss RNFL Detection

Input: A RNFL Box image
Output: Healthy RNFL or RNFL Loss box Classification

Step 1 : RNFL Box images are selected from RNFL-Loss and RNFL-Healthy dataset respectively.

Step 2: Angular Second Moment, Contrast, Correlation, Variance, Homogeneity, Sum Average, Sum Variance, Sum Entropy, Entropy, Difference Variance, Difference Entropy, Long Run Low Grey-Level Emphasis (LRLGE), Long Run High Grey-Level Emphasis (LRHGE) features are realized for individual RNFL Box images with proper tagging.

Step 3: Linear Discernment Analysis (LDA) is applied to the textural feature list to reduce the Dimensionality.

Step 4: Support Vector Machine (SVM) based on the Radial Basis Function Kernel on the reduced Feature list is applied to do the classification of Loss-RNFL section and Healthy-RNFL section.

3.2   Feature Extraction

Extraction of features is the way to verify the higher percentage of image information, such as texture, shape, and shading. Analysis of texture is the main region of visual observation for humans. Exact texture measurable examines the spatial growth of the dark aspect, enlisting neighborhood highlights at each stage in the picture and surmising many estimates from the adjacent highlights allocations. In 1973, Haralick et al. demonstrated Gray-Level Co-occurrence Matrix (GLCM) and surface highlights [19]. This system has been commonly used in apps for image examination, especially in the field of biomedicine. It involves two extraction highlights phases. GLCM exhibits how routinely each decrease estimation happens at a pixel organized at a fixed geometric position with respect to one another pixel, as a portion of the lesser estimation. In the following (Fig. 5) the texture of RNFL Healthy Box and the texture of RNFL Loss Box is depicted as follows.

Haralick Features:

Differentiation or standard deviation between the target pixel and its neighbor is an amount of strength or a dull assessment assortment. Huge principle of complexity represents enormous differences of contrast in GLCM:

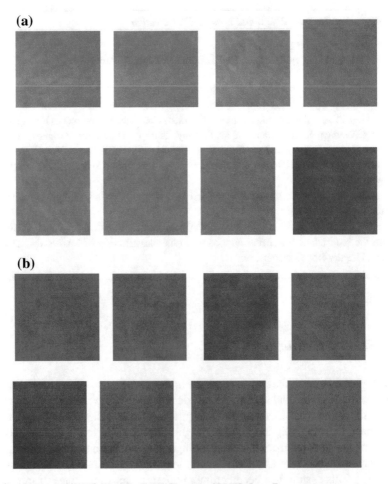

Fig. 5 **a** Texture of RNFL-Healthy Box **b** Texture of RNFL-Loss Box

Contrast:

$$\sum_{n=0}^{N_g-1} n^2 \left\{ \sum_{i=1}^{N_g} \sum_{j=1}^{N_g} p(i, j) \right\}, |i - j| = n \tag{14}$$

Homogeneity assesses how near the GLCM segment dissemination is to the GLCM border. The contrast generally reduces as homogeneity extends:

Inverse Difference Moment (Homogeneity):

$$\sum_i \sum_j \frac{1}{1 + (i - j)^2} p(i, j) \tag{15}$$

Entropy is the irregularity or level of confusion present in the image. The estimation of entropy is the greatest when all parts of the co-occurrence matrix are identical and little when components are unequal:

Entropy:

$$-\sum_i \sum_j p(i, j) \log(p(i, j)) \tag{16}$$

The ASM, mean, and entropy are most basic difference statistic texture descriptions as following:

Angular Second Moment:

$$\sum_i \sum_j p(i, j)^2 \tag{17}$$

Energy derives from the second angular moment (ASM). The ASM measures the continuity of the dark measurements in the neighborhood. The ASM appreciation will be huge at the stage where pixels are basically the same as they are.

$$Energy = ASM \tag{18}$$

When the $P_{x-y}(k)$ values are very similar or close, ASM is small. ASM is large when certain values are high and others are low:

When the values of $P_{x-y}(k)$ are focused near the origin, the mean is small and the mean is large when far from origin:

Entropy:

$$-\sum_i \sum_j p(i, j) \log(p(i, j)) \tag{19}$$

Entropy is smallest when $P_{x-y}(k)$ values are unequal and largest when values are equal.

The computation of the Haralick texture features utilizing the past conditions for the fundus images volume sequences for each eye independently was performed. For every member, the dark dimension co-event surface highlights: differentiate, homogeneity, entropy, vitality, connection, and $m1$, $m2$, $m3$, and $m4$ joined by the distinctive measurable highlights: ASM, difference, mean, and entropy were gotten for each eye.

The Run-Length features:

With the perception that, in a coarse texture, generally long gray-level runs would happen all the more regularly and that a fine surface ought to contain fundamentally short runs, Galloway proposed the utilization of a run-length grid for surface component extraction [17]. For a given picture, a run-length network $p(i; j)$ is characterized as the number of keeps running with pixels of dim level I and run-length j. Different texture features would then be able to be gotten from this run-length matrix. Here, we structure a few new run-length frameworks, which are slight however one of a kind varieties of the customary run-length lattice. For a run-length framework $p(i; j)$, let M be the number of gray levels and N be the most extreme run-length. In a later report, Dasarathy and Holder [18] depicted another four-element extraction capacities following joint statistical proportion of dim level and run-length, as pursues.

Long Run Low Gray-Level Emphasis (LRLGE):

$$\text{LRLGE} = \frac{1}{n_r} \sum_{i=1}^{M} \sum_{j=1}^{N} \frac{p(i, j) \cdot j^2}{i^2} \tag{20}$$

Long Run High Gray-Level Emphasis (LRHGE)

$$\text{LRHGE} = \frac{1}{n_r} \sum_{i=1}^{M} \sum_{j=1}^{N} p(i, j) \cdot i^2 \cdot j^2 \tag{21}$$

Dasarathy and Holder [18] tried every one of the eleven highlights on the order of a lot of cell pictures and demonstrated that the last four highlights gave better execution. These highlights are altogether founded on intuitive think, trying to catch some clear properties of run-length appropriation. For instance, the eight highlights showed in Fig. 2 are weighted-entirety proportions of the run-length focus in the eight headings, i.e., the positive and negative 0_, 45_, 90_, and 135_bearings. Two disadvantages of this methodology are: there is no hypothetical confirmation that, given a specific number of highlights, greatest surface data can be removed from the run-length framework, and huge numbers of these highlights have profoundly corresponded with one another.

4   Results and Discussion

A confusion matrix as given in the figure (Fig. 6) is an overview of outcomes of forecast on an issue of ranking. The amount of right and inaccurate projections are compiled and split down by each category with list numbers. This is the core of the confusion matrix. The confusion matrix demonstrates how, when making projections, your ranking model is incorrect. It provides us understanding not only into a classifier's mistakes, but more closely the kinds of mistakes being produced.
   Here,

- Class 1 represents the RNFL-Loss case detection
- Class 2 represents the RNFL-Healthy case detection
  Definition of the Terms:
- True Positive (TP) indicates that the observation is positive, and the predicted value is also positive.
- False Negative (FN) indicates that the observation is not positive, and the predicted value is also not positive.
- True Negative (TN) indicates that the observation is negative, and the predicted value is also negative.
- False Positive (FP) indicates that the observation is negative, but the predicted value is positive.

Accuracy:
Accuracy can be defined by the following equation:

$$\frac{TP + TN}{TP + TN + FP + FN} \tag{22}$$

However, accuracy has some constraints. For both types of mistakes, it implies equivalent expenses. Depending on the problem definition, a 99 percent precision can be outstanding, good, mediocre, bad, or awful.

Recall:
Recall can be described as the proportion of the complete amount of favorable instances properly categorized divides from the complete amount of favorable instances. High Recall shows that the category is acknowledged properly (tiny FN range). Recall can be represented by the following equation.

Fig. 6 Confusion matrix

	Class 1 predicted	Class 2 predicted
Class 1 Actual	TP	FN
Class 2 Actual	FP	TN

$$\frac{TP}{TP + FN} \tag{23}$$

Precision:

To get the accuracy score, we divided the complete amount of favorable instances properly categorized by the complete amount of favorable instances anticipated.

High precision suggests an instance marked as favorable (tiny amount of FP) is indeed negative. Precision is defined by the following equation:

$$\frac{TP}{TP + FP} \tag{24}$$

High recall, low precision:

This implies that most of the beneficial instances are acknowledged properly (small FN), but many false positives exist.

Low recall, high precision:

This demonstrates that we ignore many favorable instances (elevated FN), but those we imagine are beneficial (small FP) in fact.

F-measure:

Since we have two measurements (Precision and Recall), a metric that reflects both of them is helpful. We calculate an F-measure using Harmonic Mean instead of Arithmetic Mean as it further punishes severe scores. The F-Measure will always be closer to Precision or Recall's lower price. The F-Measure will always be nearer to the smaller value of Precision or Recall [13]. We have got around 68% accuracy. F-measure is defined by the following equation:

$$\frac{2 * Recall*Precision}{Recall + Precision} \tag{25}$$

The performance of the proposed methodology is given below.

The classification performance in Table 1 clearly shows that the linear SVM significantly outperforms the RBF SVM. When we are applying PCA as the dimensionality reduction technique for the feature list of the texture and after that by applying SVM as the classifier we have gotten the better result rather than applying the before mentioned techniques. Although SVM seems to be inherently capable of

Table 1 Performance table

Linear SVM	65%
RBF kernel based SVM	68%
PCA + linear SVM	71.5%
LDA + RBF kernel based SVM	85%
LDA + linear SVM	76.5%
PCA + RBF kernel based SVM	73%

```
Confusion Matrix :
[[92 26]
 [42 54]]

Accuracy Score :
0.6822429906542056

Report :
              precision    recall  f1-score   support

         0.0       0.69      0.78      0.73       118
         1.0       0.68      0.56      0.61        96

avg / total       0.68      0.68      0.68       214
```

Fig. 7 Testing result and accuracy score of our algorithm

handling high-dimensional data, PCA can improve performance by removing corre-lations between variables and possibly removing outliers. In contrast, experimental results reveal the benefits of using LDA in combination with RBF Kernel Based SVM. The performance on the basis of the accuracy level in the Confusion matrix in Table 1 has shown that using the RBF Kernel Based SVM Classifier after doing the LDA based feature dimensionality reduction is better rather than the others compiled classifier. Figure 7 displayed the result of our process where RNFL-Loss and RNFL-Healthy texture are accurately classified. The accurate identification of RNFL-loss and RNFL-healthy section gives scope to the analysis for Glaucoma.

In the following figure Fig. 8, the performance of the classification for RNFL-Healthy and RNFL-Loss texture box will be depicted very clearly. As per the previous discussion it is to be noted that on the generation of the Intracular Pressure (IOP), the RNFL regions are facing the structural transformation. To do this classifica-tion, methods need to be implemented to identify that transformation. The proposed methodology can identify the heterogeneity [20] transformation of the RNFL tex-ture box very accurately. The high proportional presence of Loss-RNFL texture—box indicates the Glaucomatous tendency of an eye. The optometrist can gather the infor-mation of the Glaucomatous tendency before looking into the Region Of Interest at the retina (Fig. 8).

5 Conclusion

The problem of determining glaucoma by RNFL texture analysis using fundus images is correctly determined by the proposed method. We proposed a novel approach for analysis of the RNFL image to determine the Glaucoma affected eye. Since the fundus image is less expensive than the OCT image, the analysis process is highly acceptable from the various sector of society. Again the proposed Support Vector

Fig. 8 Loss-RNFL and
Healthy-RNFL classification

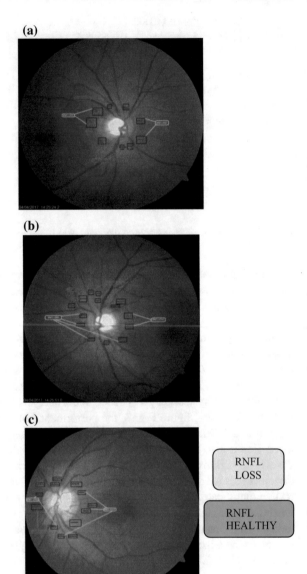

Machine (SVM) based on the Radial Basis Function Kernel performs best giving high average true positive value and least average false positive value. In this method overall accuracy of above 68% using Support Vector Machine (SVM) based on the Radial Basis Function Kernel on the reduced Feature list generated by LDA. We are looking forward to enhance the accuracy in the near future. In the future we are motivated to design an approach by which we can automatically identify proportional ratio in between Healthy-RNFL and Loss-RNFL.

Acknowledgements The research presented in this paper was partially supported by Academy of Technology, University of Calcutta, Suryoday Eye Centre in technical collaboration with L. V. Prasad Eye Institute, Hyderabad, The Calcutta Medical Research Institute and The Currae Eye Hospital. The author would like to sincerely thank Dr. Debasis Chakrabarti, M.S., Fellow Glaucoma (LVPEI) of the Currae Eye Hospital and Dr. Sailaja Sengupta, M.S., Fellow Glaucoma (LVPEI) for their guidance and giving a free hand in choosing research directions.

References

1. Swathy Ravi V. A survey on glaucoma detection methods
2. Gopinath K, Sivaswamy J, Mansoori T. Automatic glaucoma assessment from angio-OCT images
3. https://www.webmd.com/eye-health/glaucoma-eyes#2
4. Odstrcilik J, Kolar R, Tornow RP, Budai A, Jan J, Mackova P, Vodakova M. Analysis of the retinal nerve fiber layer texture related to the thickness measured by optical coherence tomography
5. Zayed N, Elnemr HA. Statistical analysis of Haralick texture features to discriminate lung abnormalities. Computer & Systems Department, Electronics Research Institute, Cairo
6. Odstrcilik J, Kolar R, Harabis V, Gazarek J, Jan J (2010) Retinal nerve fiber layer analysis via Markov random fields texture modelling. In: Proceedings of the 18th European signal processing conference, pp 1650–1654
7. Muramatsu C, Hayashi Y, Sawada A et al (2010) Detection of retinal nerve fiber layer defects on retinal fundus images for early diagnosis of glaucoma. J Biomed Opt 15, Article ID 016021
8. Tuulonen A, Alanko H, Hyytinen P, Veijola J, Seppänen T, Airaksinen PJ (2000) Digital imaging and micro texture analysis of the nerve fiber layer. J Glaucoma 9(1):5–9
9. Kolář R, Jan J (2008) Detection of glaucomatous eye via color fundus images using fractal dimensions. Radio Eng 17(3):109–114
10. Jan J, Odstrcilik J, Gazarek J, Kolar R (2012) Retinal image analysis aimed at blood vessel tree segmentation and early detection of neural-layer deterioration. Comput Med Imaging Graph 36(6):431–441
11. Bock R, Meier J, Michelson G, Nyul LG, Hornegger J (2007) Classifying glaucoma with image-based features from fundus photographs. Lect Notes Comput Sci (Springer) 4713:355–365
12. Mitra A, Roy S, Roy S, Setua SK (2018) Enhancement and restoration of non-uniform illuminated fundus image of retina obtained through thin layer of cataract. CMPB (Elsevier) 156(March):169–178. https://doi.org/10.1016/j.cmpb.2018.01
13. Mitra A, Banerjee PS, Roy S, Roy S, Setua SK (2018) The region of interest localization for glaucoma analysis from retinal fundus image using deep learning. Comput Method Program Biomed (Elsevier) 165(October):25–35. https://doi.org/10.1016/j.cmpb
14. Yogesan K et al (1998) Tele-ophthalmic screening using digital imaging devices. Aust N Z J Ophthalmol 26:S9–S11

15. Haralick RM, Shanmugam K, Dinstein I (1973) Textural features for image classification. In: IEEE transactions on systems, man, and cybernetics, vol SMC-3, no 6, pp 610–621. https://doi.org/10.1109/TSMC.1973.4309314
16. Srinivasan G, Shobha G (2008) Statistical Texture Analysis
17. Galloway MM (1975) Texture analysis using gray level run lengths. Comput Graph Image Process 4:172–179
18. Dasarathy BR, Holder EB (1991) Image characterizations based on joint gray-level run-length distributions. Pattern Recognit Lett 12:497–502
19. Zongging L et al (2009) A variational approach to automatic segmentation of RNFL on OCT data sets of the retina. In: 16th IEEE international conference on image processing (ICIP), Cairo, Egypt, Nov 2009, pp 3345–3348
20. Roy S, Bhattacharyya D, Bandyopadhyay SK, Kim T-H (2018) Heterogeneity of human brain tumor with lesion identification, localization and analysis from MRI. Inform Med Unlocked (Elsevier) 13:139–150

Chapter 14
Artificial Intelligence Based Glaucoma Detection

Prabhjot Kaur and Praveen Kumar Khosla

1 Introduction

Glaucoma is often termed as "silent robber of vision" mainly because it doesn't cause any symptoms, until the advancement of the disease to the last stage, where it causes permanent non-curable blindness. Glaucoma is caused by damage to the nerve cells in Optic Nerve Head (ONH) leading to loss of neurons responsible for transmitting signals to the brain. Glaucoma can be detected at early stages by measuring structural damage in ONH or it can be diagnosed in later stages by the use of perimetry visual field test. According to the World Health Organization (WHO), glaucoma is one of the leading causes of blindness worldwide [1]. Therefore, it becomes important to detect glaucoma at an early stage, as the damage caused is irreversible.

Computer Aided Diagnosis (CAD), expert systems, and computer vision are the hot topics these days and all of this has become possible with recent advancements in Artificial Intelligence (AI) and Machine Learning (ML) based techniques. Recent advancements in Deep Learning (DL) have also become possible due to availability of highly powerful high-speed GPUs made available from Intel (Intel Xenon processors) and NVIDIA, which have considerably reduced the training time, as compared with taken by general CPUs. Therefore, more and more complex DL based self-trained systems are becoming possible, which were considered impossible before. At the same time large amount of medical data is digitalized, more and more patient diseases data in form of images is becoming available, leading to revolution in AI-based CAD systems. AI is extensively used in medical field for image segmentation,

P. Kaur (✉) · P. K. Khosla
C-DAC Mohali, A-34 Industrial Area, Phase VIII, Mohali 160071, Punjab, India
e-mail: prabhjotkaur25@gmail.com

P. K. Khosla
e-mail: drpkhosla@gmail.com

© Springer Nature Singapore Pte Ltd. 2020
O. P. Verma et al. (eds.), *Advancement of Machine Intelligence in Interactive Medical Image Analysis*, Algorithms for Intelligent Systems,
https://doi.org/10.1007/978-981-15-1100-4_14

image enhancement, classification, and disease severity detection. In case of glaucoma detection AI-based techniques have become mature enough to perform expert analysis on data obtained from visual field tests, fundus images depicting internal eye structure comprising of optic nerve head, retinal nerve fiber layer to provide automated disease detection with high level of accuracy. Recent advances in AI include transfer learning and ensemble of various deep learning neural networks.

This paper brings out an overview of various AI-based techniques in use for glaucoma detection and also describes the general motivation behind the use of such techniques and the current challenges being faced by the use of such techniques. Further, relevant issues and various associated problems corresponding to the various automated glaucoma detection techniques are also described.

2   Clinical Diagnosis of Glaucoma

The Clinical diagnosis of glaucoma is usually performed by performing comprehensive eye examination, which may consist of measuring Intraocular Pressure (IOP), the measurement of the extent of visual field patient is able to see, out of the normal field of vision, and the measurement of the various other structural features related to eye or Optic Nerve Head (ONH) [2, 3]. Color Fundus Imaging technology (CFI) and Optical Coherence Tomography (OCT) are the two most widely used techniques for extracting structural measurements of various eye parameters, used for glaucoma detection.

There are various AI-based techniques currently in use for automated glaucoma detection using Visual field testing, OCT, and CFI technique.

2.1   Measurement of Intraocular Pressure

Measurement of eye pressure is usually performed in a test known as Tonometry. Schiotz Tonometer [4], Slit Lamp applanation Tonometer, and Air Puff based non-contact Tonometers are the few of the prominent devices used for the measurement of intraocular pressure. Schiotz Tonometer is a low cost, robust device. Schiotz Tonometer is an indentation based tonometer, which is based on the principle of measuring eye pressure from eye compression. Applanation Tonometers are more precise than indentation-based Tonometers, but indentation based Tonometers are more cost-effective and simpler in nature. Applanation Tonometers work on the principle of IOP being directly proportional to the pressure required for applanation of the cornea [5].

High intraocular pressure is usually associated with glaucoma, but it is not always true. Some people even with normal eye pressure may develop glaucoma, but having exceptionally high IOP, will certainly lead to glaucoma [4].

2.2 Visual Field Test (Perimetry)

A person suffering from glaucoma develops telescopic vision. Such functional requirement of the eye or impairment can be tested using visual fields. The visual fields and intraocular pressure are two parameters which are predominantly used to detect glaucoma and understand its progression. Perimeter or Visual Field Analyzer is the instrument used to test the visual fields. Here the patient is made to fix his chin on the instrument. He is advised not to rotate his eye. The patient has to look into the hollow sphere of the eye. The instrument randomly switches on the LED. Every time patient observes an LED, he presses a button which increments the count [2]. If instrument has switched on the LED and also the button has been pressed it's a positive case and is plotted as shown Fig. 1. If the patient has wider field of view, the test prescribed is 24–10. If he has more predominant glaucoma, 10–2 is prescribed for a narrower field of view.

The visual field test is generally performed to identify the progression of glaucoma as the vision loss once occurred due to glaucoma, cannot be reversed. Figures 1, 2, 3 and 4 are 10–2 Field of View of the same patient with a gap of 7 years. The readers can use this data to check the progression of the disease over a period of 7 years.

Standard Automated Periphery (SAP) is the currently most commonly used method to measure the visual field. Different intensities of lights are shown to the patient and a threshold sensitivity value is derived from that [6]. The visible intensity which is slightly visible to the patient is termed as threshold sensitivity. The visual field map shows such threshold sensitivity values at each of the location in the field of view. The threshold value varies from 0 to 40 dB. 0 dB indicates total blindness at the given point. In case of glaucoma the loss of vision is irreversible [7].

Various Machine Learning (ML) based clustering and classification techniques are currently in use for automated diagnosis and progression detection of glaucoma from visual field analysis tests [8, 9]. ML techniques are used for cluster formation of visual field defect patterns and further division of clusters along the axis. Patient visual field data results are compared to normal visual field data to find better measure of abnormality. There are various statistical methods in use to identify rate of change of visual field and hence to ascertain glaucoma progression from visual field. Bayesian networks [10], Support Vector Machines, Multilayer Perceptron, and Mixture of Gaussian are few of such ML-based techniques being used for analyzing visual field measurements [8].

2.3 Glaucoma Diagnosis Using Structural Features

The Glaucoma can be easily detected from identifying structural damage mainly in Optic Nerve Head (ONH) and RNFL defects from retina. There are two main imaging techniques in use for measuring structural features of eye namely OCT and CFI.

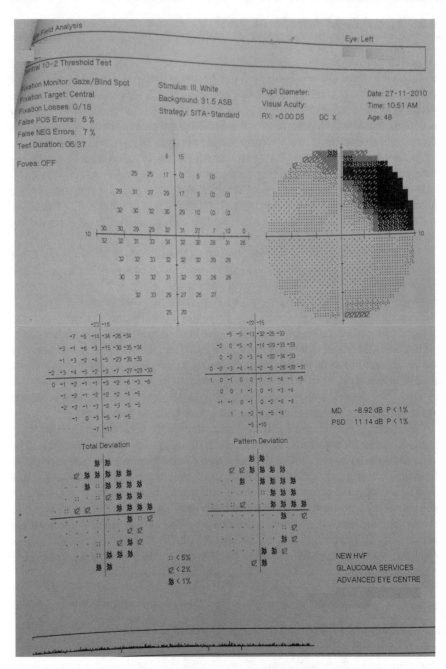

Fig. 1 Anonymised field of view of the left eye of a patient in 2010

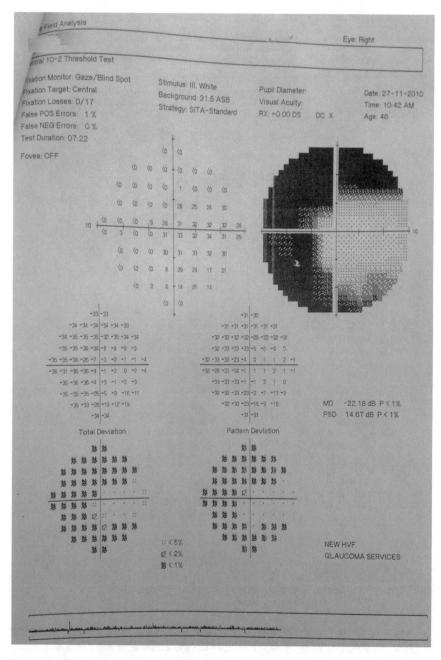

Fig. 2 Anonymised field of view of the right eye of a patient in 2010

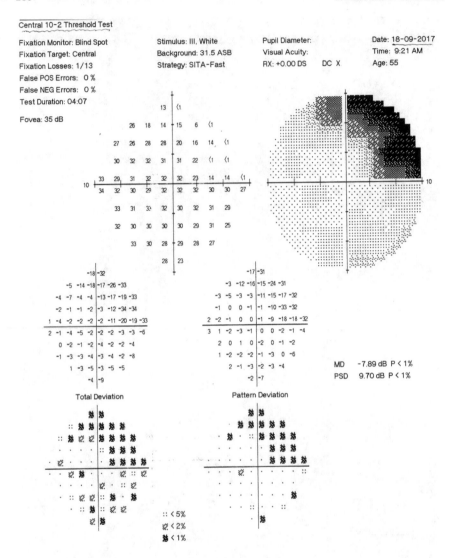

Fig. 3 Anonymised field of view of the left eye of the same patient in 2017

Glaucoma causes increase in cup size, termed as cupping in ONH. Due to loss of retinal ganglion cells, optic disc becomes excavated leading to an increase in cup size. The second important parameter to look for is, non-following of ISNT rule by the Neuroretinal Rim (NRR) [11]. The area between the cup boundary and disc outer boundary forms NRR. The NRR is generally divided into four equal regions namely, inferior, superior, nasal, and temporal. Generally in a healthy eye, the inferior reason

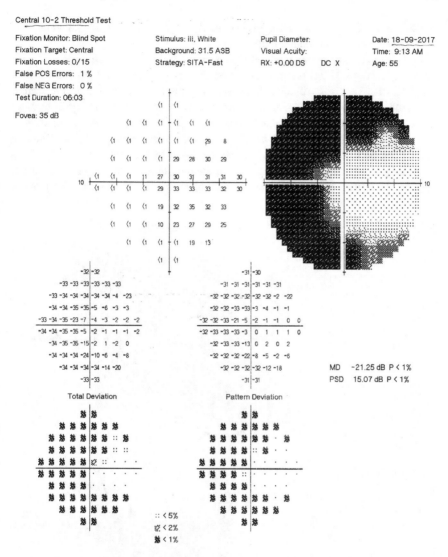

Fig. 4 Anonymised field of view of the right eye of the same patient in 2017

thickness is greater than superior reason, which is further greater than nasal reason thickness and the temporal reason is the thinnest [12]. The third important parameter to look for is CDR value of both the left and right eyes. If the CDR ratio varies by more than 0.2 value in both the eyes provided the disc size of left and right remains same, it makes the eye more susceptible to glaucoma. Another important parameter to look for is the disc size. In a normal-sized disc CDR > 0.5 makes disc susceptible to glaucoma, but if the disc size is abnormally small, CDR value of 0.3 can also be glaucomatous. On the other hand, in abnormally large-sized optic disc, having CDR value of 0.7 can be non-glaucomatous [13].

In addition to the above factors, the presence of NRR notch is an important indicator of glaucoma. The abrupt damage in NRR area is termed as NRR notch. Not all the glaucomatous discs will have NRR notching, but presence of it is a sure sign of glaucoma. Other than above, the other important parameters are baring of circumlinear vessels, bayonetting of vessels, presence of laminar dots, and nasalization of vessels [14, 15]. Figure 5 shows a healthy optic disc having healthy NRR area, having inferior region thickness of NRR greatest, followed by the superior region, which is further greater than the nasal region and the temporal region being the thinnest. Figure 6 shows a glaucomatous optic disc having hardly any NRR area, due to loss of

Fig. 5 Healthy optic disc

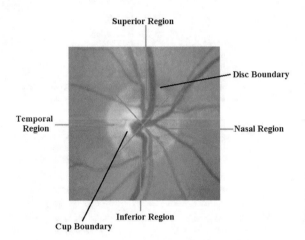

Fig. 6 Glaucomatous optic disc

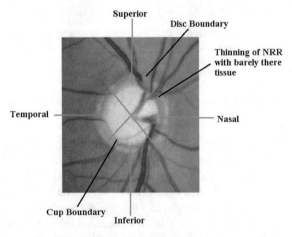

Thickness of NRR in inferior region is same as in superior , nasal and temporal in violation of ISNT rule

neurons in the optic disc due to glaucoma. This disc has an NRR rim area thickness in inferior region same as in superior, nasal, and temporal region, in violation of ISNT rule. It also shows bayonetting of vessels.

It can be clearly seen, there are a number of parameters which must be considered collectively for accurately making decisions about the eye optic disc being glaucomatous or non-glaucomatous. There are various AI-based techniques in use namely, the traditional machine learning based techniques and the recently advanced deep learning based techniques. The current AI-based techniques have shown promising results but still there are certain challenges being faced by both of the methodologies which are discussed in later sections of this book chapter.

2.4 Color Fundus Imaging Technology (CFI)

CFI technology makes use of low-powered microscope to capture an inside view of the eye known as fundus. Fundus camera is extremely useful in diagnosis of various eye diseases including glaucoma, diabetic retinopathy, microaneurysm, and retinal detachment. Fundus camera is used for capturing view of retina, blood vessels, macula, Retinal Nerve Fiber Layer (RNFL) and Optic Nerve Head (ONH). Different fundus cameras may have different angles of views ranging from 20° to 140°. A normal fundus camera has angle of view of 30° whereas a fundus camera having angle of view of 20° is considered to be a narrow view camera and a wide view camera can capture between 40° and 120° of field of view [16, 17].

Roughly there are two main types of fundus cameras available in the market, mydriatic and non-mydriatic. Non-mydriatic fundus cameras do not require pupil dilation for eye examination whereas mydriatic cameras do not work well and provide unclear view of retina, without pupil dilation. The fundus images obtained by non-mydriatic cameras are superior in quality and are much easier to operate than mydriatic cameras [18, 19]. Figure 7 shows a non-mydriatic AI-powered fundus camera "FOP-NM10", by Remido currently in use at Aravind Eye Hospital, Pondicherry, India. It can automatically screen and detect diabetic retinopathy with sensitivity and specificity of 93% and 98%, respectively. Traditional fundus cameras have good image quality but are expensive and can be operated only by trained professionals [20].

2.5 Smartphone-Based Color Fundus Imaging

Giardini et al. [21] proposed a smartphone-based ophthalmoscope in 2014 for taking colored fundus images without dilation. They used Samsung mobile having a back camera along with an adapter having an optic assembly. It provides low cost alternative for color fundus imaging. The use of such devices is very beneficial in poor countries having shortage of expertise where pictures taken from remote locations

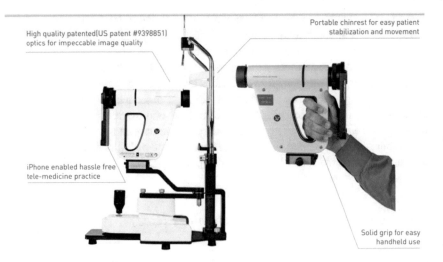

High quality patented(US patent #9398851)
optics for impeccable image quality

Portable chinrest for easy patient
stabilization and movement

iPhone enabled hassle free
tele-medicine practice

Solid grip for easy
handheld use

Fig. 7 Non-mydriatic AI-powered fundus camera "FOP-NM10", by Remido

can be sent to trained technicians for further evaluations. Lin et al. [22] proposed
the use of smartphone for capturing fundus images of children. The children were
administrated local anesthesia so that there is no eye movement. A smartphone with
30D lens was used, which captured the fundus photographs of acceptable quality
which were further used for diagnosis of retinopathy of prematurity.

2.6 Optical Coherence Tomography

OCT is the latest technique used for measuring in-depth and multidimensional information of retina. OCT is used especially to have a measure of thickness and depth
of Retinal Nerve Fiber Layer (RNFL), depth of cup and macula, which are important parameters in glaucoma diagnosis. Spectral Domain (SD-OCT) is considered
to be better than the Time Domain-OCT (TD-OCT) because of being faster and
therefore less susceptible to unintentional patient movement. OCT makes the use of
time-delay in reflection of light waves from the object, which helps in constructing
multidimensional and in-depth model of the object [23, 24].

Figure 8 provides a genuine image of OCT test details of an anonymous patient,
depicting the test result values. This figure clearly shows measurements of the various
important parameters including RNFL thickness, extent of following of ISNT rule,
RNFL symmetry, rim area, disc area, average Cup to Disc Ratio (CDR), and cup
volume, which are important for glaucoma detection. OCT helps in confirming the
glaucoma diagnosis when the case is in borderlines and other methods of glaucoma
diagnosis are nonconclusive.

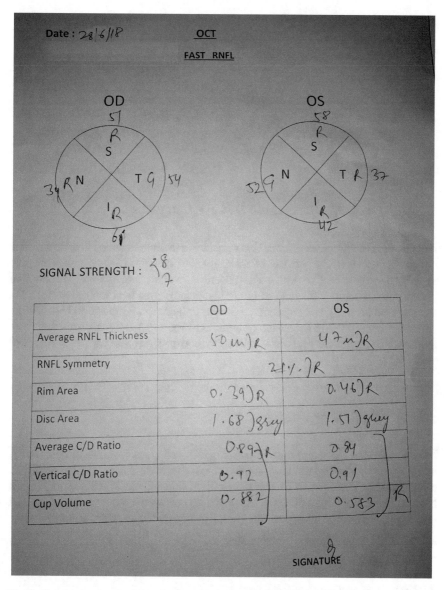

Fig. 8 Genuine image of OCT test details of an anonymous patient, depicting the test result values

3   Overview of AI-Based Methods/Techniques

The following section describes briefly the concepts of Artificial Intelligence, Machine Learning, methods of learning, Neural Networks and the Deep Learning as relevant to ophthalmic diseases detection.

Fig. 9 Relational hierarchy between various AI-based techniques

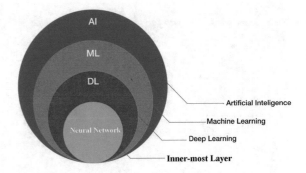

3.1 AI-Powered Ophthalmic Disease Detection

AI is an area of computer science that creates intelligent machines that behave like human beings. It is the process of making smart machines which can work like experts, do inference like humans and even more can make sense of information obtained from images. In some of the fields including medical image analysis, the accuracy achieved by such systems is similar to human experts and in some of the situations like anomaly detection systems, AI-based systems perform better than humans. Figure 9 shows relational hierarchy between various subsets of AI-based techniques.

3.2 Machine Learning

Man has a natural inherent ability to distinguish and classify things automatically, and can easily identify patterns in various kinds of information. In AI parlance this is termed as classification and pattern recognition, respectively. Machines outperform humans in pattern recognition because machines can use and process more data than humans. The machine interpretation generally follows the sequence of steps:

a. Read the Image Dataset
b. Apply preprocessing filters.
c. Region of Interest Extraction
d. Create a Model using training dataset.

  i.   Features Extraction
  ii. Features Optimization
  iii. Classification

e. Compare Training and Testing data to evaluate performance.
f. Visualization of results.

Machines continue to learn from data without reprogramming and hence called machine learning. Various methods of machine learning are given below:

Supervised Learning. It means learning by example. The machine is trained by providing it sufficient number of examples, taking care of under-fitting or over-fitting of training data. The machine learns by associating features extracted with the targeted class. Support Vector Machine (SVM) and K-Nearest Neighbor (KNN) classifier are examples of supervised learning.

Unsupervised Learning. In case of unsupervised learning there are no predefined class labels for the data. The unsupervised machine learning algorithms try to form different object groups based on characteristics of the object. Other examples include clustering in images and detection of anomaly in data. K-means clustering is one such machine learning clustering algorithm, which is extensively used for performing texture-based clustering and segmentation in image processing, where k means the number of clusters desired.

Reinforcement Learning. Reinforcement learning is based on the principle of reward and punishment. It works by the agent or reinforcement algorithm finding the best possible strategy so that reward becomes maximum. It is commonly used in robotics and gaming.

3.3 Artificial Neural Network

Artificial Neural Network (ANN) is an intelligent system that is inspired by the working of human biological neurons [25]. There are basically three types of ANNs feed forward, feedback neural ANNs known as recurrent neural networks and recursive neural networks. Perceptron is the basic building block of ANN. Figure 10 shows a simple feed-forward ANN. The simplest ANN is perceptron and is known as single-layer neural network. Perceptron takes various signals as input with associated weights. The output is determined by projecting activation function on weighted sum of inputs. The ANN is formed by collection of nodes called artificial neurons, or perceptron. The input layer in ANN receives input data. The output of input layer is connected as input with hidden layers. There may be a number of hidden layers according to need of the task. The hidden layers nodes receive weighted input from multiple layers just like in human brain where one neuron receives weighted signals from multiple neurons through synapse. If the sum of signals received becomes greater than certain threshold the neuron fires an electric signal in the brain. This threshold in ANN is determined by activation function used [26, 27].

Activation function decides the threshold value and the strength of signal being fired. The activation function helps in incorporating nonlinearity in the output of a neuron. Examples of activation functions include: Sigmoid, ReLU, Hyperbolic Tangent [28, 29].

Fig. 10 A simple
feed-forward ANN

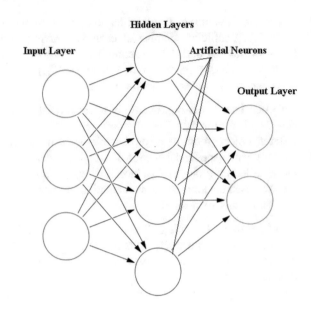

3.4 Deep Learning (DL)

Deep Learning Neural Networks (DLNN) is the most powerful concept of AI. DLNN differ from traditional ML-based techniques in respect that DLNN do not need hand-crafted feature extraction but can generate and extract features by themselves from the training data provided to them. The power of DL lies in the fact that DL algorithms can learn optimal set of features by itself and that too, much better than the handcrafted ones. For example, DLNNs may be provided with images labeled with object identification, and then it is expected to train itself to identify the objects in the test images by itself. The only constraint is that the training data should be extensive, representing all the possible scenarios.

Various deep learning models include Deep Belief Networks, Convolutional Neural Networks, and Auto Encoders [25]. The desirable characteristics of DLNNs include being spatial invariance in feature extraction, ability to learn features hierarchically, and ability to scale [30].

4 AI-Based Image Preprocessing and Enhancement Techniques in Use for Glaucoma Detection

Image preprocessing techniques consist of segmentation, image enhancement, and region of interest extraction techniques. Image enhancement techniques are very important in image processing. Since images obtained from color fundus camera may have poor quality because of insufficient illumination, movement by patient, glare or having a certain disease like cataract, prevent to have complete field of view of retina. Therefore, it becomes important to enhance the quality of images using various image enhancement methodologies, so that it can lead to accurate diagnosis. The use of preprocessing techniques is very important in case of ordinary machine learning techniques, but they can be omitted in case of deep neural network based techniques as the system can adapt itself for the noisy input images.

4.1   Histogram Equalization

Histogram in a digital image is the mapping between the number of pixels having a particular gray level intensity. Generally, histograms are normalized, before any use by dividing each of its values by the total number of pixels in the image. Histogram is used in image enrichment, segmentation as well as in reduction of image size. Histogram Equalization (HE) means dispersion of the histogram into various intensity values for better contrast of image, leading to significant image enhancement [31]. HE is particularly useful in underexposed, overexposed images, X-ray images, satellite images, infrared images, and microscopic images. HE may also lead to increase in noise in image signal, and may also lead to development of unrealistic effects in image. Other modified forms of HE include Local Contrast Enhancing HE, Adaptive HE (AHE), Contrast Limiting Adaptive HE, Multi-peak and Multi-purpose Beta Optimized HE [32]. HE is used in [33] as the preprocessing step for image enhancement in automated glaucoma detection.

5   Current State-of-the-Art AI-Based Methods/Techniques in Use for Automated Glaucoma Detection for Feature Extraction

5.1   Traditional Machine Learning Based Techniques

Radon Transformation. Radon Transformation (RT) is a widely used technique for image reconstruction. Raghavendra et al. [34], in 2018, uses a method based on Radon Transformation followed by Modified Census Transform (MCT) for compensating against change in illumination levels, for image reconstruction for glaucoma detection. RT is generated by mapping images along with certain directions at various angles to generate a signature of the image. The generated signature can be used for matching various images. A sinogram in RT is generated by sum of projections at various angles α, defined by the sampling rate. Larger angle α, results in loss of information, whereas smaller α may generate lots of data for computation [35, 36].

Census Transformation (CT) CT is a way of representing image structure which is independent of illumination variance. Therefore, the transformation, that is the local pattern obtained is independent of overall illumination changes and is usually used to overcome illumination problems. CT works by linking each pixel of the image, a binary encoding, depending upon the intensity of the neighboring pixels. The value of binary encoding is totally dependent on relative sequence of intensities, therefore making it resistant to illumination variations. CT is very well apt to detect edges and shapes, rims, and boundaries which is quite helpful in glaucoma detection. It leads to better detection of NRR boundary, cup and disc shape and size, which are important parameters for glaucoma detection in color fundus images [34, 37–39].

Wavelet Transformation Based Methods. Wavelet Transformation (WT) is generally used in image processing for various applications such as in removing noise from images and in texture analysis. In WT an appropriate frequency band is selected for signal representation. WT is a time-frequency analysis technique for feature extraction and compression [40].

Discrete Wavelet Transformation. Discrete Wavelet Transformation (DWT) is popularly used for representational of biomedical signals such as Electrocardiogram (ECG), Electromyography (EMG), and Electroglottography (EGG). DWT is also used for image compression in computer vision and for matrix eigenvalues computation. DWT is similar to Fast Fourier Transform (FFT), in the way that both are invertible and orthonormal [41]. DWT are localized in space whereas FFT are not. Due to this reason DWT has many uses including data compression, feature extraction and noise removal [42]. DWT is composed of various sets of wavelet functions. There are various types of DWT methods, such as Malaat, Haar, daubechies, Coiflet, Meyer, Morlet DWT [43]. DWT extracts features from a signal by applying following high and low pass filters [44]. Nitha Rajandran [45] used DWT-based fourteen energy features obtained by Daubecies, Symlets, and Biorthogonal DWT for glaucoma detection. These features are subjected to feature reduction and selection followed by ANN to obtain an accuracy of 94% in automated glaucoma detection using color fundus images.

Empirical Wavelet Transformation. Empirical Wavelet Transformation (EWT) is used to decompose and extract features from the image. EWT represents an image in the form of various frequency bands. The red, green, blue channel and the grayscale image are fed separately to obtain different EWT components from the image. EWT is signal-dependent [46]. Maheshwari et al. [47] in 2015, used EWT along with correntropy for feature extraction, which were further subjected to t-value feature selection algorithm for feature reduction, for automated glaucoma diagnosis. The proposed system obtained accuracy of 98.33% in threefold validation.

GLCM Based Texture Feature Extraction. GLCM (Gray Level Co-occurrence Matrix) is a machine learning based texture feature extraction method which takes care of intensity levels as well as their co-occurrence in the image. It relies on second-order statistics. The GLCM converts an image into its gray level co-occurrence matrix. GLCM records the count of co-occurrence of same intensity levels along with the location specified in terms of direction [27]. Following are the steps to compute the gray level co-occurrence matrix:

a. Convert the image into gray scale or single-channel image.
b. Calculate total intensity levels used in image.
c. Generate $k \times k$ matrix, where k is the total no. of intensity levels.
d. Calculate values of GLCM in the function of number of times each intensity value co-occurs with each other.

In the next step normalization of GLCM matrix is performed so that the GLCM matrix values obtained becomes independent of the image size. This is performed by dividing the values of the GLCM matrix with total number of matrix elements.

The main drawback of GLCM based feature extraction is that though GLCM can find texture information it fails to detect edge information.

Kavya et al. [48] uses GLCM based Feature Extraction (FE) along with Hough transformation for the region of interest extraction, extraction of Markov random field to obtain accuracy of 94% for automated glaucoma detection. Salam et al. [49] use GLCM-based feature extraction along with Gabor Wavelet transformation, Local Binary Patterns, Harlick FE, color moments FE, to obtain sensitivity of 100% and specificity of 87% for automated glaucoma detection using color fundus images.

Higher Order Spectra Features. Higher Order Spectra (HOS) consists of higher order moments of a signal. HOS features are useful in extraction of nonlinear features from an image. HOS can eliminate noise from the image, and can also preserve the phase information of the image [31]. This is with contrast to second-order statistics which is largely based on Fourier Transformation of correlation, which can extract only Gaussian and Linear properties, without any phase information. HOS is widely used method for texture features extraction from different types of images [50, 51].

5.2 Deep Learning Based Techniques

An automated system for glaucoma detection with AI-based techniques takes as an input an image. Hence, all the intensity values of the image form the input for deep neural networks. Input layers are followed by a number of hidden layers, from here comes the term "deep" in deep learning neural networks. These numbers of Hidden Layers (HLs) determine the depth of the DLNN. Various levels of features of image are extracted by these HLs. Theses HLs act in hierarchy. For example at the initial hidden layer the DNN system may extract only edges. The next hidden layers use the features extracted by previous hidden layers to extract more sophisticated features, and so on. The higher hidden layers may extract more abstract features comprising of overall shape of objects. The prominent example of deep learning is automatic feature extraction along with optimization of features also. The last layer in case of deep learning based glaucoma detection systems usually consists of a classifier to make glaucoma prediction.

There are various modifications possible in the deep learning based architectures, these consists of transfer learning neural network and ensemble of neural networks which are explained in later sections.

Convolutional Neural Network. Convolutional Neural Networks (CNNs) are the most commonly used Neural Network (NN) in various image processing applications of classification, object recognition. Besides these CNNs are also commonly used in self-driving cars, robotics, Natural Language Processing (NLP), and speech recognition. CNNs are extensively used in expert systems based on computer vision for image segmentation and classification [52, 53]. Usually CNNs consist of a large number of layers, therefore considered as part of DLNNs. Different layers in CNNs respond to different features present in image. For example, certain layers and nodes of CNN are sensitive only to edges and some are sensitive to overall texture of the object.

Types of layers in convolutional neural network:

a. Convolutional
b. Activation function
c. Pooling
d. Fully connected
e. Dropout

Chen et al. [54] used six layers based deep CNN for automated glaucoma detection. It consisted of four convolution layers and two fully connected layers. In first step Region of Interest (ROI) containing ONH is extracted from the fundus image, which is fed to the input layer of NN. Response-normalization and pooling is used to overcome the problem of over-fitting. In the end softmax classifier is used for making prediction. This system got Area Under Curve (AUC) of the receiver operating characteristic curve at 0.831 and 0.887 with ORIGA and SCES database, respectively.

Transfer Learning Based Techniques. In transfer learning a pretrained model is usually selected as the base model for newly developing deep neural network based system. Transfer learning is very commonly used in the case of image processing. It is very common to use a pretrained deep learning model from ImageNet challenge which is already pretrained for 1000-class classification. These pretrained models are used as starting point in the new model being developed followed by appropriate model tuning to suit the current needs and challenges of the task. Many of such pretrained models are AlexNet, VGG-16, GoogLeNet, ResNet-50, and ResNet in use for glaucoma detection. Raghavendra et al. [55] used CNN architecture comprising of 18 layers for automated glaucoma detection using color fundus images. The proposed system got accuracy of 98.13% in glaucoma detection using CFI. Gómez-Valverde et al. [56] proposed use of VGG-19 CNN for transfer learning and got AUC of 0.94 in automated glaucoma detection. Alan Carlos de Moura Lima et al. [57] proposed feature extraction using pretrained ResNet neural network for automated glaucoma detection. The proposed system obtains AUC of 0.957 by using Logistic Regression on RIM-ONE-re database and Logistic Regression, on the RIM-ONE-r2 dataset, with AUC of 0.957. Li et al. [58] proposed a novel transfer learning based method of glaucoma detection. The system used AlexNet, VGG-16, VGG-19, GoogLeNet, ResNet-50, and ResNet-152 for feature extraction, with the total number of 19,456 features extracted, along with SVM as classifier with AUC of 0.838.

AlexNet. AlexNet was introduced in 2012. Though being simpler then LeNet, AlexNet outdid all other CNNs, participating in ImageNet Large-Scale Visual Recognition Competition, by scoring classification error rate of 15%. There were many reasons due to which AlexNet performed better than LeNet. Firstly, it was trained on a huge dataset containing 15 million images of different objects, having around 22,000 different classes. It made use of ReLU, which was faster than traditional functions and has no problem of vanishing gradient. AlexNet in total consists of only five convolutional, three pooling and fully connected layers, and softmax as classifier.

Oxford VGG. VGG neural network was introduced in 2014 consists of small-sized convolutional filters of 3×3, therefore it has got more layers and is more deep then ZFNet. This in turn has further lead to increase in accuracy as compared with ZFNet.

VGG-19. VGG19 consists of five stacks consisting of 2 and 4 convolution operation layers, followed by a max-pooling layer. VGG-19 is widely used in automated breast cancer detection and skin diseases detection along with glaucoma detection [59–61].

GoogLeNet. GoogLeNet is a powerful CNN developed by Google in 2015. GoogLeNet has 22 layers. It was among the top performers in ImageNet Challenge of 2015 for making accurate predictions. It makes use of inception module to reduce the number of parameters. The inception module consists of convolution layer with different kernel sizes and pooling sublayers concatenated with output filter banks in a single vector. GoogLeNet consists of 22 layers in total. Because of the use of inception layer the number of parameters used by GoogLeNet is twelve times lesser than those used in AlexNet. With the increase in number of layers, it is usually assumed that accuracy will also increase due to increase in number of features extracted. But this is not always true because of Vanishing gradient problem. Further it may pose optimization difficulty due to increase in number of features extracted. Therefore, the inception model was further refined to overcome the negative effects [62, 63]. Inception-V3 and Inception ResNet are the improved versions of inception model.

Microsoft ResNet. ResNet, introduced in 2015, is a CNN-based architecture having, 152 layers in total. ResNet makes use of residual connections, which resulted in better flow of information between various layers. ResNet won ImageNet object recognition challenge of 2015 by obtaining accuracy to the tune of 97% which is assumed to be better than humans. ResNet uses lesser number of parameters than VGGNet but has more depth [62].

Ensemble Learning Based Techniques. Fu et al. [64] proposed, an ensemble-based neural network for automatic glaucoma detection. This system integrates four deep learning neural networks consisting of global image neural network, global segmentation stream, local disc region stream, and disc polar transformation stream. The first stream consists of a pretrained network, ResNet based on a CNN. ResNet-50 act as backbone model for feature extraction from fundus image. The second stream is the neural network, which segments OD region from fundus image. SGNN is inspired by U-shape convolutional network (U-Net) consisting of an encoder and decoder. It consists of two local streams to learn local details. The first local stream consists of ResNet based classification network for the local disc region. The second local stream is used for performing polar transformation.

6 Conclusion

As evident from this book chapter, the traditional machine learning techniques appear to work well in automated computer vision based glaucoma detection, but the design of such systems is a very tedious, time-consuming process and is highly dependent on the knowledge of the expert. The traditional handcrafted feature extraction techniques are very complex to design and pose a challenge to find the most appropriate techniques of feature extraction and optimization. The glaucoma detection is a complex process and is not based on the result of single parameter only, but to consider different parameters together. However, the traditional ML-based methods lack sufficient depth in representing the nonlinear functions and the depth required for complex tasks. Deep learning based techniques have got edge over the traditional ML-based techniques, in a way that DL-based techniques can adapt very well to the kind of complexity required due to automatic feature extraction and optimization. DL based systems can be trained for very high accuracy, sometimes even surpassing humans, by providing a large dataset, which sufficiently represents all the use cases. With the advent of different types of deep learning based systems, including transfer learning, ensemble learning, and hybrid approach, deep learning has revolutionized computer vision based disease detection expert systems. Artificial Intelligence has got lots of applications in medical field not only in glaucoma detection, but in detection of cancer cells and other diseases also.

References

1. Tham YC, Li X, Wong TY, Quigley HA, Aung T, Cheng CY (2014) Global prevalence of glaucoma and projections of glaucoma burden through 2040—a systematic review and meta-analysis. J Ophthalmol 121(11):2081–2090
2. Khalil T, Khalid S, Akram MU, Jameel A (2017) An overview of automated glaucoma detection. In: IEEE computing conference
3. Kaur P, Khosla PK (2019) Comparative study of recent automated glaucoma detection techniques using color fundus images. Int J Innov Technol Explor Eng
4. Ismael Cordero, National Center for Biotechnology Information. https://www.ncbi.nlm.nih.gov/pmc/articles/PMC4322748/. Accessed 29 June 2019
5. Draeger J (1967) Principle and clinical application of a portable applanation tonometer, investigative ophthalmology and visual science. ARVO J

6. American Optometric Association, "Glaucoma". https://www.aoa.org/patients-and-public/eye-and-vision-problems/glossary-of-eye-and-vision-conditions/glaucoma. Accessed 18 June 2019
7. Meng S-H, Turpin A, Lazarescu M, Ivins J (2005) Classifying visual field loss in glaucoma through baseline matching of stable reference sequences. In: Proceedings of the fourth international conference on machine learning and cybernetics, Guangzhou
8. Chan K, Lee T-W, Sample PA, Goldbaum MH, Weinreb RN, Sejnowski TJ (2002) Comparison of machine learning and traditional classifiers in glaucoma diagnosis. IEEE Trans Biomed Eng
9. Yousefi S, Goldbaum MH, Balasubramanian M, Medeiros FA, Zangwill LM, Liebmann JM, Girkin CA, Weinreb RN, Bowd C (2013) Learning from data—recognizing glaucomatous defect patterns and detecting progression from visual field measurements. In: IEEE 13th international conference on data mining
10. Ceccon S, Garway-Heath DF, Crabb DP, Tucker A (2014) Exploring early glaucoma and the visual field test: classification and clustering using Bayesian networks. IEEE J Biomed Health Inform 18(3)
11. Jonas JB, Gusek GC, Naumann GO (1988) Optic disc, cup and neuroretinal rim size, configuration and correlations in normal eyes. Invest Ophthalmol Vis Sci 29:1151–1158
12. Ruengkitpinyo W, Kongprawechnon W, Kondo T, Bunnun P, Kaneko H (2015) Glaucoma screening using rim width based on ISNT rule. In: IEEE 6th international conference of information and communication technology for embedded systems (IC-ICTES)
13. Tatham AJ, Weinreb RN, Zangwill LM, Liebmann JM, Girkin CA, Medeiros FA (2013) The relationship between cup-to-disc ratio and estimated number of retinal ganglion cells. Investig Ophthalmol Vis Sci
14. Nyul LG (2009) Retinal image analysis for automated glaucoma risk evaluation. Proc SPIE: Med Imaging Parallel Process Images Optim Tech 7497:1–9
15. Jakirlic N (2016) Optic nerve evaluation in glaucoma. California Optometric Association
16. https://www.sciencedirect.com/topics/medicine-and-dentistry/fundus-photography. Accessed 30 June 2019
17. University of British Columbia, Department of Ophthalmology and Visual Sciences. https://ophthalmology.med.ubc.ca/patient-care/ophthalmic-photography/color-fundus-photography/. Accessed 30 June 2019
18. Remidio. https://www.remidio.com/fop.php. Accessed 30 June 2019
19. Coburntechnologies. https://www.coburntechnologies.com/2016/04/12/non-mydriatic-fundus-camera/. Accessed 30 June 2019
20. National Center for Biotechnology Information. https://www.ncbi.nlm.nih.gov/pmc/articles/PMC6543857/. Accessed 28 June 2019
21. Giardini ME, Livingstone IAT, Jordan S, Bolster NM, Peto T, Burton M, Bastawrous A (2014) A smartphone based ophthalmoscope. In: 36th annual international conference of the IEEE Engineering in Medicine and Biology Society
22. Lin S-J, Yang C-M, Yeh P-T, Ho T-C (2014) Smartphone fundoscopy for retinopathy of prematurity. Tiwan J Ophthalmol
23. Bussel II, Wollstein G, Schuman JS (2014) OCT for glaucoma diagnosis, screening and detection of glaucoma progression. Br J Ophthalmol 98(Suppl-2)
24. Fujimoto JG, Pitris C, Boppart SA (2000) Optical coherence tomography—an emerging technology for biomedical imaging and optical biopsy. J PMC
25. Du X, Cai Y, Wang S, Zhang L (2016) Overview of deep learning. In: 31st youth academic annual conference of Chinese Association of Automation (YAC)
26. Pedrycz K, Pedrycz W (eds) Springer handbook of computational intelligence. Springer
27. Gad AF (2018) Practical computer vision applications using deep learning with CNNs. Apress
28. Ding B, Qian H, Zhou J (2018) Activation functions and their characteristics in deep neural networks. In: Chinese control and decision conference (CCDC). IEEE
29. Shanmugamani R. Deep learning for computer vision: expert techniques to train advanced neural networks using TensorFlow and Keras
30. Bengio Y et al (2012) Representation learning: a review and new perspectives

31. Gonzalez RC, Woods RE (2001) Digital image processing. Prentice Hall, Englewood Cliffs, NJ
32. Sargun, Rana SB (2015) A review of medical image enhancement techniques for image processing. Int J Curr Eng Technol
33. Hasikin K, Isa NAM (2014) Adaptive fuzzy contrast factor enhancement technique for low contrast and non-uniform illumination images. SIViP 8(8):1591–1603
34. Raghavendra U, Bhandary SV, Gudigar A, Rajendra Acharya U (2018) Novel expert system for glaucoma identification using non-parametric spatial envelope energy spectrum with fundus images. Bio Cybern Biomed Eng 38(1):170–180
35. D'Acunto M, Benassi A, Moroni D, Salvetti O (2014) Radon transform: image reconstruction and identification of noise and instrumental artifacts. In: 22nd signal processing and communications applications conference. IEEE
36. Kertész G, Szénási S, Vamossy Z (2017) Application and properties of the radon transform for object image matching. In: IEEE 15th international symposium on applied machine intelligence and informatics
37. Kublbeck C, Ernst A (2006) Face detection and tracking in video sequences using the modified census transformation. Image Vis Comput 24(6):564–572
38. Froba B, Ernst A (2004) Face detection with the modified census transform. In: Sixth IEEE international conference on automatic face and gesture recognition
39. Jo H-W, Moon B (2015) A modified census transform using the representative intensity values. In: International SoC design conference (ISOCC). IEEE
40. Xizhi Z (2008) The application of wavelet transform in digital image processing. In: IEEE international conference on multi media and information technology
41. Olkkonen H (2011) Discrete wavelet transforms—biomedical applications. InTech
42. Agarwal R, Raman B, Mittal A (2015) Hand gesture recognition using discrete wavelet transform and support vector machine. In: 2nd international conference on signal processing and integrated networks (SPIN)
43. Ghazali KH, Mansor MF, Mustafa MM, Hussain A (2007) Feature extraction technique using discrete wavelet transform for image classification. In: IEEE the 5th student conference on research and development-SCOReD
44. Kiran SM, Chandrappa DN (2016) Automatic detection of glaucoma using 2-D DWT. Int Res J Eng Technol (IRJET)
45. Rajandran N (2014) Glaucoma detection using DWT based energy features and ANN classifier (2014)
46. Kirar BS, Agrawal DK (2017) Empirical wavelet transform based pre-processing and entropy feature extraction from glaucomatous digital fundus images. In: International conference on recent innovations is signal processing and embedded systems. IEEE
47. Maheshwari S, Pachori RB, Acharya UR (2017) Automated diagnosis of glaucoma using empirical wavelet transform and correntropy features extracted from fundus images. IEEE J Biomed Health Inform 21(3):803–813
48. Kavya N, Padmaja KV (2017) Glaucoma detection using texture features extraction. In: IEEE 51st asilomar conference on signals, systems, and computers
49. Salam AA, Khalil T, Usman Akram M, Jameel A, Basit I (2016) Automated detection of glaucoma using structural and nonstructural features. Springerplus
50. Lagdali S, Rziza M (2017) Higher order spectra in image processing. In: 3rd international conference on advanced technologies, for signal and image processing, ATSI
51. Mendel JM (1991) Tutorial on higher-order statistics (spectra) in signal processing and system theory. Theoretical results and some applications. Proc IEEE 79(3):278–305
52. Krizhevsky A et al (2012) Imagenet classification with deep convolutional neural networks. NIPS
53. Le QV et al (2011) Building high-level features using large scale unsupervised learning. ICML
54. Chen X, Xu Y, Wong DWK, Wong TY, Liu J (2015) Glaucoma detection based on deep convolutional neural network. In: 37th annual international conference of the IEEE Engineering in Medicine and Biology Society, EMBC

55. Raghavendra U, Fujita H, Bhandary SV, Gudigar A, Tan JH, Rajendra Acharya U (2017) Deep convolution neural network for accurate diagnosis of glaucoma using digital fundus images, computerized medical imaging and graphics. ScienceDirect (Elsevier) 55:28–41

56. Gómez-Valverde JJ et al (2019) Automatic glaucoma classification using color fundus images based on convolutional neural networks and transfer learning. Biomed Opt Express 10(2)

57. de Moura Lima AC, Maia LB, Pereira RMP, Junior GB, de Almeida JDS, de Paiva AC (2018) Glaucoma diagnosis over eye fundus image through deep features. In: 25th international conference on systems, signals and image processing (IWSSIP)

58. Li A, Wang Y, Cheng J, Liu J (2018) Combining multiple deep features for glaucoma classification. In: IEEE international conference on acoustics, speech and signal processing (ICASSP)

59. Simonyan K, Zisserman A (2014) Very deep convolutional networks for large-scale image recognition. arXiv:14091556 Cs

60. Antropova N, Huynh BQ, Giger ML (2017) A deep feature fusion methodology for breast cancer diagnosis demonstrated on three imaging modality datasets. Med Phys 44(10):5162–5171

61. Liao H (2016) A deep learning approach to universal skin disease classification

62. Arif Wani M, Bhat FA, Afzal S, Khan AI (2019) Advances in deep learning studies in big data. Springer

63. Szegedy C, Liu W, Jia Y, Sermanet P, Reed S, Anguelov D, Erhan D, Vanhoucke V, Rabinovich A (2015) Going deeper with convolutions, pp 1–9

64. Fu H, Cheng J, Xu Y, Zhang C, Wong DWK, Liu J, Cao X (2018) Disc-aware ensemble network for glaucoma screening from fundus image. IEEE Trans Med Imaging

Chapter 15
Security Issues of Internet of Things in Health-Care Sector: An Analytical Approach

Pranjal Pandey, Subhash Chandra Pandey and Upendra Kumar

1 Introduction

The Internet as we know is a massive global network that allows people to communicate with each other and it can use homogenous as well as heterogeneous data [1]. The heterogeneous object includes not only communication devices but also diverse physical objects.

The Internet of Things (IoT) is a promising approach and it is pivoted on the interconnectivity of things or devices to each other as well as with users. Indeed, this approach is a cornerstone for the development of smart homes and cities. Perhaps, trustworthiness of any technology is of paramount importance form the viewpoint of users. Further, security and privacy issues play a pivotal role in this pursuit. In spite of vigorous striving of researchers, extensive literature survey revealed the fact that we are still lagging behind trustworthy security and privacy prospects. In IoT, unlike the Internet, things or devices use their computing capabilities together with the network connectivity to sense and collect data from the world around us and share the data across the Internet or cloud and utilize the analyzed and processed data for decision-making. In the world of IoT there are millions of devices connected together. Thereby services should be discoverable by the people by making use of the service discovery mechanism. Nowadays IoT application can be seen in houses, social places,

P. Pandey
Department of Electronics and Communication Engineering, Indraprastha Institute of Information Technology, Delhi, Okhla Industrial Estate, Phase 3, New Delhi, India
e-mail: pranjalpandey200@gmail.com

S. C. Pandey (✉) · U. Kumar
Department of Computer Science and Engineering, Birla Institute of Technology, Mesra, Ranchi (Patna Campus), Patna, Bihar, India
e-mail: s.pandey@bitmesra.ac.in

U. Kumar
e-mail: upendrakumarphdp@gmail.com

© Springer Nature Singapore Pte Ltd. 2020
O. P. Verma et al. (eds.), *Advancement of Machine Intelligence in Interactive Medical Image Analysis*, Algorithms for Intelligent Systems,
https://doi.org/10.1007/978-981-15-1100-4_15

and industries. For instance, IoT application in houses could be helpful in automatic ordering of online groceries and other home supplies. Cameras together with the sensors, inside the refrigerator, could keep track of the grocery supply and automatic online order is triggered in case of any shortage. Another very interesting example of the smart home IoT application is the self-programmable, Wi-Fi enabled, and sensor-embedded Nest Thermostat which learns what temperature you like and accordingly sets the schedule. Some of the more future prospects of IoT application could be like automatic parking or getting cautionary advice on your phone or wearable device if any physical danger is detected nearby. Addressing these challenges and ensuring security in IoT services and products is very essential. The data and services should be well protected from the cyber-attacks as hacking any single device could provide access to the user data. The information included is personal data of the user like location and time which needs to be secured and privatized. Security services which are needed to be paid are authentication, authorization, confidentiality, and integrity. The application domain of Internet of things could be versatile for example if we are on the way to our holiday vacation, our car could have access to the network and already detect the location of the nearest hotel. There could be many more such applications where in everyday life each object via communication could help do the house chores as well as in business and industrial field. With the excess of IoT devices coming across, the business must develop a universal approach to IoT management, including IoT security, to prevent vulnerabilities of device and related data and services.

With the help of IoT, millions of devices are going to be connected with each other through heterogeneous infrastructures, by avoiding the risk of attack. It requires very less or no human involvement which in turn renders the IoT prone to various attacks such as eavesdropping, man-in-the-middle attack (MMA), and denial of service attack (DSA). In addition, some antagonist devices can have unauthorized access to any physical device in the network. Attacks could easily damage the physical system as well as the connections by compromising the issues of security and privacy pertaining to IoT. Further, the realm of IoT is subject to resource constriction in terms of power, bandwidth, and storage. Therefore, security solutions are needed which will not masticate through the resources of IoT.

Many challenges do stand in the way of security and privacy. Main attacks on IoT system are:

- Spoofing
- False signal injection
- Replay attacks
- Eavesdropping.

The aspects of security are adversely affected by such attacks. Different aspects adversely affected are privacy, reliability, and authenticity. Addressing these challenges and ensuring security in IoT services and products is very essential. The data and services should be well protected from the cyber-attacks as hacking any single device could provide access to the user data. Cryptographic security solutions can gain trust over IoT since the interaction is between human and things and things and things, much crucial information is shared across the network that could be hacked easily if any of the connected devices is hacked. So, there must be well-defined security architecture deployed with a shared common set of protocols as heterogeneous devices are connected through the IoT network. Many layered architectures are proposed already for example three-layered architecture—perception, network, and application layer. With the impetus of application requirements, large amount of information is being shared among the devices. Moreover, IoT faces many other challenges like scalability, power, bandwidth, security, and privacy. Hence the cryptographic solutions provide the way to utilize the feature of the lightweight symmetric and asymmetric algorithm such that less execution time with optimum energy requirements can be taken care.

In this chapter authors have discussed and analyzed different issues pertaining to the security aspects of IoTs. Further, different measures have been suggested related to security issues of the IoT in different domains. This chapter is organized as follows: Sect. 2 presents the naive information related to IoT. Various security concerns are given in Sect. 3. Security architecture is discussed in Sect. 4. Further, Sect. 5 preludes layered classification. Cryptographic solutions for IoT are given in Sect. 6. Moreover, security issues of IoT in medical sector are discussed in Sect. 7. Over and above, security analysis of IoT in business sector is delineated in Sect. 8. Finally, chapter is concluded in Sect. 9.

2 Background

The term IoT was coined in Massachusetts Institute of Technology (MIT). As of now, IoT has been the buzzword since the past decades. IoT has versatile importance and applications in different areas, which are very useful for human beings, but IoT being hacked through different ways is also a big challenge for security provider agencies or users. One of the major problems is malware (malicious software) attacks on IoT Systems due to obscurity to patch end devices. Intermediate nodes are used for patching to minimize propagation of malware in IoT system. In [2], authors have compared randomized patching scheme with traffic-aware patching scheme. Since traffic management is one of the major problems of IoT, which is efficiently managed by technique of Operation Research (OR) and system thinking, but system thinking is not a very easy task to implement in IoT for security purposes [3].

The large volume of collected data are mined and used to discover some hidden information to further analyzed data to get certain result in different areas like health care, and business management [4]. In IoT environment, dozens of devices like health-care devices, house controls, house security, and camera are connected with each other. Here, service discovery comes into the picture, which is automatic detection of the devices and services offered by these devices on a computer network. To enhance the capability of IoT, it is necessary for improvement of related technology, methodology, and tools. All these related components will improve living standard of human beings due to diversity of IoT. Diversity of IoT standardized the lifestyle in different fields like development of agriculture, supply chain management system, work culture of Industry, caring in health, development of smart cities, transportation, entertainment, etc. [5]. In order to capture the full potential use of service discovery, there are many protocols like BLE (Bluetooth Low Energy), Apple Airdrop, and Multicast DNS. In order to secure the quality of services through traffic measurement, it is necessary to improve network performance through simulation for machine type communication [6]. IoT covers large domains to deliver services for smart manufacturing, environmental monitoring, and health services. It also improves quality of life and productivity of humans [7].

Security and privacy were the most concerning parts for the IoT application facing enormous challenges. IoT is extending the frontier of connectivity beyond laptops and smartphones to smart homes, wearable, smart city, connected car, connected health, and many more. A predator–prey model can help to understand the security threats of wireless nano-sensor network (WNSN) which is made of nano-IoT [8]. Inclusion of active mode and sleep mode of sensor nodes makes this work interesting.

It is not a very easy task to connect many devices with IoT, but researchers have taken it as a challenging task for future IoT. Moreover, different models are also proposed to present IoT ecosystem [9]. Such type of network systems will require massive storage device to maintain the cloud servers. In existing scenario, future of IoT greatly depends on decentralization of IoT networks [10]. The various technological challenges which need to be addressed are connectivity, security, compatibility, standards, intelligent analysis, and actions. However, deploying utmost security in all IoT devices is not feasible. Different security solutions have been architectured for homogeneity of network. Another challenge arises due to the risks involved with mission-critical operations in open and un-trusted environments [11].

3 IoT Security Concerns

Any object connected to the Internet is bound for security threats. The new vulnerability emerges and software securities degrade over time. By looking at eternally changing threat landscape from cyber espionage to phishing and from malware to ransomware, we can see most of the attacks are targeted on data. With the advent of IoT, we see more and more connected devices impacting our everyday life. By 2032 it is predicted that every personality will be connected to average 3000–5000 devices. This leads to immense security challenge for all IoT segments [12]. Adoption of ever-changing technology also opens up new avenues for attackers to target and abuse IoT. Perhaps, security architecture is not standardized and varies from vendors to vendors. Keeping different IoT insecurities, security implementation, and protection architecture, in general contain three different levels [13]:

a. Device security (Perception layer)—This is device-level protection and deals mostly with devices, mobile app, and web app integrity. IoT security threat at this level is physical threat which attacks the physical devices.

b. Communication securities (Network and transport layer)—Used for network and communication channel protection.
c. Cloud security (Application layer)—Cloud services particularly vulnerable to exploits, such as SQL injection flaws, will likely be targeted first. Cloud security will insure data protection and will protect data leakage.

3.1 Basics and Related Definitions

Perhaps, prior to dealing with deep insight of IoT, it would be incisive to prelude basic terminologies and definitions to make the further discussions explicit. Following are the terms often used while discussing the IoT:

- IoT
- Big Data
- Cloud Server
- Cloud Computing

IoT is considered as entire things capable to transmit or receive information and can have the efficacy of mutual interaction with each other. Indeed, this realm is moderately pervasive in different walks of life. The salient feature of IoT network can be precisely enumerated as:

- Ability to have data accumulation
- Data Processing
- Data Transmission
- Ability to analyze the entire network component

Researchers are visualizing that in the recent future the IoT will control the world by physical things that would be connected together through an infrastructure of its own [14]. Precisely, we can percept IoT as a gigantic network which encompasses an enormous group of things, sensors, and smart devices [15]. However, sensors are the prevailing component of this network. Other components of this network can be briefly enumerated as:

- RFID
- Smart card
- Smartphone
- Computers to share the necessary information.

Further, there is no explicit definition for the term big data [16]. However, the term big data entails a substantially large volume of data [17]. Different definitions observed through the literature survey unveil the fact that there are varying definitions of big data and is mainly a function of intent and research field. It is also pertinent to mention that the rate of data produced is not a fixed quantity in big data. However, few well known big data generators are YouTube, Facebook, and Twitter [18]. Moreover, wireless sensor networks are also considered as big data generators [19, 20].

It is the researcher's prediction that in the coming future IoT network will have maximum number of shared documents. The IoT network consists of an extremely powerful server to process the information. The cloud server elicits the information from distinct avenues such as from sensors, RFID, and other smart devices. Subsequently, this information is stored in memory [21]. The industry is considered as the genesis of cloud computing such as Google and Amazon [15]. The cloud server is of paramount importance in IoT network. It is due to the fact that IoT network consists of ordinary devices which are not able to store and process the big data.

It is an important issue to consider the trade-off between the big data and IoT. As stated above years ago Facebook was visualized as the gigantic creator of big data. However, in future IoT will undoubtedly be the greatest information exchanging network. Perhaps, in near future major big cities of the world will be transformed into large-scale smart cities [22]. It is worthy to mention that in fourth industrial revolution the apex use of IoT is observed. Moreover, the concomitant use of sensor which is an indispensable component of IoT also begun to augment [23]. Accumulation of environmental information, GIS through wireless sensors also renders the relationship between IoT and big data [24]. In [17], interdependencies between big data and IoT have been discussed with deep insight.

3.2 Key Security Issues

Social media is constantly reflecting the information sharing in different domains which in turn is increasing the pervasive demand of IoT. Further, malevolent and deceitful individuals are also increasing to deceive genuine users. The security of the individual's privacy is of utmost importance. Therefore, in order to save the privacy of innocent users and things some security measures are of prime concern [25–27]. Two substantial components in this pursuit are given below.

- Privacy
- Security

It is of profound interest to prevent the accessibility of information of one user by others [28]. In [29], it has been mentioned that the privacy information can subsequently be categorized in three streams. These are

- Infrastructure security
- Information security
- Information management

Further, there must be provision for each user which can be individual in devices to provide access of its information to others. However, there must be provision to inhibit the access and stop the privacy. It is a strenuous task to provide security while there is a huge volume and velocity of the data. Indeed, plenty of uncertainties subsist on data and information security. Researchers are persistently striving in this pursuit. But, big data security is still murkier and a matter of investigation. However, to prevent the users against these attacks is an arduous and challenging task. Perhaps, the growing technological advancement will resolve this issue soon [25]. Moreover, security threats entail sub-elements such as [29]:

- Denial of service (DoS)
- Functionalities of encryption
- Access control.

3.3   Features of Big Data

In this section, authors will prelude important features of big data as it is the most indispensable component for providing security. It is also important to mention that authors have considered mainly the current challenges [17, 25, 30–34]. Some important features of big data are enumerated in Table 1.

4   Security Architecture in IoT

IoT network architecture is scalable to manage the surge of devices and will circumscribe a vast range of technologies [35]. Architecture could be delineated as a simple skeleton for the network characteristics and pragmatic organization and composition. Different communication models often used are Thing to Application server and Thing to Human or Thing to Thing communication [36]. IoT architecture will have evolutionary growth and several architecture models which are currently under development like M2M (Machine to Machine) model from ETSI (European Telecommunications Standards Institute), ITU-T (International Telecommunication Union) model should be factored into future developments [37]. Further, it can be divided into four levels as given below.

- Perceptual Layer
- Network Layer
- Support Layer
- Application Layer

The first layer, i.e., the perceptual layer gathers all information like RFID [38] and all kinds of sensor information. Information could be object properties, environmental condition, etc. Security problems in this layer include the attacks from external networks such as denial of services. The functioning of second layer, i.e., network layer is to perform secure communication of the information from the first layer as well as the initial information processing. Security in network layer of IoT is very important. Since threats like MMA can pose danger to security. The support layer is the third layer and it is helpful in setting up the platform for the application layer. This layer does the intelligent processing of massive information. Finally, the

Table 1 Important features of big data

Serial number	Features	Descriptions
1	Volume	• Substantial challenge in BD • Not possible to define the limitations for the big data
2	Velocity	• Velocity is often fast in case of big data
3	Variability	• In general, different sort of data is transmitted in the case of big data
4	Veracity	• The data structure is substantially complex in big data • Difficult to establish confidence
5	Visualization	• Required lucid and adequate information • Key information must be available from each sender
6	Value	• Storage and maintenance of the value of data is required
7	Data acquisition	• Accumulation and storage of data received from different sources are required • Security of this accumulated data is of paramount importance
8	Data mining and cleansing	• It is an indispensable challenging task
9	Data aggregation and integration	• It is an indispensable challenging task
10	Complexity	• Complexity increases with volume • Increased data complexity • Increased computational complexity • Increased system complexity
11	Data analysis and modeling	• Required segregation of information • Fetching of data lost
12	Data interpretation	• Required for decision-making process
13	Privacy	• Most challenging task
14	Security	• Required to maintain the users' privacy and prevention from malware
15	Data governance	• With the increase of big data, companies and organizations need more thrust on data governance
16	Data and information sharing	• Coordination of information sharing is required by all organization
17	Cost	• It increases with demand
18	Data ownership	• Ownership is an indispensable parameter

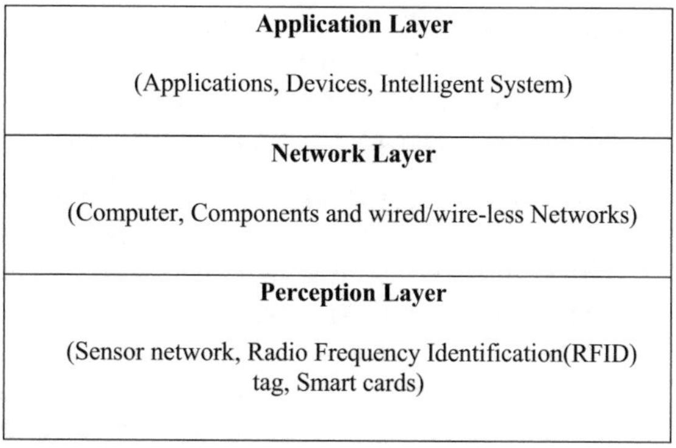

Fig. 1 Different layers in
three-layered architecture

Application Layer
(Applications, Devices, Intelligent System)
Network Layer
(Computer, Components and wired/wire-less Networks)
Perception Layer
(Sensor network, Radio Frequency Identification(RFID) tag, Smart cards)

uppermost layer is application layer. This layer caters to the users' need. Moreover, sharing of the data is the major characteristic of this layer and thus poses serious concerns for the security. Advantages and risks are discussed in [36]. The schematic representation of different layers is shown in Fig. 1.

4.1 Security Issues in Perception Layer

It is the lowest layer of the IoT architecture. The functionality of this layer is to collect information. This is done by various devices like smart cards, RFID tags, and sensor networks. This layer has ability to get the information. The security concerns in this layer lie in the fact that IoT cannot provide security to the sensing devices and information. The reason includes limited energy, fragile security capability of the sensing devices. The RFID system in this layer has security issues such as information leakage, man-in-the-middle attacks, tampering, cloning attacks, and information tracking. Furthermore, security problems faced by perception layers are congestion attack, Denial of service attacks, node replication attack, capture gateway node, and forward attack. Security issues for devices in this layer can be categorized as (a) Terminal security issue and (b) Sensor network security issue.

4.1.1 Terminal Security Issue

For collecting large amounts of real-time information, perception layer needs a lot of terminals to present this information to the user. This process requires an authentication and integrity of data to be maintained. Here, exchange of information is done through RFID technology using RFID tags, which requires no human intervention. The RFID tags are vulnerable to various attacks such as man-in-the-middle or sniffing, power analysis, viruses, cloning, and many more. At this time RFID tags do not have enough memory to store the virus, but in future the virus could be a great threat to the system. Virus programmed on the chip by the unknown source and when read, virus can transfer from the tag to the reader and then to the company's network bringing down every connected computer, RFID components, and networks. Four common types of attacks in terminal security are:

a. Unauthorized tag disabling: In these types of attacks RFID tags will become incapable temporarily or permanently. In this the attacker manipulates the behavior of the tag which can be done if one is very far away.
b. Unauthorized tag cloning: It cracks up the integrity of the information. The hacker captures the tag's identification information through the manipulation of the tags. Once the identification information of the tag is leaked, this can be used by the hacker to bypass simulated protection and plan for the further theft scheme.
c. Unauthorized tag tracking: This is a privacy attack. The hacker can trace tags by using illegitimate RFID readers. Using which, one can capture sensitive information, for instance a person's address, etc. This type of attack, management could have planned to trace their employees. Another example could be of a customer buying a product having RFID tag which guarantees them no confidentiality and thus endangers their privacy.
d. Replay attacks: This is also an eavesdropping attack. The hacker uses an illegitimate reader to imitate a tag by using a tag's response. This occurs when one side of the communication is obstructed or cut-off recorded and replayed at the later time to the receiving device in order to gain information or access.

4.1.2 Sensor Network Security Issue

There are many security concerns in sensor network. Perhaps, security need in WSN is of utmost concern and security of sensor network is substantially different from conventional network because of mobility, heterogeneity, and cost. Conventional protocols of security cannot be used in WSN. However, many innovative protocols have been designed in the recent past which visualizes the aspect of the WSN [35].

4.2 Security Issue in Network Layer

IoT is deployed in open and physically insecure with the environment which is open for attack. The layer chiefly includes computers, wired or wireless network, and also entails security issues. Further, the layer faces security issues like confidentiality, illegal access, data eavesdropping, DoS attacks, man-in-the-middle attack, virus attack, and so on. A large number of devices in this layer collect data of different formats with information having massive, multisource, and heterogeneous characteristics [35]. Transferring of large data between these nodes encounters the security issue like network congestion, thus giving way to attacks. The attacks taking place in network layer are:

a. Selective Forwarding—This attack includes a compromised node or malicious node taken over by the attacker which may send message to the incorrect pathway [39]
b. HELLO flood attack—The adversary node, which is not a legal node in the system, floods the hello request to any legitimate node. This creates congestion in the traffic by sending a useless message. The current solution of the hello flood attack is cryptographic solution which has high computational complexity [39].
c. The wormhole attack—In simple words, it's the rearrangement of the bits.
d. Sybil attack—This attack aims at fault-tolerant schemes, such as multiple routing, distributed storage, and topology maintenance. The solution to this attack which would prevent any stranger to start a Sybil attack could be Authentication and Encryption techniques.
e. Acknowledgment spoofing—In Acknowledgement flooding attack, an adversary node spoofs the acknowledgements updating the wrong information to the destined neighboring node. Acknowledgement is needed in times when routing algorithm is used.
f. Sinkhole attack—Makes selective forwarding extremely uncomplicated.

4.3 Security Issue in Application Layer

Application layer provides authenticity, integrity, and confidentiality of data. The security concerns in this layer include tampering and eavesdropping. Application layer carries out the task of traffic management and data management with the help of applications software. These attacks target the application layer to rupture the genuine service request traffic and inundate the service which ultimately creates denial of service.

5 Layered Classification

There are on the whole two categories of layered classification of attacks, which are:

- Layered classification of attacks on RFID
- Layered classification of attacks on the WSN
- Layers used for security solution of IoT

The wireless medium used in RFID network has drawbacks since it leaves the system vulnerable to attacks. These attacks are classified based on the layers in which it could be performed, which is shown in Fig. 2a. The figure classifies the RFID network attacks segregating the attacks which could be deployed to physical layer as well as the multilayer, which affects more than one layer. Understanding of security architecture gives the attacker advantage to attack a particular layer. Amongst

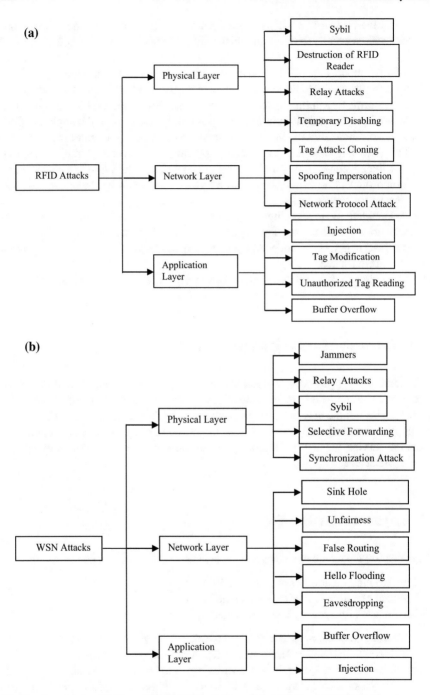

Fig. 2 a layered classification of RFID attacks, **b** layered classification of WSN attacks

various other attacks, Sybil attack, replay attack, destruction of RFID readers, temporarily disabling passive interference, active jamming, etc. could be said to have security concerns that are faced by the physical layer. Likewise, on network layer we could find attacks such as cloning, spoofing, eavesdropping, etc. Attacks which are viewed on this layer are injection, buffer overflows, unauthorized tag reading, and tag modification.

There are several attacks possible in WSN (Wireless Sensor Network) and are classified here in Fig. 2b. It depends upon which layer the attacks happen. Classifications of attack on the layer basis helps locate, identify, and mitigate the vulnerabilities due to attack and in many cases prevent these attacks. The intention of attacker in WSN layer is to either jeopardize the benefit of WSN or eavesdrop the network. There are many tools and techniques to deal with attacks in WSN security attacks.

CISCO has claimed that the IoT needs different sorts of network models for communication and processing. A seven layers model is proposed by CISCO. This model of seven layers mainly aims to secure each device and system, and every process at each level. It also provides secure communication between each level [40]. These seven layers are illustrated in Fig. 3.

6 Cryptographic Solutions for IoT

Cryptography involves creating written or generated codes that secure the information. Unauthorized readers cannot access the information since cryptography converts the data into a format that is only readable by the authorized users [40]. Cryptography can be divided into three main functions:

- Confidentiality
- Integrity
- Authentication

Confidentiality implicates that only the sender or receiver can access the data or information. Integrity does not permit an intruder to access the data during the process of transmission. It implicates the identity verification of the sender by the receiver.

6.1 Symmetric Lightweight Algorithm

These algorithms are mainly used for confidentiality purposes [12]. Advanced encryption standard (AES) is an example of this category. It is used to encrypt data and protect it against the unauthorized access. It is originally introduced by the American Institute of Standards and Technology. The cryptographic process key of varied length is used—AES-128, AES-192, or A256 depending on the length. This type of encryption of any type of data is particularly safe and effective and is

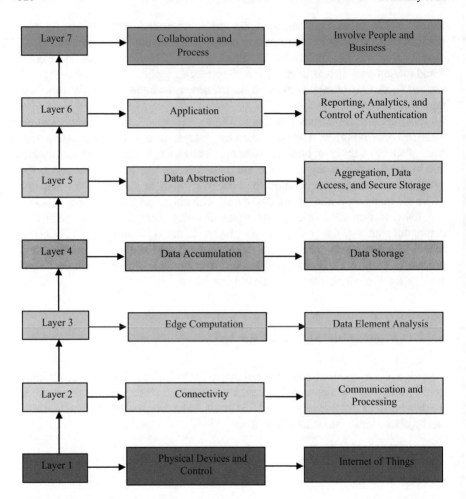

Fig. 3 Illustration of seven layers

used in various protocols and transmission technology for instance WPA2 protection of Wi-Fi networks utilizes the Advanced Encryption Standard. Today, AES is integrated into hardware of many devices to give more effective and rapid encryption and decryption than software solutions. AES is more popular due to the advantage it provides. It is freely usable, no license fees or patent restrictions.

Moreover, it has low storage, and encryption algorithm is simple to code and implement. AES is prone to the attack of MMA.

- High Security and Lightweight (HIGHT)—During Encryption and Decryption phase its keys are generated. Parallel implementation of HIGHT requires less power, fewer lines of code, and a relatively faster and efficient Radio Frequency Identification System.

- Tiny Encryption Algorithm (TEA)—This algorithm works on a constrained environment and uncomplicated. TEA is very less complex and uses basic operations like addition, shifting, and XOR. In [41], space requirement for this algorithm is given.
- PRESENT—It is ultra-lightweight algorithmic solutions. It is vulnerable to differential attack on 26 out of 31 rounds.
- RC5—RC5 is a fast symmetric block cipher suitable for hardware and software implementations. It is well known for its simplicity.

6.2 Asymmetric Lightweight Algorithm

- RSA—This algorithm basically generates a public and private key pair. Further, the public key is made open to everyone where on the other hand the private key is made secure.
- Elliptic Curve Cryptography (ECC). It is discovered in 1985.

6.3 Protocols for IoT Security

There are some essential protocols to secure IoT. These protocols are applicable to execute both symmetric as well as asymmetric cryptographic algorithms.

- HTTP: Hyper Text Transfer Protocol is a keystone of client server-based protocol for the Internet, which is used only in the client side of IoT to initiate connection but not to receive connection request.
- XMPP: Extensible Messaging and Presence Protocol defines a novel concept in IoT. It is especially applicable for voice and video calls as well as strengthening in the sense of addressing security and scalability of IoT systems.
- CoAP: It is applicable to define miserable nodes and is also capable to manipulate existing recourse with the help of interfaces. Some functions of TCP are also replicated in it, although it uses user datagram protocol.
- MQTT: Message Queue Telemetry Transport has been designed and developed for unreliable network system having low bandwidth. It controls the overall network from backend server.
- AMQP: Advance Message Queuing Protocol is applicable for queuing and routing in IoT-based system for point to point communication and security.

7 Security Issues of IoT in Medical Sector

In [42], it has been mentioned that by 2016 the approximate market of medical-related devices will be $5 billion. However, in [43], it has been given that the bulk of medical organizations are not concentrating properly on the issue pertaining to protect patient data. It is reflected in the study of [44] that the plethora of cases have occurred regarding the security in the domain of health care. Perhaps, due to the lack of effective security measure the health data is an enticing and tempting object for a malicious doer. Nowadays, smartphones equipped with sensors and other wearable devices are being used to collect the biometric data of the patients. Generally, these are connected with the apps for processing and interpreting the data and signals. The functionalities of these apps are often broadened by communicating the data to the cloud and this phenomenon required complex algorithms. The realm of security in this prospect encompasses three main ingredients. These are:

1. Confidentiality
2. Genuine use of information
3. Legal issues.

Further, three points of vulnerability pertaining to security are:

1. Devices
2. Data transmission
3. Cloud storage

It is pertinent to mention that the susceptibility of data theft from devices is less. However, one cannot defy from the chances of data theft from devices by using malware. Physical devices are also prone to be stolen or lost. Further, protection does not entail wearable sensors. But, there is provision of configuration of mobile by users. However, data can be hacked using different malware during the phenomenon of data transmission. Indeed, efficient encryption approach and method of authentication is needed to avoid the hacking of data. But there are certain pitfalls pertaining to encryption and authentication which need to be resolved. These are the reduction in the data transfer rate, difficulty of pragmatic implementation, and high energy consumption. Finally, the data is saved in the database of cloud computing architecture. However, this service is imparted by different third party vendors and thus there are more chances of attacks. Further, these databases are accessible to the Internet network so that data can be fetched from the users. It further augments the chances of attacks. In order to resolve these issues multifaceted authenticity, some access controlling technique, and complex password is needed which in turn render the system more complicated to use.

7.1 An IoT Architecture for Medical Sector

In [45], architecture for the data accumulation is proposed which needs scalable data storage devices with high performance computing (HPC) infrastructure so that efficient storage, processing and sharing of sensor data pertaining to health can take place. In this proposed architecture, logical solution monitoring is given. The solution thus obtained encompasses many parameters such as noninvasive sensors. The characteristics of these sensors are to permit the processing of voluminous data. Moreover, the solution also renders the searching and retrieval of health-related information. Figure 4 shows the general architecture for the purposes of activity monitoring based on the features given above. Furthermore, the elements of Fig. 1 are developed for the project given in [46]. This architecture lets to register the observation of persistent and nonpersistent patients and healthy people as well. Moreover, this architecture also permits the mutual communication between the family and hospitals via pragmatic application of cloud computing, big data, IoT. In this architecture IoT plays pivotal roles. In addition, this architecture incorporates the following components:

- Smart mobile phone.
- Cloud-based infrastructure.
- Analytic module.
- Access element.
- Interoperability and messaging platform.
- Website platform.

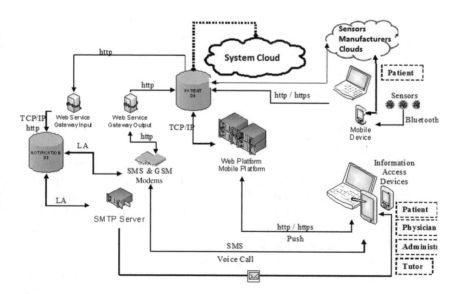

Fig. 4 Proposed architecture for patients monitoring [48]

The first element accepts the data from the sensors. The second element is required for the purposes of data storage. The third element is the need for sending the alarms to the patient and hospital. The fourth element is required to access different sensors of cloud manufacturers. Further, the fifth element is required for delivery of information for everyone involved in the system. Finally, the last element is required to permit the access of related information of the patient from the computer and from the mobile as well.

8   Security Analysis of IoT in Business Sector

IoT strategy implemented in the business sector influences the end users and is mainly designed on the basis of devices. Moreover, security features are included in an ad hoc fashion and thrust is given on threats. However, threats associated at hardware level are always there [47]. Indeed, hardware-level security issues are of the prime concern while designing the IoT architecture to cater to the need of the business sector.

8.1   Case Study

Authors are including include home automation system (Haier's system) as a case study in this chapter. The said system is treated as a smart system for controlling and reading the information obtained from different sensors installed at various places in the user's home. Different information obtained from these sensors is given below.

- Leakage of water.
- Smoke detecting sensor.
- Sensor to check the opening and closing position of the door.
- A remotely situated power switch.

Different sensors are linked by a specific protocol known as "Zigbee". The basic utility of this system is to observe the home and to receive the warning alarm on the basis of information achieved by the sensors. First of all, users are required to install a mobile app prepared by the home care system of producer. Further, this home system is linked to users' network by ethernet connectivity. Furthermore, the mobile is also linked to the same local network. Subsequently, the mobile app is opened and an account is created through the cloud service of the producer. This will enable the users to see their sensor data while the users are away from the local network.

The first phase of analysis is to analyze the hardware elements of this home security system. The processor used in this home security system supports the Linux and Android OS. The printed circuit board hardware platform and daughter board of this system is shown in Fig. 5a and b, respectively. The system also provides the capability of reading the serial data as well as the startup. At the commencement of

Fig. 5 **a** Smart care hardware platform [45], **b** smart meter CL200 daughterboard [45]

the booting the system displays "whether user wants to terminate the usual booting progression?" Moreover, at the termination of this phenomenon the user is allowed to go into a U-booting shell. This shell permits the modification of peculiar parameters of the system pertaining to boot. However, it is still possible that hackers can fetch low-level information from this system after modification in starting shell.

Subsequently after the modification of parameters booting process is initialized. After the completion of the booting process, the user enters into "rudimentary shell".

After observing the output of the system, let us suppose that the device is working on Linux. Further, in Linux environment it is indispensable to be acquainted with the type of permission users have, administration id displayed to the users that they were on the source account of the apparatus. Further, the busy box utility delineates that the user is able to run a telnet server. Further, it will permit the "TFTP" file transfer and "wget" enables the system to get the files from web. The root shell also provides the efficacy to observe the password hashes on the system.

The password can implement up to eight characters. It is pertinent to mention that getting the root password requires cracking of password hash which requires a dictionary attack. In this attack, each included password is hashed and thereafter checked for the hash under consideration. Matching of hash will provide the password otherwise checking will be continued. In addition, brute force attack is another option. However, time complexity of this approach is more.

Often, excellent hardware performance coupled with parallel processing is pragmatically used for optimizing the cracking phenomenon. In this case study, two "AMD R9290" graphics cards are used for speeding up the process. Further, it has been observed that almost 5 h is needed to obtain the root password. However, if root password is already known moving to the other layer of attack is the next disposition.

The next attack of interest to be accomplished is the remote attack on the basis of network. First of all, in this attack network scanning is performed. The network scanning enables the user to recognize "whether device possess a telnet server or not?" Next, after the linking of the device with telnet, login prompt is displayed. Further, with the help of root records, root shell can be obtained over a local network. Furthermore, type of traffic generation is observed. Moreover, in order to analyze the type of traffic generation "man-in-the-middle" is done.

Computer is required as the gateway of the system's network. Now, the internet access can be performed via the gateway. In the next step, packet sniffing is performed to observe the type of traffic the device is generating. It is worthy to mention that the firmware is being fetched on plaintext linking. Now, the updates can be fetched with the help of "wget" and file can be analyzed on it which in turn imparts the "ZIP archive". Subsequently, this archive is unzipped and lets the observation of main binary of this system with hash script needed for the device updating. On the basis of this initialized script, the device will re-setup from its own accord and further the main binary is executed. After getting this information, next phase in this analysis is to observe the way device is tackling the firmware. In this phenomenon reverse engineering and binary is needed.

Further, software is used for the binary analysis. This analysis renders the binary searching which in turn displays the way of updates handling. "MQTT" protocol is used for secured communication through the encrypted channel. Moreover, it performs as a publisher and backpropagates the sensors' related information to producer's server.

9 Conclusions

In this chapter, authors prelude an analytic view of existing key security issues pertaining to IoT security and several protocols have been reviewed. The latest research has demonstrated that IoT has huge potential for serving the industry with a titanic impact on the world, but it also poses a great threat to privacy and security. In this chapter, preoccupied architecture which connects physical world to the network has been discussed. IoT being the upcoming technology isstill in its early phase of development. Indeed, it seems that it will encompass a wide spectrum in the future. Many subdivisions of smart city, connected industry, smart energy, connected car, smart retail, and smart agriculture are controlled by IoT using RFID system. Numerous energy-constrained devices and sensors are being continuously communicating with each other via IoT technology but the security of which must not be compromised.

IoT cannot be used if it's not safe, therefore security attacks and countermeasures have been mentioned in this chapter. Various security challenges are laid before and to prevent quite a lot of attacks, one needs lightweight cryptographic security solutions, which adds on to high computational complexity. The implementations show promising results making the algorithm suitable to be adopted in IoT applications. However, it is intuitive to ponder that there is no unique solution which can cater to all the needs. Moreover, the IoT solution varies from domain to domain. Therefore, it is of paramount importance to comprehend all the necessities of an IoT implementation prior to designing an IoT solution. Designing a viable and pragmatic security solution in the realm of IoT can be considered as a future research endeavor.

References

1. Evans D (2011) The internet of things: how the next evolution of the internet is changing everything. CISCO 1–11
2. Cheng S-M, Chen P-Y, Lin CC, Hsiao H-C (2017) Traffic-aware patching for cyber security in mobile IoT. IEEE Commun Mag 29–35
3. Ryan PJ, Watson RB (2017) Research challenges for internet of things. System (MDPI) 1–32
4. Ammar M, Russello G, Crispo B (2018) Internet of things: a survey on the security of IoT frameworks. J Inf Secur Appl (Elsevier) 8–27
5. Ray PP (2018) A survey on internet of things architectures. J King of Saud Univ Comput Inf Sci (Elsevier) 291–319
6. Al-Shammari BKJ, Al-Aboody N, Al-Raweshidy HS (2018) IoT traffic management and integration in the QoS supported network. IEEE Internet Things J 352–370
7. Bertino E, Choo K-KR, Georgakopolous D, Nepal S (2016) Internet of things: smart and secure service delivery. ACM Trans Internet Technol 16(4), Article 22

8. Keshri AK, Mishra BK, Mallick DK (2018) A predator–prey model on the attacking behaviour of malicious objects in wireless nanosensor networks. Nano Commun Netw (Elsevier) 15:1–16

9. Al-Karaki JN, Kamal AE (2004) Routing techniques in wireless sensors networks: a survey. IEEE Wirel Sens Netw 11(6):6–28

10. Xiwen S (2012) Study on security issue of internet of things based on RFID. In: Fourth international conference on computational and information sciences, pp 566–569

11. Wood A, Fang L, Stankovic J, He T (2006) SIGF: a family of configurable secure routing protocols for wireless sensor networks. ACM, SASN, Alexandria, Virginia, USA, pp 1–14

12. Lu T, Wu M, Ling F, Sun J, Duh HY (2010) Research on the architecture of internet of things. In: 3rd international conference on advanced computer theory and engineering, pp 484–487

13. Batra I, Luhach AK (2016) Analysis of lightweight cryptographic solutions for internet of things. Indian J Sci Technol 1–7

14. Rahman AFA, Daud M, Mohamad MZ (2016) Securing sensor to cloud ecosystem using internet of things (IoT) security framework. In: Proceedings of the international conference on internet of things and cloud computing. ACM

15. Hashem IAT et al (2015) The rise of "big data" on cloud computing: review and open research issues. Inf Syst 47:98–115

16. Perera C, Ranjan R, Wang L (2015) Big data privacy in internet of things era. Internet Things (Mag) 32–39

17. Chen CP, Zhang C-Y (2014) Data-intensive applications, challenges, techniques and technologies: a survey on big data. Inf Sci 275:314–347

18. Khan N et al (2014) Big data: survey, technologies, opportunities, and challenges. Sci World J

19. Ding X, Tian Y, Yu Y (2016) A real-time big data gathering algorithm based on indoor wireless sensor networks for risk analysis of industrial operations. IEEE Trans Industr Inf 12(3):1232–1242

20. Kang Y-S et al (2016) MongoDB-based repository design for IoT generated RFID/sensor big data. IEEE Sens J 16(2):485–497

21. Cai H et al (2017) IoT-based big data storage systems in cloud computing: perspectives and challenges. IEEE Internet Things J 4(1):75–87

22. Sun Y et al (2016) Internet of things and big data analytics for smart and connected communities. IEEE Access 4:766–773

23. Khan M et al (2017) Big data challenges and opportunities in the hype of industry 4.0. In: 2017 IEEE international conference on communications (ICC). IEEE

24. Chen M, Mao S, Liu Y (2014) Big data: a survey. Mob Netw Appl 19(2):171–209

25. Jakobik A (2016) Big data security. Comput Commun Netw 12:241–261

26. Cheng C et al (2017) Securing the internet of things in a quantum world. IEEE Commun Mag 55(2):116–120

27. Dubey A, Srivastava S (2016) A major threat to big data: data security. In: Proceedings of the second international conference on information and communication technology for competitive strategies. ACM

28. Tole AA (2013) Big data challenges. Database Syst J 4:31–41

29. Ye H et al (2016) A survey of security and privacy in big data. In: 2016 16th international symposium on communications and information technologies (ISCIT). IEEE

30. Jung JJ (2017) Computational collective intelligence with big data: challenges and opportunities, pp 87–88

31. Gani A et al (2016) A survey on indexing techniques for big data: taxonomy and performance evaluation. Knowl Inf Syst 46(2):241–284

32. Jin X et al (2015) Significance and challenges of big data research. Big Data Res 2(2):59–64

33. Sivarajah U et al (2017) Critical analysis of big data challenges and analytical methods. J Bus Res 70:263–286

34. Bertino E, Ferrari E (2018) Big Data security and privacy. In: A comprehensive guide through the Italian database research over the last 25 years. Springer International Publishing, pp 425–439

35. Chuankun WU (2010) A preliminary investigation on the security architecture of the internet of things. Bull Chin Acad Sci 25(4):411–419
36. Poudel S (2016) Internet of things: underlying technologies, interoperability, and threats to privacy and security. Berkeley Tech LJ 31:997
37. Minerva R, Biru A, Rotondi D (2015) Towards a definition of the internet of things (IoT). IEEE Internet Initiat 14–16
38. Bhabad MA, Bagade ST (2015) Internet of things: architecture, security issues and counter-measures. Int J Comput Appl (0975-8887) 125:1–4
39. Karlo C, Wangner D (2003) Secure routing in wireless sensor networks: attacks and counter measures. Adhoc Netw (Elsevier) 293–315
40. Derbez P, Fouque PA (2014) Exhausting Demirci-Selçuk meet-in-the-middle attacks against reduced-round AES. In: International workshop on fast software encryption, pp 504–558
41. Nyberg K (2015) Links between truncated differential and multidimensional linear properties of block ciphers and underlying attack complexities, pp 162–179
42. Angela McIntyre JE (2013) Gartner, market trends: enter the wearable electronics market with products for the quantified self, July
43. Ponemon Institute LLC (2015) Fifth annual benchmark study on privacy & security of health-care data
44. Symantec: internet security threat report (2015). http://www.symantec.com/security_response/publications/threatreport.jsp
45. Wurm J, Hoang K, Arias O, Sadeghi A-R, Jin Y (2014) Security analysis on consumer and industrial IoT devices. http://jin.ece.ufl.edu/papers/ASPDAC16_IOT.pdf
46. Gachet D, Aparicio F, de Buenaga M, Ascanio JR (2014) Big data and IoT for chronic patients monitoring. In: Hervás R, Lee S, Nugent C, Bravo J (eds) Ubiquitous computing and ambient intelligence. Personalization and user adapted services. LNCS, vol 8867. Springer, Heidelberg, pp 416–423
47. Hernandez G, Arias O, Buentello D, Jin Y (2014) Smart nest thermostat: a smart spy in your home. In: Black Hat USA
48. Teresa Villalba M, de Buenaga M, Gachet D, Aparicio F (2016) Security analysis of an IoT architecture for healthcare. ICST Institute for Computer Sciences, Social Informatics and Telecommunications Engineering. https://doi.org/10.1007/978-3-319-47063-4-48

Printed in the United States
By Bookmasters